Apoptosis: Involvement of Oxidative Stress and Intracellular Ca^{2+} Homeostasis

Apoptosis: Involvement of Oxidative Stress and Intracellular Ca^{2+} Homeostasis

Edited by

Gines Maria Salido
University of Extremadura, Spain

and

Juan Antonio Rosado
University of Extremadura, Spain

Springer

Editors
Dr. Gines Maria Salido
University of Extremadura
Dept. of Physiology
Avda. Universidad s/n,
10071 Caceres
Spain
gsalido@unex.es

Dr. Juan Antonio Rosado
University of Extremadura
Dept. of Physiology
Avda. Universidad s/n,
10071 Caceres
Spain
jarosado@unex.es

ISBN 978-90-481-8215-2 e-ISBN 978-1-4020-9873-4

DOI 10.1007/978-1-4020-9873-4

Printed on acid-free paper

9 8 7 6 5 4 3 2 1

springer.com

Preface

Apoptosis or programmed cell death is a physiological mechanism required for an adequate organic function. It is required, as well as mitosis, for an adequate tissular development, i.e. metamorphosis and fetal formation. Apoptosis is also necessary to remove cells that represent a threat to the integrity of the organism, such as infected cells, immunocompetent cells when the immune response decreases or DNA damaged cells. In contrast to necrosis, an uncontrolled form of cell death that leads to cell lysis, inflammation and, potentially, health complications, apoptosis, is a process where cells play an active role in their own death; for this reason apoptosis is often referred to as cell suicide. There are several reasons that make a cell undergo apoptosis, or commit suicide, including oxidative stress, DNA damage, accumulation of unfolded or misfolded proteins, occupation of death receptors by extracellular signals and lack of stimulation from other cells.

Apoptosis: Involvement of Oxidative Stress and Intracellular Ca^{2+} Homeostasis, presents a concise synthesis of recent developments in the understanding of the programmed cell death or apoptotic mechanism in different cell types, from electrically excitable cells, such as neurons, to non-excitable cells, including vascular endothelial cells, blood cells or germ cells. Particular attention is given to the involvement of apoptosis in the development of human diseases, such as different forms neurodegenerative and immune diseases. This comprehensive text integrates the most innovative and current findings from several related disciplines, including physiology, pathology, cell biology and immunology.

The text contains two main sections; the two first chapters describe general aspects of the apoptotic events and their relationship with oxidative stress and disorders in intracellular Ca^{2+} homeostasis. This section is followed by a more specialized description of the apoptotic events in a number of electrically excitable and non-excitable cells focused on the particular aspects of apoptosis in these cells and the relationship with diseases. All of the contributors are at the forefront of scientific discovery, and the reviews presented examine the most exciting and innovative aspects of their particular areas of expertise. An extensive glossary is included to help non-specialists with the most common terms. This book is designed for scientists and academics in the medical sciences, graduate and undergraduate education in cell biology, biochemistry, physiology, pharmacology, public health and experimental pathology.

Acknowledgements

We wish to thank our colleague Dr. Bernardo Santano for his invaluable help in the editing of the chapters of the book.

Cáceres, Spain G.M. Salido
Cáceres, Spain J.A. Rosado

Contents

Contributors

Noemi N. Atucha Department of Physiology, University of Murcia, Campus Espinardo, 30100 Murcia, Spain, ntma@um.es

Maria L. Campo Department of Biochemistry, Molecular Biology and Genetics, University of Extremadura, Avda Universidad s/n, 10071 Caceres, Spain, mlcampo@unex.es

Javier Garcia-Estañ Department of Physiology, University of Murcia, Campus Espinardo, 30100 Murcia, Spain, jgestan@um.es

Javier González-Gallego Department of Biomedical Sciences; Centro de Investigación Biomédica en Red de Enfermedades Hepáticas y Digestivas (CIBEREHD) and Institute of Biomedicine, University of León, Campus de Vegazana s/n, 24071 León, Spain, jgonga@unileon.es

Antonio González Mateos Department of Physiology, University of Extremadura, Avda Universidad s/n, 10071 Caceres, Spain, agmateos@unex.es

Jose A. Pariente Department of Physiology, University of Extremadura, Avda Elvas s/n, 06071 Badajoz, Spain, pariente@unex.es

Fernando J. Peña Veterinary Teaching Hospital, Laboratory of Spermatology, University of Extremadura, Avda Universidad s/n, 10071 Caceres, Spain, fjuanpvega@unex.es

Juan A. Rosado Department of Physioligy, University of Extremadura, Avda Universidad s/n, 10071 Caceres, Spain, jarosado@unex.es

Gines M. Salido Department of Physiology, University of Extremadura, Avda Universidad s/n, 10071 Caceres, Spain, gsalido@unex.es

Jose A. Tapia Department of Physiology, Laboratory of Spermatology, University of Extremadura, Avda Universidad s/n, 10071 Caceres, Spain, jatapia@unex.es

Maria J. Tuñón Department of Biomedical Sciences; Centro de Investigación Biomédica en Red de Enfermedades Hepáticas y Digestivas (CIBEREHD) and Institute of Biomedicine, University of León, Campus de Vegazana s/n, 24071 León, Spain, mjtung@unileon.es

Chapter 1
Oxidative Stress, Intracellular Calcium Signals and Apoptotic Processes

G.M. Salido

Abstract Apoptosis, an essential physiological process that is required for the normal development and maintenance of tissue homeostasis, is mediated by active intrinsic mechanisms, although extrinsic factors can also contribute. Aerobic metabolism induces the production of reactive oxygen species (ROS), which are able to induce oxidative stress that promotes cellular apoptosis. The mechanisms of ROS-induced modifications in ion transport pathways involves oxidation of sulphydryl groups located in the ion transport proteins, peroxidation of membrane phospholipids, inhibition of membrane-bound regulatory enzymes and modification of the oxidative phosphorylation and ATP levels. Alterations in the ion transport mechanisms lead to changes in a second messenger system, primary Ca^{2+} homeostasis. Ca^{2+} disregulation induces mitochondrial depolarization, which further augments the abnormal electrical activity and disturbs signal transduction, causing cell dysfunction and apoptosis. Control of ROS levels in cells is important, because cellular dysfunction triggered by ROS is a major factor contributing to the development of many diseases. Available evidences show that ROS can induce increases in cytosolic free Ca^{2+} concentration ($[Ca^{2+}]_c$) by release of the divalent cation from internal stores and impairment of Ca^{2+} clearance systems. In fact, $[Ca^{2+}]_c$ increase is a constant feature of pathological states associated with oxidative stress and apoptosis.

Keywords Apoptosis · Homeostasis · Reactive oxygen species · Calcium · Oxidative stress

---

G.M. Salido (✉)
Department of Physiology, University of Extremadura, Avda Universidad s/n, 10071 Cáceres, Spain
e-mail: gsalido@unex.es

G.M. Salido, J.A. Rosado (eds.), *Apoptosis: Involvement of Oxidative Stress and Intracellular Ca^{2+} Homeostasis*, DOI 10.1007/978-1-4020-9873-4_1,

1.1 Introduction

Oxygen is life's molecular tragedy because it always gets us all. While it is required to run the high performance engine provided by respiration, life is a continuous battle against oxygen damage. If one acquiesces to these notions, it is not difficult to accept that evolution has provided several check-and-balance mechanisms to assure the stability and efficiency of respiration, while minimizing the harms of oxidation. Nevertheless, some of the electrons that should reduce oxygen gas to harmless water leak out from the respiratory system and reduce oxygen to the particularly dangerous reactive oxygen species (ROS). ROS, such as superoxide radical anion (O_2^-), singlet oxygen ($^1O_2$), hydrogen peroxide (H_2O_2), hydroxyl radical (•OH), and hypochlorous acid (HOCl), are thus the ancillary products of oxidative metabolism in all aerobic organisms.

The oxidative stress due to the endogenous production of ROS by mitochondria is normally counteracted by endogenous antioxidant systems that most mammalian cells have developed, including glutathione, ascorbic acid and enzymes such as superperoxide dismutase, glutathione peroxidase, and catalase. When the antioxidant machinery is overwhelmed by ROS production, the resulting oxidative damage can lead to cell death. Oxidative stress is known to activate cell death by using different execution pathways, namely apoptosis or necrosis. In general, while high levels of oxidative stress can cause necrosis (i.e.: plasma membrane rupture and release of lysosomal and granular contents to the medium), lesser degrees of oxidative stress induce apoptosis (a highly coordinated process that implies breakdown of the cell into multiple spherical bodies that retain membrane integrity).

Despite the fact that ROS can damage cells by oxidizing membrane phospholipids, proteins, and nucleic acids and that there are many pathologies which have been attributed to ROS-induced cell dysfunction, it is now recognized that these species may also act as normal intracellular messengers (Rhee 2006). In the end, oxygen is life's molecular paradox.

1.2 Intracellular Calcium Signals

The calcium ion (Ca^{2+}) is an almost universal intracellular messenger (Case et al. 2007) controlling a diverse range of cellular processes such as contraction and secretion, gene transcription and cell growth. In most cell types, Ca^{2+} has its major signalling function when cytosolic free Ca^{2+} concentration ($[Ca^{2+}]_c$) increases in response to a wide variety of hormones and neurotransmitters. But Ca^{2+} signalling is more than a simple increase in $[Ca^{2+}]_c$; spatial patterning, which includes amplitude, frequency and duration of the Ca^{2+} signal, determines its intracellular function (Berridge et al. 2000; Yano et al. 2004). Usually, the detection of dynamic changes in $[Ca^{2+}]_c$ is achieved by introducing a fluorescent Ca^{2+} indicator into the cell, being fura-2/AM the most widely used (Grynkiewicz et al. 1985). Because introduction of the indicator into the cytosol inevitably perturbs the time-course of $[Ca^{2+}]_c$ by acting as a buffer, thus altering the quantity to be measured, mathematical modelling can be used as an alternative approach (Borst and Abardanel 2007). In addition, chimeric proteins containing aequorins with different targeting sequences and Ca^{2+} affinities permit simultaneous and independent monitoring of $[Ca^{2+}]$ in different subcellular domains of the same cell (Manjarres et al. 2008).

Ca^{2+}-mobilising cellular agonists increase $[Ca^{2+}]_c$ by means of two mechanisms: the release of Ca^{2+} from intracellular stores and the entry of extracellular Ca^{2+} through plasma membrane (PM) channels (Fig. 1.1). The endoplasmic reticulum (ER) is the most investigated organelle and probably represents the major Ca^{2+} store in most cell types. The ER expresses at least two major types of Ca^{2+} release channels, namely inositol 1,4,5-trisphosphate (IP_3R) and ryanodine (RyR) receptors (Pozzan et al. 1994; Ashby and Tepikin 2002; Bootman et al. 2002). Several other cellular organelles also store Ca^{2+} and act as physiological agonist-releasable Ca^{2+} compartments. A number of studies have provided evidence of the importance of mitochondria in cellular Ca^{2+} homeostasis (Gilabert et al. 2001; Collins et al. 2002; Villalobos et al. 2002; Gonzalez et al. 2003). In addition, a role for the nuclear envelope, the Golgi apparatus, secretory granules and lysosomes, has recently received support (Pinton et al. 1998; Yoo 2000; Gerasimenko et al. 2003; Lopez et al. 2005).

In electrically excitable cells, such as neurons and muscle cells, Ca^{2+} entry mostly occurs through voltage-operated Ca^{2+} channels (VOCs); however, in non-electrically excitable cells, where VOCs are not present Ca^{2+} influx mainly occurs

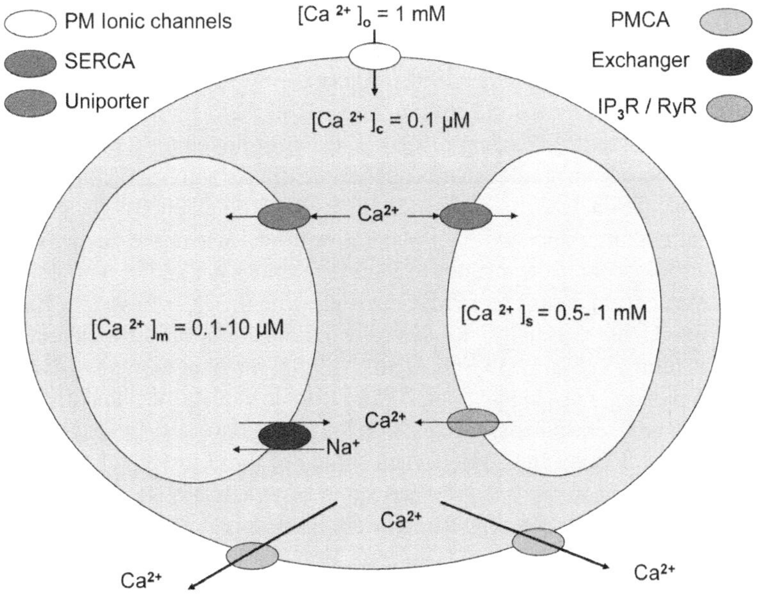

Fig. 1.1 Schematic diagram depicting the major ROS-sensitive Ca^{2+}-handling mechanism (See also Plate 1 in the Color Plate Section on page 223)

through receptor-operated channels (ROCs), second messenger-operated channels (SMOCs) or store-operated channels (SOCs).

VOCs are Ca^{2+} permeable channels that are briefly activated by changes in the membrane potential (Tsien et al. 1995). These channels are mainly found in electrically excitable cells, where they open in response to membrane depolarisations to allow Ca^{2+} to enter the cell (McCleskey 1994). ROCs belong to a heterogeneous family of channels that are especially relevant in secretory cells and neurons and are activated by a number of cellular agonists inducing a rapid Ca^{2+} entry, indicative of a direct coupling between the receptor and a Ca^{2+} permeable channel (Sage 1992). SMOCs are Ca^{2+} channels activated by a second messenger, such as inositol phosphate or Ca^{2+} itself. These channels have been described mainly in non-electrically excitable cells. For instance, in endothelial cells a Ca^{2+} channel activated by Ca^{2+} and inositol 1,3,4,5-tetrakisphosphate has been found (Lückhoff and Clapham 1992); in human platelets, thrombin activates a store-independent (non-capacitative) Ca^{2+} entry, which is mediated by protein kinase C (PKC) (Rosado and Sage 2000a). In addition, some transient receptor potential channels (TRPCs) have been reported to be activated by diacylglycerol (DAG) analogues in different non excitable cells (Ma et al. 2000).

The major mechanism for Ca^{2+} entry in non-electrically excitable cells is store-operated Ca^{2+} entry (SOCE) through SOCs, controlled by the filling state of the intracellular Ca^{2+} stores. It is not yet clear how store depletion is communicated to the plasma membrane, but a number of hypotheses have been suggested. They can be divided into those which propose a role for a diffusible messenger and those which propose a direct interaction between proteins in the ER and PM (conformational coupling). Diffusible messengers include cyclic GMP, small GTP-binding proteins, a product of cytochrome P450, tyrosine kinases and a yet unknown Ca^{2+} influx factor (CIF) (Parekh and Penner 1997). Alternatively, the conformational coupling model suggests an interaction between the IP_3R in the membrane of the ER and a Ca^{2+} channel in the PM (Berridge 1995). The conformational coupling has recently received support from studies that propose a *de novo* conformational coupling for the activation of SOCE (Rosado et al. 2000b; Redondo et al. 2003). The *de novo* conformational coupling appears as an integrative model where messenger molecules and the actin cytoskeleton interact to facilitate a physical and reversible coupling between elements in the ER and PM. Consistent with this, proteins of the Ras family and tyrosine kinases, initially considered as members of the diffusible messenger hypothesis for the activation of SOCE, are essential for actin reorganisation induced by store depletion (Rosado and Sage 2000b; Rosado et al. 2000a) and proteins classically involved in exocytosis, such as SNAP-25 appear to be involved in Ca^{2+} entry (Redondo et al. 2004a). In addition, remodelling of the cytoskeletal cortical barrier has been suggested to facilitate the activation of SOCE by CIF (Xie et al. 2002). Finally, a secretion-like coupling based on the insertion of preformed Ca^{2+} channels in the PM has also been reported (Yao et al. 1999; Patterson et al. 1999).

Ca^{2+} removal from the cytosol is carried out by several Ca^{2+} pumps and exchangers which reintroduce Ca^{2+} into the internal stores or extrude it out of the cell (Meldolesi and Pozzan 1998). Ca^{2+} uptake into the intracellular stores mostly occurs against a concentration gradient, since $[Ca^{2+}]_c$ is lower than Ca^{2+} concentration in the stores ($[Ca^{2+}]_s$), which at rest is in a high-micromolar to low-millimolar concentration range. Released Ca^{2+} is returned to the stores by the sarcoendoplasmic reticulum Ca^{2+}-ATPase (SERCA). SERCA proteins are encoded by three differentially expressed genes in mammals (SERCA 1, 2, and 3; Wu et al. 1995; Bobe et al. 2005). SERCA 1 gene products are expressed in fast-twitch skeletal muscles. SERCA 2a protein is expressed in cardiac and slow-twitch striated muscles while SERCA 2b is ubiquitously expressed. SERCA 3, which is expressed in some non-muscle tissues, is alternatively spliced, generating mRNAs that encode three protein isoforms. SERCA has a high affinity for Ca^{2+} (0.1–0.4 μM), which suggests that SERCA is likely to be activated by an increase in $[Ca^{2+}]_c$ and inhibited by an increase in $[Ca^{2+}]_s$ (Carafoli 1991). Several pharmacological tools, such as thapsigargin, 2,5-di(tert-butyl)-1,4-benzohydroquinone (TBHQ) and cyclopiazonic acid, have been developed to investigate the role of SERCA in Ca^{2+} signalling. Among them, the most widely used is thapsigargin, which binds to all SERCAs although with different affinity (Cavallini et al. 1995) and causes an irreversible inhibition of their activity by blocking the ATPase in the Ca^{2+}-free state (Wictome et al. 1992). A similar effect is induced with TBHQ, although with lower potency, and some isoforms seem to be insensitive to this inhibitor, which has been used to identify distinct intracellular Ca^{2+} stores (Cavallini et al. 1995) that, in turn, activate different Ca^{2+} entry mechanisms (Rosado et al. 2004a).

Perhaps the major mechanism for the removal of cytosolic Ca^{2+} is the extrusion of Ca^{2+} to the extracellular medium against a concentration gradient. Ca^{2+} efflux is mainly carried out by two different transporters, the plasma membrane Ca^{2+}-ATPase (PMCA) and the Na^+/Ca^{2+} exchanger. The PMCA is an ATPase highly sensitive to vanadate and lanthanum (Pariente et al. 1999; Lajas et al. 2001; Pedersen 2007). Molecular biology studies have revealed the expression of at least four PMCA isoforms in humans: PMCA1-4, although its number is increased by the existence of alternative splice variants (Strehler and Zacharias 2001; Bobe et al. 2005). The structure of the PMCA consists of ten transmembrane domains and five extracellular regions, with the NH_2 and COOH termini located in the cytosolic site of the membrane (Guerini 1998; Strehler and Zacharias 2001). PMCA activity is regulated by several messenger molecules including Ca^{2+}/calmodulin, protein tyrosine kinases, PIP_2, protein serine/threonine kinases and by proteases like calpain (Strehler and Zacharias 2001). Agonists might also either increase or inhibit the PMCA activity by activating these intracellular pathways (Rosado and Sage 2000c; Pariente et al. 2001).

On the other hand, the Na^+/Ca^{2+} exchanger is a bidirectional electrogenic ion transporter that couples the movement of Na^+ in one direction with the transport of Ca^{2+} in the opposite direction. The Na^+/Ca^{2+} exchanger modulates $[Ca^{2+}]_c$ by either removing Ca^{2+} from the cytosol (forward mode) or by transporting Ca^{2+} inside the cell (reverse mode). Three different Na^+/Ca^{2+} exchangers have been described. Two of them are electrogenic, the K^+-independent Na^+/Ca^{2+} exchanger, which catalyses the countertransport of either 3 or 4 Na^+ for 1 Ca^{2+}, and the K^+-dependent Na^+/Ca^{2+} exchanger, which catalyses the exchange of 4 Na^+ by 1 Ca^{2+}

and 1 K^+ (Blaustein and Lederer 1999). In addition, an electroneutral Na^+/Ca^{2+} exchanger has been described in mitochondria (Matsuda et al. 1997).

Finally, mitochondria are relevant components of the Ca^{2+} signalling machinery. Localised in the vicinity of the Ca^{2+} channels, mitochondria sequester Ca^{2+} modulating the Ca^{2+} signals (Gonzalez and Salido 2001; Parekh 2003). Ca^{2+} enters the mitochondria by a high capacity and low affinity uniporter that requires local high $[Ca^{2+}]_c$ to function (Berridge et al. 2000). Despite this high threshold, Ca^{2+} transport by mitochondria can be activated at smaller $[Ca^{2+}]_c$ (Xiong et al. 2004). Efflux of Ca^{2+} occurs by means of two different exchangers that countertransport Ca^{2+} for either Na^+ or H^+, or through a permeability transition pore showing a reversible low conductance state, that allows mitochondria to participate in Ca^{2+} signalling, and an irreversible high conductance state that collapses the mitochondrial membrane potential (ψ_m) leading to the activation of apoptosis (Ichas et al. 1997; Berridge et al. 2000; Jeong and Seol 2008)

1.3 Mechanisms of Ros-Induced Modifications in Ca^{2+} Movements

Among the metabolic pathways that are known to produce ROS in mammalian cells, the electron transport system of mitochondria is one that probably has received more attention from the scientific community. The electron transport chain is the mechanism, localized in the mitochondrial inner membrane, responsible for the generation of cellular energy. Although it is a very efficient mechanism, under normal conditions, a small leakage of single electrons occurs and causes the production of superoxide (O_2^-) and hydrogen peroxide (H_2O_2), which in the presence of iron can produce hydroxyl radical (•OH). Superoxide has difficulties to cross lipid membranes but can oxidize proteins present in the organelles where it is produced (Thannickal and Fanburg 2000). In addition, the presence of the enzymes superoxide dismutases in the cytosol, mitochondria and extracellular space rapidly dismutate O_2^- to H_2O_2. Hydrogen peroxide, a small molecule with biological diffusion properties across biomembranes similar to water (Antunes and Cadenas 2000), is maintained in physiological intracellular concentrations by the action of the naturally occurring enzymes catalase and glutathione peroxidase, which metabolize H_2O_2 to H_2O and O_2.

Albeit the cellular antioxidant machinery is able to maintain intracellular ROS concentration in a physiological range, a disturbance in the prooxidant/antioxidant balance (i.e. oxidative stress) can induce cell damage and apoptosis (Chandra et al. 2000). This is due at least in part to the ability of ROS to interact with cell signalling pathways by way of modifications of key thiol groups (SH groups) on proteins that possess regulatory functions, including Ca^{2+}-channel forming proteins and transporters (Hool and Corry 2007).

According to this, many proteins are targets of such oxidative attack due to the presence of reactive cysteine residues which have a sulfhydryl group in their side chain and the proton is labile, which makes it a chemical hot spot for a wide variety of biochemical interactions. The exposure of cysteines on the protein surface is a functional necessity to prevent redox changes to spread through the entire protein molecule (Biswas et al. 2006). SH groups can react with ROS forming intramolecular disulfide bonds (Gilbert 1990) and it is known that Ca^{2+} channels contain many cysteines, although not all of these residues will be susceptible to oxidation by exogenous sulfhydryl reagents. Another target for ROS interaction with channel proteins is methionine residues, although the role of methionines in redox regulation of many of the ion channels and transporters has yet to be determined (Hool and Corry 2007).

It is generally reported that ROS cause an increase in $[Ca^{2+}]_c$ of different cell types, including smooth (Roveri et al. 1992; Krippeit-Drews et al. 1995) and skeletal (Favero et al. 1995) muscle cells, mesangial cells (Meyer et al. 1996), blood mononuclear cells (Korzets et al. 1999), pancreatic β-cells (Krippeit-Drews et al. 1999), neurons (Whittemore et al. 1995), cardiomyocytes (Wang et al. 1999) and renal tubular cells (Ueda and Shah 1992). It seems that this increase can be due to both Ca^{2+} release from intracellular stores such as the ER, and to Ca^{2+} influx from the extracellular medium through the PM (Fig. 1.1). However, the effect of ROS on Ca^{2+} signalling can vary from stimulative to repressive, depending on the type of oxidants, their concentrations, and the duration of exposure. For example, in human aortic endothelial cells, low concentrations (1–10 μM) of H_2O_2 exhibit no effect on $[Ca^{2+}]_c$, while 100 μM H_2O_2 induced intracellular oscillations (Hu et al. 1998). In pancreatic acinar cells, our research group and others have shown that the sulphydryl group oxidising agents thimerosal (Thorn et al. 1992), vanadate (Pariente et al. 1999) and phenylarsine oxide (Lajas et al. 1999) are able to mobilise Ca^{2+} from intracellular stores, and this effect is reversible in the presence of the thiol-reducing agent dithiothreitol. Additionally, it has been shown that thimerosal is able to mobilise Ca^{2+} from intracellular stores in pancreatic acinar cells (Thorn et al. 1992) and HeLa cells (Bootman et al. 1992) by sensitising the IP_3R to the endogenous level of IP_3, whereas in skeletal muscle cells thimerosal was also able to produce Ca^{2+} release through RyR (Abramson et al. 1995). However, our results using the membrane-permeable IP_3R blocker, xestospongin C, demonstrate that H_2O_2 releases Ca^{2+} from a non-mitochondrial and agonist-sensitive Ca^{2+} pool in mouse pancreatic acinar cells by an IP_3R independent mechanism (Pariente et al. 2001). Notwithstanding, in cardiac-derived fibroblasts, pretreatment with xestospongin C reduced the Ca^{2+} release evoked by H_2O_2 (Colston et al. 2002).

Alternatively, the existence of a redox sensor in the agonist-sensitive Ca^{2+} stores in human platelets has been proposed (Redondo et al. 2004c). The redox sensor in the agonist-releasable pool might consist of hyperreactive sulphydryl groups present in the IP_3R. These groups are highly sensitive to oxidation by agonist-generated ROS (Granados et al. 2004; Rosado et al. 2004b) or when platelets are exposed to exogenous ROS. Consistent with this, we have previously shown that ROS induce concentration-dependent Ca^{2+} release from agonist-sensitive Ca^{2+} stores by oxidation of sulphydryl groups in IP_3R but independently of IP_3 generation. Blockade of either the IP_3 turnover by lithium or PLC by the specific inhibitor U-73122 is unable to prevent ROS-induced Ca^{2+} release from the agonist-sensitive pool (Redondo et al. 2004c). Similarly, tert-butyl hydroperoxide-induced Ca^{2+} release (Sakaida et al. 1991) occurs without any requirement for PLC activation in hepatocytes (Rooney et al. 1991).

It has been also reported that •OH, generated by hypoxanthine/xanthine oxidase, releases Ca^{2+} from thapsigargin-insensitive (Bielefeldt et al. 1997), but ryanodine-sensitive, Ca^{2+} stores (Klonowski-Stumpe et al. 1997; Weber et al. 1998).

ROS have been shown to activate receptor-operated Ca^{2+} entry (ROCE) through TRPM2 in granulocytes by enhancing NAD concentration (Heiner et al. 2003). In contrast, oxidative stress inhibits ROCE in endothelial cells, suggesting that high concentrations of ROS might have adverse effects on ROCE.

ROS, such as H_2O_2, play a concentration-dependent effect in the activation of SOCE. SOCE has been reported to be reduced by treatment with H_2O_2 at concentrations $\geq 100\,\mu M$ through the activation of PKC, which leads to membrane depolarization and increased Ca^{2+} extrusion (Törnquist et al. 2000). Our observations in human platelet indicate that exposure to low H_2O_2 concentrations favours Ca^{2+} release from intracellular stores and subsequently SOCE, showing a positive correlation between Ca^{2+} release and entry at $10\,\mu M$ and $100\,\mu M$ H_2O_2. However, the ability of H_2O_2 to induce SOCE decreases at higher concentrations, so that it induces a small amount of Ca^{2+} entry despite the extensive depletion of the intracellular Ca^{2+} stores. In addition, 1 mM H_2O_2 reduces both the activation and maintenance of SOCE stimulated by agonists (Redondo et al. 2004b).

The role of other ROS, such as superoxide anion (O_2^-) on SOCE has also been described. In vascular endothelial cells, incubation with the O_2^--generating system xanthine oxidase/hypoxanthine resulted in an increased intracellular Ca^{2+} release and SOCE in response to bradykinin and ATP in a time- and concentration-dependent manner (Graier et al. 1998). In contrast, it has been reported that high O_2^- concentrations reduce SOCE in PLB-985 cell lines and neutrophilic granulocytes from peripheral blood, which further supports that high concentrations of ROS impair the activation of SOCE.

The correlation between Ca^{2+} release and entry, and the similar actin reorganization induced by low concentrations of H_2O_2 and physiological agonists, suggests that endogenous H_2O_2 production might play a role in the activation of SOCE under physiological conditions. A physiological role of ROS on the activation of SOCE has been demonstrated in different cell types, such as mast cells, where inhibition of ROS production by diphenyleneiodonium impairs SOCE stimulated by FcεRI cross-linking (Suzuki et al. 2003). In endothelial cells, enzymatically produced non-toxic H_2O_2, rather than O_2^- or •OH induces Ca^{2+} release from thapsigargin sensitive stores and activates SOCE, at least partially by activating PLC (Volk et al. 1997). In addition, in human platelets, Ca^{2+} store depletion, induced by physiological agonists or by pharmacological tools, stimulates the production of H_2O_2 in the micromolar range ($\leq 100\,\mu M$). Generated H_2O_2 stimulates actin filament reorganisation and, subsequently, the activation of $pp60^{src}$ by a PKC-dependent mechanism, are required for the coupling between naturally expressed TRPC1 and IP_3R type II and the subsequent activation of SOCE in these cells (Rosado et al. 2004b). The role of H_2O_2 as a messenger molecule release after store depletion was confirmed by the stimulation of Ca^{2+} entry by 10 μM H_2O_2 in the absence of Ca^{2+} release from the intracellular stores (Rosado et al. 2004b; Redondo et al. 2005). Ca^{2+} mobilisation induced by low concentrations of H_2O_2 clearly differ from pathological "oxidative stress" associated with a progressive increase in $[Ca^{2+}]_c$.

Despite the fact that it has been reported that H_2O_2 can affect the activity of SERCA pumps, the mechanisms that underlie this process remain unclear. Several authors have described that metal-catalyzed oxidation result in SERCA inhibition by direct disulfide bonds oxidation, this is the case in platelet (Redondo et al. 2004c), but in other cell types, such as skeletal-muscle cells (Moreau et al. 1998) and myocardic H9c2 (Ihara et al. 2005), SERCA inhibition is induced by an independent oxidative mechanism. Other ROS like hydroxyl radicals and peroxynitrite can also reduce the activity of SERCA (Moreau et al. 1998; Gutiérrez-Martín et al. 2004).

ROS can also modify the activity of PMCA. A number of studies have shown that H_2O_2 produces an alteration on the ability of PMCA to extrude Ca^{2+} from the cytosol (Ermak and Davies 2002; Zaidi and Michaelis 1999), but the underlying mechanism still remains unclear. Different hypotheses include modulation by direct disulfide bounds oxidation or by the activity of intermediate oxide-sensitive proteins such as calmodulin (Gonzalez and Salido 2001; Chen et al. 2005). However, other oxidants, such as peroxynitrite ($ONOO^-$), induce loss of activity by direct changes on the PMCA structure in neurons and other cell types (Chen et al. 2005). PMCA inhibition by ROS has very important physiological consequences and may be targets of oxidative stress in the aging brain. In fact, it has been shown that a reduction in PMCA activity may contribute to age-related alterations in neuronal $[Ca^{2+}]_c$ regulation (Zaidi and Michaelis 1999), and could be the cause for other diseases related to Ca^{2+} homeostasis.

The mechanisms of mitochondrial Ca^{2+} uptake and release can be modified by several ROS (Pariente et al. 2001; Gonzalez et al. 2005), but at the same time the $[Ca^{2+}]_c$ can increase the production of ROS when the complex I activity is altered (Votyakova and Reynolds 2005). The relationship between Ca^{2+} mitochondrial ($[Ca^{2+}]_m$) and ROS production is not clear. An increase in $[Ca^{2+}]_m$ by uptake from the cytosol is able to induce ROS production on the electron mitochondrial transport chain. Oxidants can also affect Ca^{2+} mitochondrial pool allowing the release of Ca^{2+} from this organelle, by inducing changes on ψ_m (Brookes et al. 2004; Gonzalez et al. 2005). It is not clear if ROS-induced changes on ψ_m are responsible for the opening of the mitochondrial pore or if this opening is produced by direct changes in the pore structure. However, it is clear that in pancreatic acinar cells, ROS, such as H_2O_2, produce changes on ψ_m, which have been proposed as the basis for free radical injury to cells (Brookes et al. 2004). Additionally, it has been reported that in pancreatic acinar cells, the Ca^{2+}-mobilizing agonist CCK induces increase in $[Ca^{2+}]_m$, depolarisation of ψ_m and increases in FAD autofluorescence. These changes in mitochondrial activity induced by CCK were completely blocked in the presence of H_2O_2 (Granados et al. 2005).

1.4 Calcium and Apoptosis

The involvement of Ca^{2+} in programmed cell death has been clearly documented from many previous studies (Orrenius et al. 2003). Although a precise description of the mechanisms that undergo apoptosis will be described in Chapter 2, a brief description of those in which Ca^{2+} participates directly will be summarized here.

Apoptosis can be triggered by two main cellular pathways: the extrinsic (consisting of cell surface TNF-related family of receptors, their inhibitory counterparts and cytoplasmic adapter or death inhibitory molecules) and the intrinsic pathway, for which ER and mitochondria are the central organelles governed by proapoptotic and antiapoptotic proteins belonging to Bcl-2 family (Bernardi and Rasola 2007). Although commonly viewed as separate pathways and capable of functioning independently, cross-talk can occur between these pathways at multiple levels, depending on the repertoire of apoptosis-modulating proteins expressed. In both cases, the executer of the apoptotic process is a family of cysteine proteases that cleave their substrates at aspartic acid residues, named caspases.

Alterations in intracellular Ca^{2+} homeostasis are commonly observed during apoptosis (Fig. 1.2), including enhanced Ca^{2+} entry and Ca^{2+} release from ER and mitochondria, thus promoting sustained increases in $[Ca^{2+}]_c$ (McConkey and Orrenius 1997; Nicotera and Orrenius 1998). Ca^{2+} release from the ER can occur through IP_3R and/or RyR, both of which function as Ca^{2+} release channels in the ER membrane. It has been demonstrated that depletion of the ER Ca^{2+} stores by activation of RyR/Ca^{2+} release channel can directly induce apoptosis in cultured Chinese hamster ovary cells (Pan et al. 2000). In addition, capacitative Ca^{2+} influx through Ca^{2+} release-activated Ca^{2+} channels is apoptogenic (Jiang et al. 1994).

There is accumulating evidence that mitochondrial Ca^{2+} uptake promotes apoptosis in different ways, making this organelle a key component of the Ca^{2+}-regulated

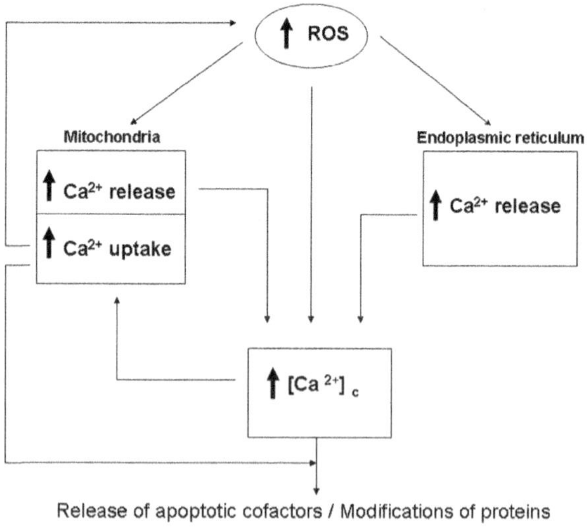

Fig. 1.2 Role of ROS evoking cell activation of apoptosis

amplification loop of apoptosis (Rizzuto et al. 2003). In a recent paper Lao and Chang (2008) show that mitochondrial Ca^{2+} spikes, observed during UV- or TNFα-induced apoptosis in HeLa cells, were synchronous with cytosolic Ca^{2+} spikes. This synchrony was due to the fact that both $[Ca^{2+}]_c$ spikes and $[Ca^{2+}]_m$ spikes were caused by the release of Ca^{2+} from ER through the IP_3R. Increases in $[Ca^{2+}]_m$ might cause opening of the permeability transition pore (PTP) or generation of reactive oxygen species (ROS), which could lead to mitochondrial dysfunction resulting in cytochrome *c* release from mitochondria and others apoptotic mediators (Brookes et al. 2004; Bernardi and Rasola 2007) to the cytosol to activate the caspase cascades (Wang 2001). The existence of certain apoptotic pathways, in which an early Ca^{2+} signal is activated upstream of cytochrome *c* release (Pu et al. 2002) is worth mentioning.

For the coordinate release of cytochrome *c* from mitochondria throughout the cell, it has been suggested that a feed-forward cycle of calcium release from IP_3R and cytochrome *c* release from mitochondria provides a molecular mechanism. Cytochrome *c* binding to IP_3R adjacent to mitochondria would sensitize IP_3R to increased Ca^{2+} release, causing mitochondrial and cytosolic Ca^{2+} overload and further cytochrome *c* release. Therefore, small amounts of cytochrome *c* released early in apoptosis would first function at ER membranes to alter Ca^{2+} handling. Subsequent global cytochrome *c* release from mitochondrion would supply the cytosolic cytochrome *c* necessary for caspase activation and completion of the apoptotic cascade (Boehning et al. 2003).

Moreover, fluctuations in $[Ca^{2+}]_c$ also affect multiple enzymes, including Ca^{2+}-activated proteases, calcineurin, endonucleases, phospholipases, nitric oxide synthase and transglutaminases. These enzymes control the breakdown of various cellular constituents, some of which are directly associated with apoptosis (Verkhratsky 2007).

Acknowledgments This work was supported by grant BFU2007-60104 funded by the Spanish Ministry of Science and Innovation.

References

Abramson JJ, Zable AC, Favero TC et al. (1995) Thimerosal interacts with the Ca^{2+} release channel ryanodine receptor from skeletal muscle sarcoplasmic reticulum. J Biol Chem 270:29644–29647

Antunes F, Cadenas E (2000) Estimation of H_2O_2 gradients across biomembranes. FEBS Lett 475:121–126

Ashby MC, Tepikin AV (2002) Polarized calcium and calmodulin signaling in secretory epithelia. Physiol Rev 82:701–734

Bernardi P, Rasola A (2007) Calcium and cell death: the mitochondrial connection. In: Carafoli E and Brini M (eds) Calcium signalling and disease. Subcellular biochemistry, vol 45. Springer, Berlin, Heidelberg, New York, pp 481–506

Berridge MJ (1995) Capacitative calcium entry. Biochem J 312:1–11

Berridge MJ, Lipp P, Bootman MD (2000) The versatility and universality of calcium signalling. Nat Rev Mol Cell Biol 1:11–21

Bielefeldt K, Whiteis CA, Sharma RV et al. (1997) Reactive oxygen species and calcium homeostasis in cultured human intestinal smooth muscle cells. Am J Physiol 272:G1439–G1450

Biswas S, Chida AS, Rahman I (2006) Redox modifications of protein-thiols: emerging roles in cell signaling. Biochem Pharmacol 71:551–564

Blaustein MP, Lederer WJ (1999) Sodium/calcium exchange: its physiological implications. Physiol Rev 79:763–854

Bobe R, Bredoux R, Corvazier E et al. (2005) How many Ca^{2+} ATPase isoforms are expressed in a cell type? A growing family of membrane proteins illustrated by studies in platelets. Platelets 16:133–150

Boehning D, Patterson RL, Sedaghat L et al. (2003) Cytochrome *c* binds to inositol (1,4,5) trisphosphate receptors, amplifying calcium-dependent apoptosis. Nat Cell Biol 5:1051–1061

Bootman MD, Berridge MJ, Roderick HL (2002) Calcium signalling: more messengers, more channels, more complexity. Curr Biol 12:563–565

Bootman MD, Taylor CW, Berridge MJ (1992) The thiol reagent, thimerosal, evokes Ca^{2+} spikes in HeLa cells by sensitizing the inositol 1,4,5-trisphosphate receptor. J Biol Chem 267:25113–25119

Borst A, Abardanel HDI (2007) Relating a calcium indicator signal to the unperturbed calcium concentration time-course. Theor Biol Med Mod 4:1–13

Brookes PS, Yoon Y, Robotham JL et al. (2004) Calcium, ATP, and ROS: A mitochondrial love-hate triangle. Am J Physiol Cell Physiol 287:C817–C833

Carafoli E (1991) Calcium pump of the plasma membrane. Physiol Rev 71:129–153

Case RM, Eisner D, Gurney A et al. (2007) Evolution of calcium homeostasis: from birth of the first cell to an omnipresent signalling system. Cell Calcium 42:345–350

Cavallini L, Coassin M, Alexandre A (1995) Two classes of agonist-sensitive Ca^{2+} stores in platelets, as identified by their differential sensitivity to 2,5-di-(tert-butyl)-1,4-benzohydroquinone and thapsigargin. Biochem J 310:449–452

Chandra J, Samali A, Orrenius S (2000) Triggering and modulation of apoptosis by oxidative stress. Free Rad Biol Med 29:323–333

Chen B, Mayer MU, Squier TC (2005) Structural uncoupling between opposing domains of oxidized calmodulin underlies the enhanced binding affinity and inhibition of the plasma membrane Ca-ATPase. Biochemistry 44:4737–4747

Collins TJ, Berridge MJ, Lipp P et al. (2002) Mitochondria are morphologically and functionally heterogeneous within cells. EMBO J 21:1616–1627

Colston JT, Chandrasekar B, Freeman GL (2002) A novel peroxide-induced calcium transient regulates interleukin-6 expression in cardiac-derived fibroblasts. J Biol Chem 277:23477–23483

Ermak G, Davies KJ (2002) Calcium and oxidative stress: from cell signaling to cell death. Mol Immunol 38:713–721

Favero TG, Zable AC, Abramson JJ (1995) Hydrogen peroxide stimulates the Ca^{2+} release channel from skeletal muscle sarcoplasmic reticulum. J Biol Chem 270:25557–25563

Gerasimenko JV, Maruyama Y, Yano K et al. (2003) NAADP mobilizes Ca^{2+} from a thapsigargin-sensitive store in the nuclear envelope by activating ryanodine receptors. J Cell Biol 163: 271–282

Gilabert JA, Bakowski D, Parekh, A (2001) Energized mitochondria increase the dynamic range over which inositol 1,4,5-trisphosphate activates store-operated calcium influx. EMBO J 20:2672–2679

Gilbert HF (1990) Molecular and cellular aspects of thiol-disulfide exchange. Adv Enzymol Relat Areas Mol Biol 63:69–172

Gonzalez A, Granados MP, Salido GM et al. (2003) Changes in mitochondrial activity evoked by cholecystokinin in isolated mouse pancreatic acinar cells. Cell Signal 15:1039–1048

Gonzalez A, Granados MP, Salido GM et al. (2005) H_2O_2-induced changes in mitochondrial activity in isolated mouse pancreatic acinar cells. Mol Cell Biochem 269:165–173

Gonzalez A, Salido GM (2001) Participation of mitochondria in calcium signalling in the exocrine pancreas. J Physiol Biochem 57:331–339

Graier WF, Hoebel BG, Paltauf-Doburzynska J et al. (1998) Effects of superoxide anions on endothelial Ca^{2+} signaling pathways. Arterioscler Thromb Vasc Biol 18:1470–1479
Granados MP, Salido GM, Pariente JA et al. (2004) Generation of ROS in response to CCK-8 stimulation in mouse pancreatic acinar cells. Mitochondrion 3:285–296
Granados MP, Salido GM, Pariente JA et al. (2005) Effect of H2O2 on CCK-8-evoked changes in mitochondrial activity in isolated mouse pancreatic acinar cells. Biol Cell 97:847–856
Grynkiewicz G, Poenie M, Tsien RY (1985) A new generation of Ca^{2+} indicators with greatly improved fluorescence properties. J Biol Chem 260:3440–3450
Guerini D (1998) The significance of the isoforms of plasma membrane calcium ATPase. Cell Tissue Res 292:191–197
Gutiérrez-Martín Y, Martín-Romero FJ, Iniesta-Vaquera FA et al. (2004) Modulation of sarcoplasmic reticulum Ca^{2+}-ATPase by chronic and acute exposure to peroxynitrite. Eur J Biochem 271:2647–2657
Heiner I, Eisfeld J, Luckhoff A (2003) Role and regulation of TRP channels in neutrophil granulocytes. Cell Calcium 33:533–540
Hool LC, Corry B (2007) Redox control of calcium channels: from mechanisms to therapeutic opportunities. Antioxid Redox Signal 9:409–435
Hu Q, Corda S, Zweier JL et al. (1998) Hydrogen peroxide induces intracellular calcium oscillations in human aortic endothelial cells. Circulation 97:268–275
Ichas F, Jouaville LS, Mazat JP (1997) Mitochondria are excitable organelles capable of generating and conveying electrical and calcium signals. Cell 89:1145–1153
Ihara Y, Kageyama K, Kondo T (2005) Overexpression of calreticulin sensitizes SERCA2a to oxidative stress. Biochem Biophys Res Comm 329:1343–1349
Jeong S-Y, Seol D-W (2008) The role of mitochondria in apoptosis. BMB Rep 41:11–22
Jiang S, Chow SC, Nicotera P et al. (1994) Intracellular Ca^{2+} signals activate apoptosis in thymocytes: studies using the Ca^{2+}-ATPase inhibitor thapsigargin. Exp Cell Res 212:84–92
Klonowski-Stumpe H, Schreiber R, Grolik M et al. (1997) Effect of oxidative stress on cellular functions and cytosolic free calcium of rat pancreatic acinar cells. Am J Physiol 272:G1489–G1498
Korzets A, Chagnac A, Weinstein T et al. (1999) H_2O_2 induces DNA repair in mononuclear cells: evidence for association with cytosolic Ca^{2+} fluxes. J Lab Clin Med 133:362–369
Krippeit-Drews P, Haberland C, Fingerle J et al. (1995) Effects of H_2O_2 on membrane potential and $[Ca^{2+}]_i$ of cultured rat arterial smooth muscle cells. Biochem Biophys Res Commun 209: 139–145
Krippeit-Drews P, Krämer C, Welker S et al. (1999) Interference of H_2O_2 with stimulus-secretion coupling in mouse pancreatic beta-cells. J Physiol 514:471–481
Lajas AI, Pozo MJ, Camello PJ et al. (1999) Phenylarsine oxide evokes intracellular calcium increases and amylase secretion in isolated rat pancreatic acinar cells. Cell Signal 11:727–734
Lajas AI, Sierra V, Camello PJ et al. (2001) Vanadate inhibits the calcium extrusion in rat pancreatic acinar cells. Cell Signal 13:451–456
Lao Y, Chang DC (2008) Mobilization of Ca^{2+} from endoplasmic reticulum to mitochondria plays a positive role in the early stage of UV- or TNFα-induced apoptosis. Biochem Biophys Res Commun 373:42–47
Lopez JJ, Camello-Almaraz C, Pariente JA et al. (2005) Ca^{2+} accumulation into acidic organelles mediated by Ca^{2+}- and vacuolar H^+-ATPases in human platelets. Biochem J 390:243–252
Lückhoff A, Clapham DE (1992) Inositol 1,3,4,5-tetrakisphosphate activates an endothelial Ca^{2+}-permeable channel. Nature 355:356–358
Ma HT, Patterson RL, van Rossum DB et al. (2000) Requirement of the inositol trisphosphate receptor for activation of store-operated Ca^{2+} channels. Science 287:1647–1651
Manjarres IM, Chamero P, Domingo B et al. (2008) Red and green aequorins for simultaneous monitoring of Ca^{2+} signals from two different organelles. Pflugers Arch – Eur J Physiol 455:961–970
Matsuda T, Takuma K, Baba A (1997) Na^+-Ca^{2+} exchanger: physiology and pharmacology. Jpn J Pharmacol 74:1–20

McCleskey EW (1994) Calcium channels: cellular roles and molecular mechanisms. Curr Opinion Neurobiol 4:304–312
McConkey DJ, Orrenius S (1997) The role of calcium in the regulation of apoptosis. Biochem Biophys Res Commun 239:357–366
Meldolesi J, Pozzan T (1998) The endoplasmic reticulum Ca^{2+} store: a view from the lumen. Trends Biochem Sci 23:10–14
Meyer TN, Gloy J, Hug MJ et al. (1996) Hydrogen peroxide increases the intracellular calcium activity in rat mesangial cells in primary culture. Kidney Int 49:388–395
Moreau VH, Castilho RF, Ferreira ST et al. (1998) Oxidative damage to sarcoplasmic reticulum Ca^{2+}-ATPase AT submicromolar iron concentrations: evidence for metal-catalyzed oxidation. Free Radic Biol Med 25:554–560
Nicotera P, Orrenius S (1998) The role of calcium in apoptosis. Cell Calcium 23:173–180
Orrenius S, Zhivotovsky B, Nicotera P (2003) Regulation of cell death: the calcium-apoptosis link. Nat Rev Mol Cell Biol 4:552–565
Pan Z, Damron D, Nieminen AL et al. (2000) Depletion of intracellular Ca^{2+} by caffeine and ryanodine induces apoptosis of Chinese hamster ovary cells transfected with ryanodine receptor. J Biol Chem 275:19978–19984
Parekh AB (2003) Mitochondrial regulation of intracellular Ca^{2+} signaling: more than just simple Ca^{2+} buffers. News Physiol Sci 18:252–256
Parekh AB, Penner R (1997) Store depletion and calcium influx. Physiol Rev 77:901–930
Pariente JA, Camello C, Camello PJ et al. (2001) Release of calcium from mitochondrial and nonmitochondrial intracellular stores in mouse pancreatic acinar cells by hydrogen peroxide. J Membr Biol 179:27–35
Pariente JA, Lajas AI, Pozo MJ et al. (1999) Oxidizing effects of vanadate on calcium mobilization and amylase release in rat pancreatic acinar cells. Biochem Pharmacol 58:77–84
Patterson RL, van Rossum DB, Gill DL (1999) Store-operated Ca^{2+} entry: evidence for a secretion-like coupling model. Cell 98:487–499
Pedersen PL (2007) Transport ATPases into the year 2008: A brief overview related to types, structures, functions and roles in health and disease. J Bioenerg Biomembr 39:349–355
Pinton P, Pozzan T, Rizzuto R (1998) The Golgi apparatus is an inositol 1,4,5-trisphosphate-sensitive Ca^{2+} store, with functional properties distinct from those of the endoplasmic reticulum. EMBO J 17:5298–5308
Pozzan T, Rizzuto R, Volpe P et al. (1994) Molecular and cellular physiology of intracellular calcium stores. Physiol Rev 74:595–636
Pu Y, Luo KQ, Chang DC (2002) A Ca^{2+} signal is found upstream of cytochrome *c* release during apoptosis in HeLa cells. Biochem Biophys Res Commun 299:762–769
Redondo PC, Harper AG, Salido GM et al. (2004a) A role for SNAP-25 but not VAMPs in store-mediated Ca^{2+} entry in human platelets. J Physiol 558:99–109
Redondo PC, Jardín I, Hernández-Cruz JM et al. (2005) Hydrogen peroxide and peroxynitrite enhance Ca^{2+} mobilization and aggregation in platelets from type 2 diabetic patients. Biochem Biophys Res Comm 333:794–802
Redondo PC, Lajas AI, Salido GM et al. (2003) Evidence for secretion-like coupling involving pp60src in the activation and maintenance of store-mediated Ca^{2+} entry in mouse pancreatic acinar cells. Biochem J 370:255–263
Redondo PC, Salido GM, Pariente JA et al. (2004b) Dual effect of hydrogen peroxide on store-mediated calcium entry in human platelets. Biochem Pharmacol 67:1065–1076
Redondo PC, Salido GM, Rosado JA et al. (2004c) Effect of hydrogen peroxide on Ca^{2+} mobilisation in human platelets through sulphydryl oxidation dependent and independent mechanisms. Biochem Pharmacol 67:491–502
Rhee SG (2006) Cell signalling: H2O2, a necessary evil for cell signaling. Science 312:1882–1883
Rizzuto R, Pinton P, Ferrari D et al. (2003) Calcium and apoptosis: facts and hypotheses. Oncogene 22:8619–8627

Rooney TA, Renard DC, Sass EJ et al. (1991) Spatial and temporal organization of calcium signalling in hepatocytes. J Biol Chem 266:12272–12282
Rosado JA, Graves D, Sage SO (2000a) Tyrosine kinases activate store-mediated Ca^{2+} entry in human platelets through the reorganization of the actin cytoskeleton. Biochem J 351:429–437
Rosado JA, Jenner S, Sage SO (2000b) A role for the actin cytoskeleton in the initiation and maintenance of store-mediated calcium entry in human platelets. Evidence for conformational coupling. J Biol Chem 275:7527–7533
Rosado JA, Lopez JJ, Harper AG et al. (2004a) Two pathways for store-mediated calcium entry differentially dependent on the actin cytoskeleton in human platelets. J Biol Chem 279:29231–29235
Rosado JA, Redondo PC, Salido GM et al. (2004b) Hydrogen peroxide generation induces pp60src activation in human platelets: evidence for the involvement of this pathway in store-mediated calcium entry. J Biol Chem 279:1665–1675
Rosado JA, Sage SO (2000a) Protein kinase C activates non-capacitative calcium entry in human platelets. J Physiol 529:159–169
Rosado JA, Sage SO (2000b) Farnesylcysteine analogues inhibit store-regulated Ca^{2+} entry in human platelets: evidence for involvement of small GTP-binding proteins and actin cytoskeleton. Biochem J 347:183–192
Rosado JA, Sage SO (2000c) Regulation of plasma membrane Ca^{2+}-ATPase by small GTPases and phosphoinositides in human platelets. J Biol Chem 275:19529–19535
Roveri A, Coassin M, Maiorino M et al. (1992) Effect of hydrogen peroxide on calcium homeostasis in smooth muscle cells. Arch Biochem Biophys 297:265–270
Sage SO (1992) Three routes for receptor-mediated Ca^{2+} entry. Curr Biol 2:312–314
Sakaida I, Thomas AP, Farber JL (1991) Increases in cytosolic calcium ion concentration can be dissociated from the killing of cultured hepatocytes by tert-butyl hydroperoxide. J Biol Chem 266:717–722
Strehler EE, Zacharias DA (2001) Role of alternative splicing in generating isoform diversity among plasma membrane calcium pumps. Physiol Rev 81:21–50
Suzuki Y, Yoshimaru T, Matsui T et al. (2003) Fc epsilon RI signaling of mast cells activates intracellular production of hydrogen peroxide: role in the regulation of calcium signals. J Immunol 171:6119–6127
Thannickal VJ, Fanburg BL (2000) Reactive oxygen species in cell signalling. Am J Physiol 279:L1005–L1028
Thorn P, Brady P, Llopis J et al. (1992) Cytosolic Ca^{2+} spikes evoked by the thiol reagent thimerosal in both intact and internally perfused single pancreatic acinar cells. Pflügers Arch 422:173–178
Törnquist K, Vainio PJ, Björklund S et al. (2000) Hydrogen peroxide attenuates store-operated calcium entry and enhances calcium extrusion in thyroid FRTL-5 cells. Biochem J 351:47–56
Tsien RW, Lipscombe D, Madison D et al. (1995) Reflections on Ca^{2+}-channel diversity, 1988–1994. Trends Neurosci 18:52–54
Ueda N, Shah SV (1992) Role of intracellular calcium in hydrogen peroxide-induced renal tubular cell injury. Am J Physiol 263:F214–F221
Verkhratsky A (2007) Calcium and cell death. In: Carafoli E and Brini M (eds) Calcium signalling and disease. Subcellular biochemistry, vol 45. Springer, Berlin, Heidelberg, New York, pp 465–476
Villalobos C, Nuñez L, Montero M et al. (2002) Redistribution of Ca^{2+} among cytosol and organella during stimulation of bovine chromaffin cells. FASEB J 16:343–353
Volk T, Hensel M, Kox WJ (1997) Transient Ca^{2+} changes in endothelial cells induced by low doses of reactive oxygen species: role of hydrogen peroxide. Mol Cell Biochem 171:11–21
Votyakova TV, Reynolds IJ (2005) Ca^{2+}-induced permeabilization promotes free radical release from rat brain mitochondria with partially inhibited complex I. J Neurochem 93:526–537
Wang X (2001) The expanding role of mitochondria in apoptosis. Genes Dev 15:2922–2933

Wang X, Takeda S, Mochizuki S et al. (1999) Mechanisms of hydrogen peroxide-induced increase in intracellular calcium in cardiomyocytes. J Cardiovasc Pharmacol Ther 4:41–48

Weber H, Roesner JP, Nebe B et al. (1998) Increased cytosolic Ca^{2+} amplifies oxygen radical-induced alterations of the ultrastructure and the energy metabolism of isolated rat pancreatic acinar cells. Digestion 59:175–185

Whittemore ER, Loo DT, Watt JA et al. (1995) A detailed analysis of hydrogen peroxide-induced cell death in primary neuronal culture. Neuroscience 67:921–932

Wictome M, Henderson I, Lee AG et al. (1992) Mechanism of inhibition of the calcium pump of sarcoplasmic reticulum by thapsigargin. Biochem J 283:525–529

Wu KD, Lee WS, Wey J et al. (1995) Localization and quantification of endoplasmic reticulum Ca^{2+}-ATPase isoform transcripts. Am J Physiol 269:C775–C784

Xie Q, Zhang Y, Zhai C et al. (2002) Calcium influx factor from cytochrome P-450 metabolism and secretion-like coupling mechanisms for capacitative calcium entry in corneal endothelial cells. J Biol Chem 277:16559–16566

Xiong J, Camello PJ, Verkhratsky A et al. (2004) Mitochondrial polarisation status and $[Ca^{2+}]_i$ signalling in rat cerebellar granule neurones aged in vitro. Neurobiol Aging 25:349–359

Yano K, Petersen OH, Tepikin AV (2004) Dual sensitivity of sarcoplasmic Ca^{2+}-ATPase to cytosolic and endoplasmic reticulum Ca^{2+} as a mechanism of modulating cytosolic Ca^{2+} oscillations. Biochem J 383:353–360

Yao Y, Ferrer-Montiel AV, Montal M et al. (1999) Activation of store-operated Ca^{2+} current in Xenopus oocytes requires SNAP-25 but not a diffusible messenger. Cell 98:475–485

Yoo SH (2000) Coupling of the IP3 receptor/Ca^{2+} channel with Ca^{2+} storage proteins chromogranins A and B in secretory granules. Trends Neurosci 23:424–428

Zaidi A, Michaelis ML (1999) Effects of reactive oxygen species on brain synaptic plasma membrane Ca^{2+}-ATPase. Free Radic Biol Med 27:810–821

Chapter 2
Apoptosis

M.L. Campo

Abstract Proliferation and cell death are the two sides of the same coin, rimmed by cellular homeostasis. The regulatory networks controlling the life and death decisions on the cellular level are more complex than we previously thought. The strict regulation of responses to external stimuli maintains tuned the signalling cascades, while unbalance is involved in a number of pathological conditions, ranging from neurodegeneration to neoplasic transformation. Apoptosis is a well-conserved physiological pathway whose basic tenets appear common to all metazoans. Key components regulate the commitment step and/or participate in effecting cell demise. Two main trails lead to apoptosis: the death receptor or extrinsic pathway and the mitochondrial or intrinsic pathway. The later is a rapid and strong way to execute the process. Breaches of mitochondria integrity result in the release of proapoptotic factors like cytochrome *c*. Tough this research area is rapidly developing many issues remain shrouded in uncertainties. The relationship between both mitochondrial membranes is uncertain and controversial. Large pores are involved, though their possible interplay is unclear. Recently the work on mitochondrial cristae remodelling has elucidated a novel checkpoint for apoptosis, which may determine sensitivity to apoptosis *in vivo*, during adult animal life. These issues will be documented in the present chapter.

Keywords Apoptosis · Death receptors · Caspases · Mitochondria · Bcl-2 proteins

2.1 Introduction: Living with Death

That life and death go together, require one another, and therefore are the two sides of the same coin, are not new insights. Cell biology and biochemistry reinforce this lopsided view, and continuously show us that the equilibrium of the disorder

M.L. Campo (✉)
Department of Biochemistry and Molecular Biology and Genetics, University of Extremadura, Avda Universidad s/n, 10071 Cáceres, Spain
e-mail: mlcampo@unex.es

G.M. Salido, J.A. Rosado (eds.), *Apoptosis: Involvement of Oxidative Stress and Intracellular Ca^{2+} Homeostasis*, DOI 10.1007/978-1-4020-9873-4_2,

in an organism i.e. life, can be easily inclined towards its own collapse i.e. death. Cells may die because they become old and defective, because they are surplus to the requirements of the tissue, or because they incur some damage. The certainty is that all cells are actually programmed to die. Yet cells are kept alive by a battery of signals that prevent them from executing their dying programs. As a result, cells survive, proliferate, differentiate, and perform their functions. These crucial surviving signals come from a variety of intra and extra cellular sources and may be derived from physical contact or from secreted factors. When such signals become blocked or absent or if other overriding signals are activated, cells undergo their predetermined death. In this scenario, life is anything but the absence of death. The paradox comes out when we realize that at the same time life has learned to cope with death, constantly relies on it.

As consequence, the clockworks for orderly growth and orderly death are very much alike; afterwards controlled death plays a pivotal role in the development and growth of complex organisms. Large number of cells die during embryonic development, e.g. during the sculpting processes that shape organs (Penaloza et al. 2006). Also throughout the life span of adult organisms, cells die on a vast scale to counterbalance cell division, in order to provide the body with the cells required at different stages, or to eliminate old, damaged or harmful cells in homeostatic cell turnover, e.g. activated immune system. Importantly, the unbalance between cell division and cell death leads to developmental abnormalities, degenerative diseases, or neoplasic transformations.

2.2 Cell's Ways to Die

2.2.1 Necrosis

Traditionally two distinct types of dying cells, characterized by their different appearance, have been described (Galluzzi et al. 2007). Pathogens and severe traumas e.g. burn, cuts, or compression injuries, may provoke necrotic cell death. During the necrosis type of death, overwhelming stress becomes incompatible with cell survival. In this case, clumps of cells in a tissue simultaneously swell. There is no evidence of continued metabolic activity. Nuclear DNA condenses in an irregular fashion, particularly at the margins, and the cellular constituents start to disintegrate rapidly in a random and uncontrolled fashion. Vital internal constituents quickly leak out of the cell, triggering inflammation by cells of the innate immune system. Recent evidence suggests that the inflammatory response is provoked by the release of a full spectrum of molecules collectively called alarmins, whose specific identity still awaits precise definition (Oppenheim and Yang 2005). What they share is common is their ability to stimulate pattern-recognition receptors on macrophages, dendritic cells and natural killer cells (Zitvogel and Kroemer 2008). This equips cells of the innate immune system with the ability to activate T cells and to initiate immune responses (Chen et al. 2007; Trinchieri and Sher 2007). The system is thus flooded with inflammatory lymphocytes to stop infection and damage to the tissue is being induced. At this point cellular debris is engulfed and removed by macrophages.

2.2.2 Apoptosis

The second process of death was coined apoptosis (in Greek the falling away of leaves of deciduous plants) and goes back to 1972, when a newly defined type of cell death was formulated by Kerr, Wyllie, and Currie, (Kerr 2002; Kerr et al. 1972). Originally it was view as being non-degenerative in nature, though an impressive body of ensuing research has proven otherwise (Hail et al. 2006). Apoptosis is not just an ordered and active process, but also a silent process in the way that dismantles the cell but does not indiscriminately propagate to the surrounding cells. At the cellular level, it is characterized by an initial shrinkage of the cells and the ensuing break of the cell-to-cell contacts. The cells round up and their internal membranes and organelles get more concentrated in the cytoplasm, which then looks darker. Notably the organelles remain intact and normal looking very late along the process, suggesting metabolic activity remains substantial for some considerable time. Cytoplasmic constituents do not leak from the cell; therefore, an inflammatory response does not follow. In the nucleus, chromatin condensation is extreme and often generates crescent shaped areas contouring the nuclear membrane or pycnosis. This highly distinctive event is not seen under any other circumstances. Endonucleases precisely cleave the DNA between nucleosomes, giving fragments of 180 (or multiples) base pares (Williams et al. 1974). Although less noticeable, other intracellular networks like the Golgi, endoplasmic reticulum and mitochondria also suffer substantial fragmentation (Frank et al. 2001; Lane et al. 2002). Whilst the process of DNA cleavage is going on, the nucleus begins to break into fragments and the cell likewise splits into a number of small intact pieces or apoptotic bodies that exclude vital dyes. Cytoplasmic budding and fragmentation is followed by phagocytic removal, a process in which migrating macrophages or healthy neighbouring epithelial cells engulf cell's fragments. This event is particularly remarkable when it is considered that phagocytes are normally engaged in recognizing and removing foreign or "non-self" entities. As the result, apoptotic bodies are enclosed in a membrane-bound vesicle in a cell called phagosome. Ultimately, the host cell or phagosome and its contents are gradually degraded and, in many cases, a new cell takes the place of the old one in a matter of hours. In some cellular systems, especially in cell culture, the time-lapse sequence just outlined may not rigorously trail all the steps. It is not unusual that the chromatin condensation results in a single dense spherical ball at one end of the original nucleus, or small cells like thymocytes and neutrophils remain a single apoptotic body. Also in contrast to *in vivo* situations, cells undergoing apoptosis *in vitro* eventually lose plasma membrane integrity. In the absence of phagocytes, the apoptotic signals (e.g. the externalization of phosphatidylserine on their plasma membrane) that encourage their expeditious removal by these cells could not take place.

2.2.3 Autophagy

Besides necrosis and apoptosis, we should introduce here a third significant player in cell death decisions, whose importance is growing very rapidly (Klionsky 2007). If apoptosis is crucial in development and normal physiology as well as in a wide range of diseases, so is autophagy (Cuervo 2004; Hara et al. 2006; Karantza-Wadsworth et al. 2007). Autophagy, or more precisely macroautophagy, is a degradative process involving sequestration of parts of the cytoplasm in double-membrane vesicles called autophagic vesicles or autophagosomes, that fuse with lysosomes forming the autophagolysosome (Thorburn 2008). In the autophagosome, the engulfed cytoplasmic material is hydrolyzed and the resulting amino acids and other macromolecular precursors can be recycled. This limited self-eating is a mechanism to provide cells with metabolic substrates to meet their energetic demands under stressful conditions, such as nutrient deprivation. It is also the primary degradation mechanism for worn out proteins and favours the selective elimination of damaged (and potentially dangerous) organelles, thus maintaining their quality control. In these instances, macroautophagy operates as a pro-survival mechanism. Though it has been described that macroautophagy can protect cells by preventing them for undergoing apoptosis (Lum et al. 2005), things are not so straightforward, and its role is currently very much in debate, as it can also do the opposite i.e. it can kill the cells (Wang et al. 2008). In fact autophagic cell death is also known as type II programmed cell death to distinguish it from apoptosis or type I programmed cell death (Tsujimoto and Shimizu 2005). Significantly, one of the best examples where the demise of the cell was shown to require the autophagic machinery occurs in cells with profound defects in the apoptosis machinery (Shimizu et al. 2004). Also, autophagic cell death can be caused by increased reactive oxygen species (ROS) resulting from autophagic degradation of catalase (Yu et al. 2006). The characteristic cellular morphology associated with apoptosis is mainly due to the early degradation of structural proteins like the cytoskeleton, whilst organelles are preserved until late in the process. In contrast, during autophagic cell death, the large accumulation of autophagic vesicles is associated with early degradation of organelles, while the cytoskeleton remains intact and functional until late in the process (Cuervo 2004).

Regardless of the substantial morphological differences between apoptosis and autophagy (Galluzzi et al. 2007), increasing evidences indicate that macroautophagy engages in a complex relationship with apoptosis, and that close connections exist between the specific stimuli and the regulatory machinery controlling both of these processes. That is, autophagy regulators can control apoptosis and *vice versa* (Wang et al. 2007; Levine et al. 2008; Thorburn 2008). Thus although the morphological studies would lead us to conclude that apoptosis and macroautophagy are two distinct ways of dying, perhaps they are just two different facets of the same integrated cell death mechanism. Even more, many examples begin to show that there are not really two different programmed cell death mechanisms (apoptosis and autophagy) and a third less regulated mechanism (necrosis). Instead, cell death can be regarded as a one integrated mechanism is which the three processes work together and regulate each other in a continuum, with not clear cut distinction among them (Thorburn 2008). Within this framework, even if a commitment to die is not the same as execution, a cell that has passed the commitment stage but is blocked from undergoing apoptosis, will die by another route (Zakeri and Lockshin 2008). In addition, what we see when cell demise is studied depends on which facet of the mechanism is most prominent at each particular time. Nevertheless, assuming this double distinction between infringed death i.e. necrosis, or committed death i.e. apoptosis and autophagy, we can assert that while infringed death is a chaotic and in a way a passive process, committed death is an ordered, programmed and in any case active process that at the molecular level looks remarkably like life.

2.3 Cell's Fate: Shooting off Apoptosis

It is now clear that healthy cells and cells that have been exposed to a damaging agent have genetic and molecular programs to receive instructions (signals) and to respond to these signals. The response may be a sequence of events that lead to cell cycle arrest and repair or alternatively to apoptosis. Cells can be infringed with several genotoxic and non-genotoxic insults that result in a rapid increase of p53. In fact, every type of DNA damage (i.e. radiation, reaction with oxidative free radicals, drugs, virus infection, etc.) is reported to the p53 protein and its pathway. In addition damage to components involved in the proper handling of the genetic material (such as the mitotic spindle), hypoxia, oncogenic activation, ribonucleotide depletion or exposure to nitric oxide, also increase the cellular levels of this protein (Pluquet and Hainaut 2001). Hence, the p53 pathway involves hundreds of genes and their products that respond to a wide variety of stress signals and p53 is arguably the most intensively studied protein today. It plays a key role in monitoring the genetic fidelity of cells, for which it has been appointed as the "guardian of the genome". Its response to DNA-damage or checkpoint failure gives rise to a series of anti-proliferative responses. The importance of p53 in the detection of genetic defects in the early pre and post implantation of mammalian embryos has been demonstrated, as it induces cells with teratogenic damage to die via apoptosis and the defective embryos to abort. Thus, one of the most important functions of p53 is its ability to induce apoptosis, while disruption of this route can promote tumour progression and chemo resistance (Soussi 2005; Meulmeester and Jochemsen 2008). The transcriptional program that p53 switches on is remarkably flexible, as it varies with the nature of the activating stimuli, the cell type and the duration of the activation signal. This flexibility may allow cells to mount alternative responses to p53 activation, such as cell cycle arrest or apoptosis (Espinosa 2008). How p53 is involved in making this decision remains uncertain (Levine et al. 2006). Though, for an individual cell, the choice may be governed by the degree to which p53 levels are raised in response to DNA damage, with low levels of p53 turning on repair genes and higher levels being required to turn on apoptosis genes or turn off survival genes. Besides its ability to promote apoptosis through transcription dependent mechanisms, p53 may also be able to activate apoptosis independent of transcriptional regulation (Fuster et al. 2007; Palacios et al. 2008).

To ensure normal cell growth, p53 levels and activity are tightly regulated (Meulmeester and Jochemsen 2008; Kuribayashi and El-Deiry 2008). In most cells and under normal conditions, p53 is ubiquitously present at extremely low or undetectable levels. The protein has a very short life (6–30 min) and its levels are predominately regulated by a quick proteolytic turn over. Upon diverse forms of cellular stress (Pluquet and Hainaut 2001) the steady state levels and transcriptional activity of p53 are rapidly (within two hours), and considerably increased. The stabilization and activation of p53 are a result of hindered inhibition by its negative regulators, e.g. Mdmx (also known as Mdm4) and Mdm2, while other activators such as HIPK2 and DYRK2 could enhance the p53 response. Once p53 is increased and activated, it gains the ability to bind to p53-responsive DNA sequence elements in the genome. Transcription of more than 150 genes are positive or negatively regulated by p53. There appear to be some p53-regulated genes that are transcribed in response to many different types of stress signals and in all tissues responding to the stress (e.g. MDM-2, GADD-45, or those coding for p21 and cyclin G), while others are either stress- or tissue specific (e.g. PTEN, TSC-2). What regulates these differences also remains unclear (Levine et al. 2006). The functions of the p53 response genes fall into several categories. A set of genes and their products are clearly involved in cell cycle arrest and cellular senescence (e.g. p21, 14-3-3 sigma, GADD-45). In cells beyond repair, a second set of p53-regulated genes involved in apoptosis are implicated. As we will discuss below, both the extrinsic and the intrinsic pathways of apoptosis are stimulated. In the extrinsic pathway p53 regulates Fas production (a secreted protein), as well as DR5/killer, the trail receptor and a membrane protein. The intrinsic pathway is populated with many p53-regulated genes of which the proteins Bax, Noxa and Puma may work in different cell types. Now the main concern is to comprehend the details of how p53 initiates apoptosis employing its transcriptional program. Although a large number of genes regulated by p53 during induction of apoptosis are known, no single target gene has been identified whose altered expression alone can sufficiently explain p53 apoptosis, and whose genetic deficiency phenocopies p53 deficiency *in vivo* (Moll et al. 2005). As an additional mode, and besides its classic transcription-dependent activities, evidence is mounting that transcription-independent activities of p53 are also important for its proapoptotic function. It has been demonstrated that a fraction of induced p53 rapidly moves out of the nucleus and promotes apoptosis upon its interaction with anti and proapoptotic members of the B-cell CLL/lymphoma (Bcl) family of mitochondrial permeability regulators. Several research groups have now confirmed the existence of a direct p53-mediated mitochondrial death (Fuster et al. 2007; Marchenko and Moll 2007; Palacios et al. 2008; Wolff et al. 2008). After all, understanding the rules that regulate the p53-mediated output is nowadays a high priority for this research filed.

2.4 Cell's Annihilation: Apoptosis Coming off

Once apoptosis is triggered, the effectors' phase of cell's annihilation is in nearly all cases a no return pathway. The morphological transformations indicative of apoptosis are endorsed by what takes place during this phase, and its interruption merely deregulates the process, but does not confer cells survival. Studies conducted over the past two decades have revealed a complex web of molecules that execute and regulate apoptosis. Interestingly, during recent years, a paradigm shift has occurred as it becomes clear that proteins involved in apoptosis also exhibit cell death-unrelated functions (Galluzzi et al. 2008). Indeed, it seems plausible that evolution has not "invented" proapoptotic factors *ex nihilo* but rather has "appropriated" molecules that already had a function in vital processes (such as adaptation to stress) into the service of programmed cell death.

At the proteome level, most of the ultrastructural features of apoptosis are the consequence of hundreds of proteins undergoing restricted proteolysis during the so-called effector phase. However, up-to-date the relevance of the majority of these proteolytic events remains unclear. A series of cystein proteases from the caspase family mediate the cleavage of these many proteins (Lüthi and Martin 2007). In fact, nearly all stimuli that trigger apoptosis seem to do so by initiating events that culminate in caspase activation, albeit in somewhat different ways. In mammals, three main routs to apoptosis-associated caspase activation have been firmly established (Danial and Korsmeyer 2004). These include the extrinsic or death-receptor mediated pathway, the intrinsic or mitochondrial-mediated pathway, and the granzyme B pathway. In addition, at least two more routes have been proposed but they are yet to be well characterized.

2.4.1 The Extrinsic Pathway

The extrinsic pathway is initiated by extracellular stimuli, such as withdrawal of growth factors that incite the engagement of transmembrane death receptors (a subset of the TNF receptor family, including TNFR1, Fas/CD95, the TRAIL receptors-1 and -2, and probably de death receptor 3) with their cognate ligands (Fig. 2.1). This provokes the direct activation of these receptors through trimerization and assembly of a large death-inducing signalling complex (DISC) on the cytoplasmic side of the plasma membrane (Debatin and Krammer 2004). Although the constituents of DISC have not been fully identified, one adapter protein, the Fas-associated death domain or FADD, appears to be the obligated component, which recruits and mediates the auto-activation of the initiator caspase, procaspase 8 (and possibly of caspase 10). Active caspase 8 then proteolytically cleaves and activates caspases 3, 6 and 7 (Fuentes-Prior and Salvesen 2004), leading to further caspase activation events that culminate in substrate proteolysis and cell death. As an alternative, the extrinsic pathway can be activated by the so-called dependency receptors, which deliver a death signal in the absence of their ligands, through yet unidentified mediators.

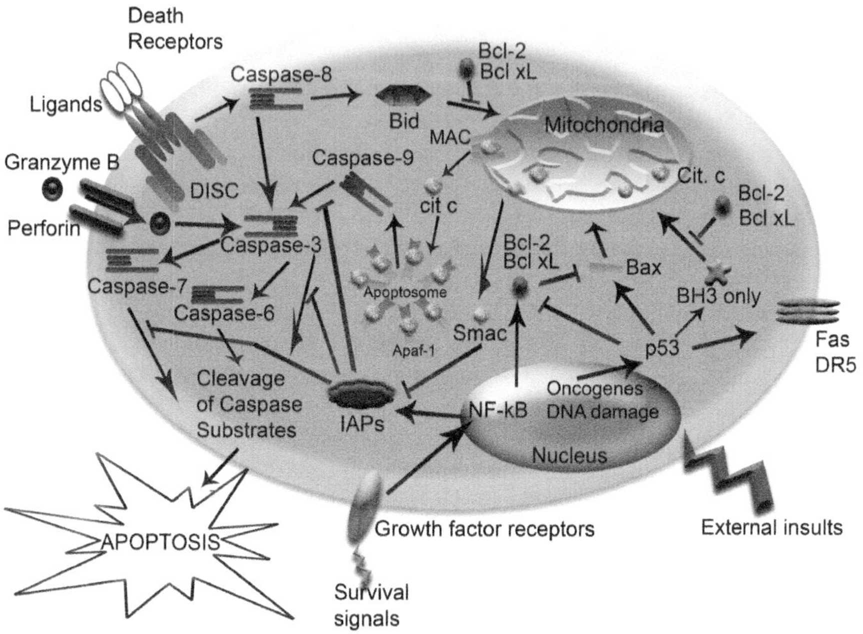

Fig. 2.1 Schematic view of the apoptosis intrinsic and extrinsic pathways. In the extrinsic pathway, engagement of death receptors with their cognate ligands provokes the recruitment of adaptor proteins, such as the Fas associated death domain protein (FADD), which recruits and aggregate several molecules of caspase 8. The death-inducing signalling complex (DISC) thus formed in the cytoplasmic face of the plasma membrane, promotes the auto processing and activation of procaspase 8 (and possibly of caspase 10), which in turn is able to cleave effector caspase 3, 6, and 7. Caspase 8 can also proteolytically activate Bid. Truncated Bid represents the main link between the extrinsic and intrinsic apoptotic pathways, because it favours the aggregation and insertion of Bax and Bak into the mitochondrial outer membrane. The resulting apoptosis induced channel (MAC) promotes mitochondrial membrane permeabilization. Mitochondrial outer membrane permeabilization is the hallmark of the intrinsic pathway. In this pathway, several intracellular signals, including DNA damage and endoplasmic reticulum (ER) stress, converge on mitochondria to induce outer membrane permeabilization, which causes the release of proapoptotic factors from the intermembrane space. Among these, Cyt *c* induces the apoptosis protease-activating factor 1 (APAF-1) and ATP/dATP to assemble the apoptosome, a molecular platform which promotes the proteolytic maturation of caspase 9. Active caspase 9, in turn, cleaves and activates the effector caspases, which finally lead to the apoptotic phenotype. Second mitochondria-derived activator of caspase/direct IAP binding protein with a low pI (Smac/DIABLO) and the Omi stress-regulated endoprotease/high temperature requirement protein A2 (Omi/HtrA2), promote apoptosis indirectly, by binding to and antagonizing members of the IAP (inhibitor of apoptosis protein) family. The granzyme B-dependent route of caspase activation involves the delivery of this protease into the target cell through specialized granules that are released from cytotoxic T lymphocytes or natural killer cells. These granules also contain the pore forming protein perforin, which permits the entry of granzymes. Granzyme B can process Bid as well as caspase 3 and 7 to initiate apoptosis (See also Plate 2 in the Color Plate Section on page 224)

In the absence of ligand, these receptors trigger cell death, thus generating a state of cellular dependence from their ligands. The prototype dependency receptors are the netrin-1 receptors DCC (deleted in colorectal cancer) and UNC5H-1, -2 and -3 (Mehlen and Bredesen 2004).

2.4.2 The Intrinsic Pathway

The intrinsic pathway relies on mitochondrial membrane permeabilization (MMP) to liberate pro-apoptotic factors like cytochrome *c* from the intermembrane space of this organelle (Kroemer et al. 2007) (Fig. 2.1). For instance, during conditions of cell stress, antiapoptotoic members of the Bcl-2 family of proteins (e.g., Bcl-2 and Bcl-xL), residing in the outer mitochondrial membrane, can be destabilized through decreased expression, or by the induction of pro-apoptotic Bcl-2 family members (e.g. Bax, Bad, and Bak). The ratio of pro-apoptotic to anti-apoptotic family members becomes greater. This promotes the assembly of Bax-Bak oligomers within the mitochondrial outer membranes. These oligomers compromise the permeability of this membrane, allowing the formation of proteinaceous outer membrane channels (Pavlov et al. 2001). As consequence, pro-apoptotic factors from the inter-membrane space of mitochondria are release into the cytoplasm. Among these, cytochrome *c* (Cyt *c*) directly activates the apoptosis protease-activating factor (Apaf-1) and, in the presence of dATP or ATP, seeds the formation of a large multimeric complex, the "apoptosome". The apoptosome recruits and mediates the auto-activation of the initiator caspase, procaspase 9, which goes on to activate caspases 3, 6 and 7 (Fuentes-Prior and Salvesen 2004).

Bcl-2 family members are not the only agents capable to promote the MMP needed in order to release pro-apoptotic factors from the intermembrane space of mitochondria. Proteases (Johnson 2000; Suzuki et al. 2004) as well as other agents (e.g. ceramide, reactive oxygen species (ROS), and Ca^{2+}) can also accomplish this function (Kroemer et al. 2007). MMP can even commit a cell to die when caspases are not activated. This caspase-independent death (Chipuk and Green 2005; Kroemer and Martin 2005) can occur because of an irreversible loss of mitochondrial function as well as because of the mitochondrial release of caspase-independent death effectors including apoptosis inducing factor (AIF) (Susin et al. 1999) and endonuclease G (Marchetti et al. 1996). Both of them translocate from the cytosol to the nuclear compartment where they favour DNA fragmentation and chromatin condensation.

At the molecular level, the cross talk between the extrinsic and intrinsic mechanisms of apoptosis takes place at the activation of caspases 3, 6 and 7 induced by both pathways (Fig. 2.1). In addition, one physiological target of the activated caspase 8 is Bid, a BH3-only Bcl-2 relative protein, which lacks a transmembrane region (see below). In some cells, in response to death receptor ligands, caspase 8 induces the cleavage of Bid to yield a truncated carboxy-terminal fragment (tBid) that translocates from the cytosol to the outer mitochondrial membrane. Oligomers of tBid can also trigger mitochondrial outer membrane permeabilization followed by the release of pro-apoptotic factors. In this way, Bid also mediates the cross talk between the extrinsic and the intrinsic forms of apoptosis. At the death receptor level, the synergy between the extrinsic and intrinsic pathways has been extensively investigated. Whereas in some type of cells (named type 1), the binding of ligands to death receptors causes the activation of effector caspases, without the necessity of MMP, in others (type 2) the complex signalling cascade outlined above (caspase 8 activation → cleavage and activation of Bid → MMP → Cyt *c*-dependent caspase 3 activation) critically depends on MMP (Krammer 2000; Krammer et al. 2007).

2.4.3 The Granzyme B Pathway

The granzyme B-dependent route of caspase activation involves the delivery of this protease into the target cell through specialized granules that are released from cytotoxic T lymphocytes (CTL) or natural killer (NK) cells. CTL and NK granules contain numerous granzymes as well as a pore forming protein, perforin. When CTL or NK cells are activated by their antigen receptor, the lytic granules move to cluster and align along the immunological synapse (Bossi and Griffiths 2005). The granule membrane fuses with the killer cell plasma membrane, releasing its contents, including perforin and granzymes, into the synapse. Granzymes bind to the target cell membrane by electrostatic interactions (granzymes are very positively charged with pIs between 9–11, and the cell surface is negatively charged) (Kurschus et al. 2005), and also by specific receptors, such as the cation-independent mannose-6-phosphate receptor (Motyka et al. 2000). However, specific receptors are not required for binding and cytotoxicity (Kurschus et al. 2005). Entry of granzymes into the target cell cytosol is generally mediated by perforin, but how perforin accomplishes this is still unclear (Pipkin and Lieberman 2007). The original model of granzyme entry through plasma membrane pores formed by perforin is generally no longer considered valid. A revised model posits that perforin makes microscopic holes in the plasma membrane that cause a Ca^{2+} influx, which triggers a cellular plasma membrane response and rapid endocytosis of granzymes and anything else bound to the cell surface (Keefe et al. 2005). In fact, entry is dynamic-dependent (Veugelers et al. 2004) and results in the formation of giant endosomes containing both granzymes and perforin (Keefe et al. 2005). Within minutes, the granzymes escape (through perforin pores in the endosome?) and find their way into the cytosol. According to some models, granzymes are primarily translocated through repairable membrane pores of finite size and not by the disruption of endocytosed vesicles (Kurschus et al. 2008).

Recent work suggests that this homologous family of serine esterases can activate at least three distinct pathways of cell death (Chowdhury and Lieberman 2008). Granzyme B activates the caspase apoptotic pathway by cleaving caspase 3 (Adrain et al. 2005). However, there is good evidence that it can also activate other pathways of cell death (particularly in the mitochondrion and the nucleus) that remain to be worked out. Thus, human granzyme B, but not the mouse enzyme, activates cell death by directly cleaving the key caspase substrates, Bid and the inhibitor of caspase activated DNase (ICAD), to activate the same mitochondrial and DNA damage pathways, respectively, as the caspases (Casciola-Rosen et al. 2007; Cullen et al. 2007). As consequence, caspase inhibitors have little effect on human granzyme B-mediated cell death and DNA fragmentation, whereas the same inhibitors significantly block the action of the mouse enzyme. The granzyme B (and caspase) mitochondrial pathway leads to ROS generation, loss of MMP, with release of cytochrome *c* and other proapoptotic molecules from the mitochondrial intermembrane space. Human granzyme B activates this pathway directly by cleaving Bid, whereas mouse enzyme activates it indirectly. Other important target of granzymes is the nucleus. Granzymes A and B rapidly translocate to the nucleus (Jans et al. 1998), where proteolytic cleavage of key substrates is important to induce programmed cell death by both them. Nuclear translocation of the granzymes may be mediated by importin-α (Blink et al. 2005). We are just beginning to understand how granzymes, other than B, activate cell death, as laboratories have begun to express active recombinant forms of many of these enzymes. Granzyme A activates cell death that has all the morphological features of apoptosis but is completely caspase-independent and involves novel mitochondrial and DNA damage pathways that are just beginning to be unravelled (Martinvalet et al. 2008). In addition, granzyme C (in mouse) and granzyme H (in humans) also activate caspase-independent cell death with a pronounced mitochondrial phenotype. There is some evidence that granzyme M may activate autophagy (Chowdhury and Lieberman 2008).

2.4.4 Other Caspase Dependent and Independent Pathways

Others less defined pathways of apoptosis are now being investigated. For quite some time two more non-caspase proteases, besides granzymes, have been implicated as effectors of apoptosis (Johnson 2000). The cathepsin family consists of cysteine, aspartate, and serine proteases. Cathepsin B and cathepsin L, both cysteine proteases, and cathepsin D, an aspartate protease, are most frequently linked to apoptosis (Johnson 2000; Turk et al. 2001). These proteases are localized in lysosomes and/or endosomes, but they translocate to the cytoplasm during apoptosis (Johnson 2000; Mathiasen and Jäättelä 2002). Cathepsins can cleave a number of substrates including Bcl-2 family members, p53, cyclin D, c-Fos, and c-Jun.30 (Turk et al. 2001). Furthermore, cathepsin activity is reportedly associated with MMP (Boya et al. 2003), chromatin condensation (Zang et al. 2001), the degradation of the intracellular matrix, (Johnson 2000), the processing of procaspases, (Yamashima 2004) and the externalization of phosphatidyl serine (PS) on the plasma membrane of apoptotic cells (Johnson 2000; Boya et al. 2003). One mechanism recently implicated in cathepsin-mediated MMP and caspase-independent apoptosis involves the novel lysosome-associated apoptosis inducing protein LAPF, which promotes lysosomal membrane permeabilization and cathepsin release in fibrosarcoma cells (Chen et al. 2005; Li et al. 2007).

The calpain family of cysteine proteases resides in the cytosol. Both μ-calpain and m-calpain have been linked to apoptotic processes, (Johnson 2000; Takano et al. 2005) and certain human diseases that are marked by excessive cell loss (e.g., Alzheimer's and Parkinson's disease) are directly linked to aberrant calpain activity. Calpains are activated by anomalous increases in intracellular free Ca^{2+} concentration (Johnson 2000). While some studies suggest that caspase activity is absent during calpain-mediated apoptosis (Takano et al. 2005), there are others that imply enhanced proteolytic activity during apoptosis via a feed-forward amplification loop that involves caspases (Wood and Newcomb 1999). Furthermore, a "calpain-cathepsin cascade" (Yamashima 2004) has also been proposed to integrate and enhance proteolytic activity during apoptosis.

Finally, it has been described that DNA damage may also signal through the activation of caspase 2. The molecular mechanism of procaspase 2 activation in the course of apoptosis remains poorly defined. Some reports suggest that caspase 2 is implicated in cytochrome *c* release and is essential for drug-induced apoptosis in several human cell lines (Guo et al. 2002; Robertson et al. 2002). Based on these data caspase 2 was proposed to be the initiator caspase in cytotoxic stress-induced apoptosis acting upstream of mitochondria. At the same time, others suggested that activation of caspase 2 occurs downstream of mitochondria (O'Reilly et al. 2002). In addition, caspase 2 was revealed to be a proximal mediator in the heat-shock induced apoptosis (Tu et al. 2006). The role of caspase 2 in death receptor signalling is also contradictory. Caspase 2 was shown to be processed in the course of death receptor–mediated apoptosis (Juo et al. 1999). It was reported that this activation is mediated by caspase 3 and, accordingly, processing of caspase 2 takes place after activation of caspase 3. In contradiction to these data, caspase 2 was shown to be necessary for optimal death receptor-mediated cleavage of Bid (Wagner et al. 2004). In addition, it has been reported that caspase 2 is activated at the DISC but does not play a role in initiating death induced apoptosis (Lavrik et al. 2006).

There are multiple pathways by which apoptosis can be accomplished and, in addition to caspases, we have sketched several examples of effector mechanisms that promote apoptotic signatures in dying cells. Perhaps the initial paradigm of caspases as the hub and only effectors of apoptosis is merely the consequence of the intensive research carried out over the past 15 years on these particular proteases. As more and more actors are coming into play, new effector mechanisms start gaining relevance. At present, these new caspase-independent mechanisms await a more thoroughgoing characterization.

2.5 Cell's Weaponry at Apoptosis

Understanding in detail the workings of the roads to cell death is an excellent way to comprehend important mechanisms in normal cellular metabolism and some of the strategies cells have at their disposal to deal with outside stress. Unavoidably, research in this area goes through the study and minute characterization of the intermediaries involved in each one of these roads. They are the true coordinators and executioners of cell's dismantling, and constitute the cell's weaponry at the apoptosis battle. Based on the available literature a brief review and discussion of their structure, physiological role, molecular mechanisms, regulation and cross talk are sketched, in relation to that of mammalian species.

2.5.1 Caspases: The Swords

The death machinery can be activated by diverse stimuli, and as central components has a group of highly conserved proteolytic enzymes called caspases (cyteinyl aspartate-specific proteases). Significantly, and irrespective to the actual route to apoptosis, many pathways converge at the activation of some of the major effector caspases. Caspases are like the cell's contingent of slender swords that in turn carry out much of the observed proteolysis and produce the alterations that we recognize as the apoptotic phenotype.

2.5.1.1 Structure and Activation

The caspases form a caspase-cascade system that plays the central role in the induction, transduction and amplification of intracellular apoptotic signals for cell fate determination, regulation of immunity, and cellular proliferation and differentiation. So far, 15 mammalian caspases have been reported (Degterev et al. 2003; Eckhart et al. 2005). They are grouped into two major sub-families, namely inflammatory caspases (comprising of caspases 1, 4, 5, 11, 12, 13, and 14) and apoptosis related caspases (comprising of caspase 2 (Bergeron et al. 1998), caspases 3, 6, 7, 8 (Fernandes et al. 1996), caspase 9 (Duan et al. 1996), caspase 10, and caspase 15 (Eckhart et al. 2005; Eckhart et al. 2006)). The apoptotic caspases are further subdivided into two sub-groups, initiator caspases and executioner caspases. Caspases 2, 8, 9, and 10 trigger apoptosis and are known as upstream or initiator caspases. They activate the executioner or downstream caspases comprising of caspases 3, 6, 7 and 15, which ultimately execute apoptotic cell death (Alnemri et al. 1996; Rupinder et al. 2007).

Caspases use a cysteine residue as the catalytic nucleophile and share a stringent specificity for cleaving their substrates after aspartic acid residues in target proteins (Alnemri et al. 1996). Structurally, their signature motif is a highly homologous protease domain. This domain can be further subdivided into two subunits, a large subunit of approximately 20 kDa and a small subunit of approximately 10 kDa (Fuentes-Prior and Salvesen 2004). Procaspases also contain a prodomain or amino terminal peptide of variable length. The initiator apoptotic caspases have long prodomains of over 100 amino acids, while the prodomains of effector caspases are usually less than 30 amino acids. The long prodomains contain a distinct motif, notably the death domain (DD). The DD is a member of TNF-receptor family (Creagh and Martin 2001; Chowdhury et al. 2008) and has two subdomains: the death effector domain (DED) and the caspase-recruitment domain (CARD) that recruits caspases to the plasma membrane before activation. Procaspases-8 and -10 possess two tandem

DEDs in their prodomain (Sprick et al. 2002). Among apoptosis related caspases, the CARD is found in procaspases 2 and 9 (Fuentes-Prior and Salvesen 2004). DED and CARD comprise 6–7 antiparallel amphipathic α-helices that interact with other proteins by electrostatic or hydrophobic interactions (Hofmann 1999), and play important roles in procaspase activation. Indeed, all caspases exist within the cell as inactive latent precursor zymogens or proforms. Distinct intracellular or extracellular stimuli lead to the formation of death-inducing signalling complexes (DISC). DED and CARD domains are responsible for the recruitment of initiator caspases into these complexes. This appears to facilitate proteolytic autoactivation of the apical caspases, because "inactive" caspase zymogens possess low but detectable catalytic activity that is sufficient to process other caspases in circumstances where sustained close proximity between the zymogens is achieved (Boatright et al. 2003). The presence of Asp at the maturation cleavage sites is consistent with the ability of caspases to autoactivate or to be activated by other caspases, as part of an amplification cascade. Upon activation, the prodomain is released from the proenzyme by cleaving and Asp-X bond between it and the large subunit. Similarly, the large and small catalytic subunits, p20 and p10 are separated *via* a second cleavage event at an Asp-X bond between these two domains (Fuentes-Prior and Salvesen 2004; Riedl and Shi 2004). Caspases with prodomains of less than 30 amino acid residues, i.e., caspase 3, 6, and 7, are activated by proteolytic processing by other caspases or granzyme B (Degterev et al. 2003; Fuentes-Prior and Salvesen 2004). Short prodomain caspases absolutely require processing for activation whereas caspase 9 and other caspases with long prodomains can have significant activity even in the unprocessed state (Stennicke et al. 1999). The mature catalytic caspases are present in a tetramer containing two large and two small subunits as heterodimer complex $(p20, p10)_2$ (heterodimers of homodimers), and therefore possess two active sites which are positioned at opposite ends of the molecule.

2.5.1.2 Substrates

Caspases bind to their substrates through interactions of the active site cleft with amino acid motifs in the substrates, that are specific for individual caspases or groups of closely related caspases (Thornberry et al. 1997). The major factors that decide substrate specificity are located inside the small subunit (Nicholson 1999), and detailed information on the binding of caspases with their substrates is now available (Chowdhury et al. 2008). Caspases typically recognize tetrapeptide (P_4-P_3-P_2-P_1) motifs in their substrates (e.g. DEVD, YVAD, DEAD), the scissile bond occurring between the P_1 amino acid residue and the adjacent C-terminal amino acid in the peptide chain (Earnshaw et al. 1999). Caspases have an absolute requirement for aspartate at the P_1 position; however, it is the residue at the P_4 position that is the most critical in determining the substrate specificity of the individual caspases. Nevertheless, substrate specificity is an area not well studied in many species. The common occurrence of sequences that match the preferred inherent substrate specificity of caspases in intracellular proteins, would suggest a multitude (somewhere in the order of several hundred) of substrates *in vivo* (Nicholson 1999).

Indeed, the list of proteins that are cleaved by caspases either *in vivo* or *in vitro* is ever growing. The studies conclude that caspases cleave key structural components of the cytoskeleton and nucleus, as well as numerous proteins involved in signalling pathways (Fischer et al. 2003). These substrates are classified as cytoplasmic proteins (actin, gelsolin, α-fodin: nonerythroid spectrin, adherens, junction components β-catenin, plakoglobin, intermediate-filament protein keratin 18, rabaptin-5: an endosome fusion protein, ser-pin: plasminogen activator inhibitor-2), nuclear proteins (lamin-A, B; lamin B receptor, nuclear mitotic apparatus protein: NuMA, RNA-binding and ribonucleoprotein-associated proteins, chromosome scaffold binding protein), DNA metabolism and repair proteins (PARP, DNA-PKcs, DNA-replication protein, DNAtopoisomerases, RNA-polymerase), protein kinases (PKC and its isoforms; MAPK, ERK, Akt, Wee1), signal transduction pathway proteins (prointerleukin cytokines, phospholipases), cell cycle and cell proliferative proteins (p21, p27, pRB, ubiquitinated proteins), human genetic disease related proteins and other apoptotic related proteins (Earnshaw et al. 1999). It is clear that these substrates represent an eclectic group of proteins and it is likely that only a relative minor subset of caspase substrates are "innocent bystanders" and that their proteolysis contributes little to the process. For this reason, the consequences of many of the substrate cleavage events that take place during apoptosis are still the subject of speculation at present. In addition, different sets of caspases are activated that cleave different substrates in different cells based on the consequences of physiological differences between the cells, caspases and type of stress. Interestingly, the substrate specificity for other vertebrate species is the same as for mammalian caspase substrates (Takle and Andersen 2007).

2.5.1.3 Regulation

The pivotal role of caspase activation in virtually all cell death events suggests that strategies that prevent or limit their activation should exist. Indeed, in normal cells, the expression, processing, activation and inactivation of caspases are precisely controlled by different mechanisms with different inhibitors or activators. In mammals, the transcriptional and post-translational regulation of procaspase genes varies among different cell types. However, detailed studies are lacking (Chowdhury et al. 2008). In contrast, the regulation of action of caspases is widely studied. The action of caspases is regulated on several levels, including blockade of activation of caspases at the DISC as well as inhibition of enzymatic caspase activity. At least three distinct types of regulators of caspases, namely, IAPs, FLIP and calpain have been described.

The inhibitor of apoptosis proteins (IAPs) are a family of cellular proteins, originally identified in insect cells infected by the baculovirus (Salvesen and Duckett 2002). The IAPs include 8 mammalian family members (Deveraux and Reed 1999) with highly conserved and differential expression patterns in various tissues. In human, six IAP relatives have been identified: NAIP (neuronal apoptosis inhibitory protein), c-IAP1/HIAP-2, c-IAP2/HIAP-1, XIAP/hILP (X-linked mammalian inhibitor of apoptosis protein), survivin and BRUCE (a conserved

528 kDa peripheral membrane protein of the trans-Golgi network) (Deveraux and Reed 1999). A hallmark of IAPs is the presence of one to three copies of a novel domain of 70–80 amino acids termed baculovirus IAP repeat (BIR), which represents a zinc binding fold that can bind to the surface of caspases, blocking the catalyzing grooves of these enzymes (Shi 2002). Some mammalian IAPs possess, in addition to their BIR domains, a C-terminal RING finger, which confers ubiquitin-ligase activity. The linker region between BIR1 and BIR2 domains selectively targets caspases 3 and 7, whereas the third BIR domain (BIR3) interacts with caspase 9 (Johnson and Jarvis 2004). Thus, IAPs specifically inhibit the activity of the effector caspases 3 and 7, as well as the initiator caspase 9. The IAPs do not bind or inhibit caspase 8, but they do bind to and inhibit its substrate procaspase 3, thus providing protection from Fas/caspase 8-induced apoptosis (Deveraux et al. 1998). In the mitochondrial pathway, caspase inactivation is regulated by XIAP, c-IAP1, and c-IAP2, which bind directly to procaspase 9, preventing its processing and activation induced by cytochrome *c* (Deveraux et al. 1998). The overexpression of IAP proteins inhibit apoptosis induced by Bax and other proapoptotic Bcl-2 family proteins (Deveraux and Reed 1999). Not all BIR-containing proteins are inhibitors of caspases or apoptosis. Survivin, which just contains one BIR domain, may act as a regulator of mitosis rather than apoptosis. In addition, BIR domains are also homing sites for apoptotic proteins, such as the mitochondrial-derived proteins Smac/DIABLO and HtrA2/Omi, which counteract the anti-apoptotic function of IAPs.

IAPs are not the only natural inhibitors of caspases, and FLIP, baculoviral p35, calpain, or Ca^{2+} are other regulators. FLIP proteins are inactive caspase homologues, whose catalytic sites are damaged and are dominant negative modulators of the caspase cascade (Thornberry and Lazebnik 1998). There are two major groups of FLIPs: viral (v-FLIP), encoded by the gamma-herpes virus, and the mammalian homologues or cellular FLIPs (c-FLIP) (Golks et al. 2005). Both v- and c-FLIPs possess two tandem DEDs at their N termini that facilitate their recruitment to the DISC. Under conditions of overexpression, all isoforms inhibit activation of procaspase 8 at the DISC by blocking its processing (Krueger et al. 2001). At the same time, there is increasing evidence that at the DISC, when present at low concentrations, the isoform c-FLIPL facilitates the cleavage of procaspase 8 by forming a heterodimer with this procaspase (Micheau et al. 2002).

The baculoviral p35 protein is a pan-caspase inhibitor that targets most caspases by forming an inhibitory complex with them (Xu et al. 2001). Baculovirus p35 is an effective inhibitor of caspases 1 to 8 (Stennicke et al. 2002). Another pan-caspase inhibitor is ser-pin CrmA, derived from cowpox virus. This protein covalently binds to the active center of caspases (Renatus et al. 2000). It is a strong inhibitor of caspases 1 and 8, and a weak inhibitor of caspases 3 and 6. Nevertheless, the mechanisms through which the members of the caspase family interact with each other, and how they interact with other proapoptotic and anti-apoptotic factors, are still uncertain.

Calpain is a kind of Ca^{2+}-dependent cysteine protease. Calpain and caspase 3 share many common substrates, including fodrin, Ca^{2+}-dependent protein kinase and ADP ribosyltransferase/PARP (Wang 2000). In cell death induced by endoplasmic reticulum (ER) stress, the role of calpains is particularly important due to the perturbed Ca^{2+} homeostasis. In the brain cells of rats suffering from unilateral hypoxia-ischemia, m-calpain first cleaves procaspase 3 into 29 kDa fragments to facilitate its further cleavage and activation (Blomgren et al. 2001). Cisplatin, a kind of anticancer agent, can cause ER stress and apoptosis. During this process, the activation of procaspase 12 by cisplatin is dependent on Ca^{2+} and calpain (Mandic et al. 2003). Calpain can also cleave Bcl-xL thereby converting to a proapoptotic molecule from an antiapoptotic one (Nakagawa and Yuan 2000).

2.5.2 Death Receptors: The Fuses

Death receptors are cell surface cytokine receptors belonging to the tumour necrosis factor/nerve growth factor (TNF/NGF) super family (Ashkenazi and Dixit 1998). These receptors transmit apoptotic signals initiated by specific death ligands, activating the caspase cascade within seconds. They are transmembrane proteins with a C-terminal intracellular tail, a membrane-spanning region, and an extracellular ligand-binding N-terminal domain. Death receptors are characterized by a significant homology in a region containing one to five cysteine-rich repeats in their extracellular domains, and in the 60–80 amino acid cytoplasmic sequence known as death domain (DD) which typically recruits downstream apoptotic proteins (Gupta 2003).

Signal transduction by death receptors is initiated by the trimerization of the receptor triggered upon juxtaposition of the intracellular domains that follows the engagement of the ligand to the receptor's extracellular domain (Fig. 2.1). This event leads to recruitment of different adapter proteins, which provide the link between the receptor and the cell effector caspases (e.g. caspase 8 and caspase 10). Adapter proteins generally have no enzymatic activity of their own, but are able to associate with receptors through homophilic interactions of the receptor's DD and an analogous DD on the adapter itself. Adapter proteins may also contain a death-effector domain (DED) that mediates the recruitment of caspases through the association with a correspondent DED on a caspase-recruitment domain (CARD) in the prodomain of the inactive initiator caspases. The resulting complex is called the death-inducing signalling complex (DISC). The proximity of several caspase molecules recruited to the receptor results in self-processing and activation of the caspase, likely through a mild proteolytic activity of the procaspase itself. The activated initiator caspase subsequently starts a cascade of caspase activation by processing and activating the effector caspases (e.g. caspase 3, caspase 6 and caspase 7), which are directly or indirectly responsible for the execution of cell death. Alternatively, initiator caspases may cleave other substrates, which induce mitochondrial dysfunction (e.g. cleavage of Bid), and activate the effector caspases through the release of the pro-apoptotic mitochondrial factors. After the apoptotic signal is triggered, the DISC is internalized where it dissociates at low pH. Death receptors may thus act like the fuses of dynamite cartridges inserted in the cell plasma membrane, that get fired by their precise matches, that is their specific extracellular ligands.

Currently, the characterized death receptors include Fas (also called CD05, APO-1), TNFR-1 (also called p55, CD120a), DR3 (also called APO-3, WSL1, TRAMP, LARD). DR4 (also called TRAIL-R1), DR5 (also called APO-2, TRAIL-R2, killer), and death receptor 6 (DR6). Their signalling cascades are somehow different. Yet, the understanding of the complex pathways originating from each one of them seems still to be incomplete.

2.5.2.1 Signalling by Fas

Fas or CD95 is a 45–54 kDa glycosilated cell-surface protein ubiquitously expressed in various tissues, but particularly abundant in thymus, liver, heart, kidney, and activated mature lymphocytes, as well as virus-transformed lymphocytes. Although the membrane-bound form is largely predominant, several soluble forms have been described, the function of which is still unclear (Cascino et al. 1995). An anti-apoptotic effect has been proposed for these soluble forms of the receptor that may antagonize Fas mediated cytoxicity by binding to and inactivating Fas ligand (FasL). In addition Fas localizes in the cytosol, in particular, in the Golgi complex and the trans-Golgi network (Augstein et al. 2002). Translocation of Fas-containing vesicles to the cell surface has been observed upon stimulation, providing an effective mechanism to regulate the plasma membrane density of the death receptor, and avoiding spontaneous activation (Sodeman et al. 2000). Fas mediated apoptosis can also be modulated by glycosilation of the receptor (Peter et al. 1995), as well as at the transcriptional level, by directly regulating its expression. A binding site for the transcription factor NF-κB has been described, and a p53-responsive element is also located within the *fas* gene.

FasL is a 40 kDa transmembrane protein with homotrimeric structure expressed on the surface of activated T cells and natural killer cells. It plays and important role in the maintenance of the peripheral T- and B-cell homeostasis, and in the killing of harmful cells, such as virus-infected cells or cancer cells, and killing of inflammatory cells at immune privileged sites such as eyes. Engagement of Fas with its ligand (or with Fas monoclonal antibodies), results in trimerization of the receptor followed by recruitment of the adaptor molecule FADD (Fas- associated protein with death domain) (Rupinder et al. 2007). FADD is a ubiquitously expressed, 28 kDa cytosolic protein with a C-terminal death domain, and a DED at the N-terminus. FADD associates with Fas receptor through its DD, while DED is required for self-association and binding procaspase 8. Recruitment and accumulation of procaspase 8 at the DISC results in spontaneous activation of this caspase via autoproteolytic cleavage, and initiation of the apoptosis signal (Gupta 2003). Procaspase 10 has also been demonstrated to be recruited and activated at the Fas DISC, with similar kinetics as procaspase 8. Both enzymes can initiate apoptosis independently of each other.

The signal downstream of DISC formation differs between cell types. In type I cells, large amounts of caspase 8 are activated at the DISC and closely followed by rapid cleavage of caspase 3. Overexpression of the anti-apoptotic protein Bcl-2 or Bc-xL does not prevent activation of caspase 8 or caspase 3 in these cells, nor does inhibit apoptosis, suggesting a mitochondrial independent activation of a caspase cascade. In contrast, DISC formation in type II cells is strongly reduced, and activation of caspases, including caspase 8, occurs mainly downstream of mitochondria, because both caspase activation and apoptosis can be blocked by Overexpression of Bcl-2 or Bcl-xL. Notably, Fas triggers the activation of mitochondria in both type I and type II cells, and in both cell types, the apoptogenic activity of mitochondria is blocked by overexpression of Bcl-2 or Bcl-xL. However, only in type II cells, and not in type I cells, overexpression of Bcl-2 or Bcl-xL blocks apoptosis. Therefore, only in type II cells are mitochondria essential for the execution of the apoptotic program, whereas in type I cells, mitochondrial dysfunction likely functions as an amplifier of the apoptosis signal.

A fine regulation of Fas signalling is required to avoid triggering unnecessary cell death and to ensure proper functioning of the apoptotic machinery. FLIP, the DED containing proteins, regulate Fas mediated apoptosis (Rupinder et al. 2007). The DED of FLIP binds to Fas-FADD complexes and inhibits the recruitment and activation of procaspase 8 and therefore acts as anti-apoptotic molecule (Gupta 2003).

2.5.2.2 Signalling by Tumour Necrosis Factor (TNF)

The TNF/TNF-receptor signalling system consists of two distinct receptors, TNF-R1 and TNF-R2, and three ligands, the membrane-bound TNF-α, the soluble TNF-α, and the soluble lymphocyte-derived cytokine TNF-β. Both receptors are transmembrane proteins with similar structural features, including an N-terminus, disulfide-rich, extracellular domain that recognizes TNF, a transmembrane helix, and a cytoplasmic tail. However, only TNF-R1, and not TNF-R2, possesses and intracellular DD, and therefore is likely to be the sole mediator of the apoptosis signal in most cell types, as well as *in vivo* studies.

Intracellular signals originating from the TNF-R1 are extremely complex and can lead to multiple, even opposite, cell responses, from cell proliferation to inflammation to cell death. Most cells treated with TNF-α, however, do not undergo apoptosis unless protein of RNA synthesis is blocked, which suggests the predominance of survival signals over death signals under normal circumstances, and the requirement of neosynthesized proteins to suppress the apoptotic stimulus. The expression of these anti-apoptotic proteins is likely to be controlled by the activity of the transcription factor NF-κB.

Engagement of TNF-R1 by TNF-α results in trimerization and conformational changes in the receptor's intracellular domain, resulting in rapid recruitment of several cytoplasmic DD-containing adapter proteins, again via homophilic interaction with the DD of the receptor. The unstimulated TNF-R1 has been found to be associated with the recently isolated SODD (silencer of death domains), which effectively prevents self-aggregation of the DD and spontaneous initiation of signalling. Upon stimulation, SODD promptly dissociates from TNF-R1, allowing the adapter protein TRADD (TNFR-associated protein with death domain) to bind to the clustered-receptor DD. TRADD functions as a docking protein that recruits several signalling molecules to activate the receptor, such as FADD, TRAF-2 (TNF-associated factor-2), RIP (receptor-interacting protein), and RAIDD (RIP- associated ICH-1/CED-

3-homologous protein with death domain). These proteins have no enzymatic activity, except for RIP, which possesses serine-threonine kinase activity. The role of its kinase activity in apoptosis, however, remains to be established; though a role in mediating death receptor-induced nonapoptotic cell death has been proposed. FADD binds and activates caspase 8, promoting apoptosis through a pathway similar, though probably not identical, to that triggered by Fas. RIP binds RAIDD, which engages a death pathway by recruitment of caspase 2 through homophilic interaction between homologous sequences in its amino-terminal domain and the prodomain of caspase 2. RIP also associates with TRAF-2, stimulating pro-survival pathways and regulating the immune response.

This apoptotic pathway is under the regulatory control of number of gene products. Apoptosis mediated via FADD is regulated by FLIP (Flice-inhibitory protein, casper, CASH, MRIT) that appears to inhibit initiator caspase. FLIP also inhibits death signal by TNFR-1 via interactions with DED of FADD. Recent studies have reported that FLIP also inhibits apoptosis by activating the survival signals. Reports have suggested that TNFR-2 lacks a cytoplasmic death domain, instead interaction between TNFα and TNFR-2 results in the binding of TRAF-1 and TRAF-2 to the cytoplasmic portion of TNFR-2. It is reported that TNFR-2 potentiates TNFα-induced apoptosis by serving as high affinity trap of TNFα and delivers it to TNFR-1 (Gupta 2003).

2.5.2.3 Signalling by TRAIL Receptors

Since the identification of the new members of the TNF family, TNF-related apoptosis-inducing ligand, TRAIL (also called APO-2L), five different cognate receptors have been shown to bind it. Two of them, TRAIL-R1 (also called DR4) and TRAIL-R2 (also called DR5 or Killer or TRICK2) are considered actual death receptors, as their engagement with TRAIL results in apoptosis. The structure of TRAIL-R1 and TRAIL-R2 is consistent with that of the other death receptors with the DD being essential for transducing the apoptotic signal. TRAIL-R1 is expressed in most human tissues, including spleen, thymus, liver, peripheral blood leukocytes, activated T cells, small intestine, and some tumour cell lines (Oldenhuis et al. 2008). TRAIL-R2 expression has a ubiquitous distribution both in normal tissue and tumour cell lines, but is particularly high in spleen, peripheral blood leukocytes, and activated leukocytes. Two of the other TRAIL receptors TRAIL-R3 (also called DcR1 or TRID or LIT) and TRAIL-R4 (DcR2/TRUNDD), are so-called decoy receptors. Although quite similar to TRAIL-R1 and TRAIL-R2 in their extracellular and transmembrane region, TRAIL-R3 and TRAIL-R4 lack a death domain, or possess a non-functional death domain, respectively. Therefore, binding of TRAIL to these receptors fails to trigger apoptosis. TRAIL-R3 and TRAIL-R4 transcripts are most ubiquitously expressed in healthy human tissues, but not in most cancer cell lines (Oldenhuis et al. 2008). The preferential distribution of the decoy receptors in normal tissues, together with their ability to compete with the death receptors for binding to TRAIL, has been though to account for the highest resistance of normal cells to TRAIL-induced apoptosis. However, recent studies suggest intracellular regulation of TRAIL signalling more likely accounts for the resistance of healthy cells to TRAIL-mediated apoptosis, rather than the surface density of decoy receptors. The fifth identified receptor for TRAIL is the soluble osteoprotegerin receptor (OPG). OPG also can act as a decoy receptor, because it efficiently binds TRAIL, but does not induce apoptosis (Truneh et al. 2000).

Like the other ligands of the TNF family, TRAIL is also a transmembrane protein. Its carboxi-teminal extracellular domain shows significant homology with the other members of this family. However, the cytosolic amino-terminus is considerably shorter and not conserved among species. The biological active form of TRAIL is a homotrimer. TRAIL seems to trigger apoptosis more specifically in tumour cell lines rather than in normal cells, although the reason for this differential sensitivity has not yet been explained.

Like Fas, activated TRAIL-R1 and TRAIL-R2 recruit FADD and caspase 8 and caspase 10 to their respective DISCs, and they seem to trigger apoptosis through a pathway similar to that activated by Fas (Peter 2000). Like Fas, TRAIL-induced apoptosis is effectively inhibited by overexpression of FILP, which appears to be the key factor in determining cell sensitivity to TRAIL-induced apoptosis. Several reports also show that TRAIL, in addition to inducing cell death promotes activation of NF-κB, via TRAF-2, and JNK through distinct, independent pathways. Because TRAF-2 does not directly associated with any of the TRAIL receptors, it is conceivable that this pathway requires a still-unidentified adapter protein other than TRADD or FADD.

2.5.2.4 Signalling by Death Receptors 3 and 6

Death receptor 3 (DR3) presents a high homology with TNF-R1, particularly in its DD. A natural ligand of DR3 has been identified and named Apo3L. The responses and signalling triggered either by overexpression of DR3 or binding to Apo3L, resemble that mediated by TNF-R1. Based on the different expression of the ligands and receptors, however, it has been suggested that, despite the similarities in the signalling mechanism, DR3 and TNF-R1 may have distinct biological roles (Wang et al. 2001).

Death receptor 6 (DR6) has been identified as a novel death receptor, based on its characteristic extracellular cystein-rich domain and the intracellular DD (Pan et al. 1998). However, unlike the other death receptors, the DD in not localized at the C-terminus, but it locales adjacent to the transmembrane domain. DR6 is highly expressed in thymus, spleen and lymphoid cells. The signalling pathway originating from this receptor has been poorly defined yet. DR6 engages a cell-death pathway in mammalian cells different by those initiated by other known death receptors, as it seems not to associate with DD-containing adaptor molecules such us TRADD, FADD, RAIDD, or RIP. Upon overexpression, DR6 is a potent inducer of JNK and NF-κB activation. The ability to activate the JNK pathway and the predominant expression in lymphoid organs suggest that DR6 may have a role in the immune system.

In summary, multiple death receptors are expressed differentially in mammalian cells. Their role in maintaining tissue homeostasis through regulation of cell death and survival is crucial, as demonstrated by the apparent redundancy of their signalling pathways. Keeping in mind that the ultimate fate of the cell depends upon the cell context, simultaneous signalling events, and other stimuli, death receptors may be particularly important in immune regulation and tissue injury in disease states. For these reasons, they have already been targeted for the treatment of several human inflammatory and autoimmune diseases. Purposeful induction of apoptosis by death receptors may ultimately be useful also in cancer therapy. Nevertheless, much remains to be learned about the complexity of death-receptor signalling, both, in health and disease.

2.5.3 Bcl-2 Proteins: The Missiles and the Shields

A critical feature of metazoan apoptosis is the permeabilization of mitochondria. In the mitochondrial pathway of apoptosis, permeabilization of the outer membrane triggers the efficient release of pro-apoptotic factors (e.g. cytochrome *c*, Smac/DIABLO (Second mitochondria-derived activator of caspase/Direct IAP-bind protein with low pI)) from the mitochondrial intermembrane space (Chipuk and Green 2008). These factors cooperate with cytosolic adaptor proteins (e.g. APAF-1 (Apoptotic protease activating factor 1)) to initiate the formation of the apoptosome and the subsequent activation of the initiator and executioner caspases, normally caspase 9 and 3 respectively (Fig. 2.1). Even if this pathway is inhibited, a caspase-independent cell death process ensures cellular demise. This process may utilize less tractable mechanisms, such as ROS (reactive oxygen species), loss of mitochondrial function or released of other mitochondrial intermembrane proteins space proteins such as apoptosis inducing factor (AIF) or endonuclease G. In both situations, caspase-dependent and caspase-independent cell death, outer membrane permeabilization occurs. In order for a mitochondrion to undergo outer membrane permeabilization, a coordinated effort between numerous Bcl-2 (B-cell CLL/Lymphoma 2) families of proteins must be engaged. Following apoptotic stimuli, these Bcl-2 proteins are targeted to the outer membrane, where they insert, aggregate, and form membrane-spanning pores for the release of the intermembrane space pro-apoptotic factors. Thus, a set of Bcl-2 proteins act like guided missiles launched against mitochondrion to break through its outer membrane.

However, the Bcl-2 family of proteins consists of both pro-apoptotic, but also anti-apoptotic members. While the pro-apoptotic members serve as sensors of death signals and executors of the death program, the anti-apoptotic members inhibit the initiation of the death program. In fact, Bcl-2, the prototype of the Bcl-2 family proteins, was cloned from the t(14;18) breakpoint in human follicular lymphoma, and one key early discovery that introduced a new paradigm for carcinogenesis, was that overexpression of Bcl-2 inhibits cell death (Vaux et al. 1988). A number of proteins were soon discovered that share sequence homology with Bcl-2. While some of them actually promote apoptosis, others engage in anti-apoptotic activity. Thus, a second set of Bcl-2 proteins may act like the missiles' shields that intercept the discharge of their counterpart family members.

2.5.3.1 Structure and Activation

One of the key features of the Bcl-2 family of proteins is that their members share sequence homology in four domains (e.g. the BH1, 2, 3, and 4 domains), although not all members have all the domains (Youle and Strasser 2008). Mutagenesis studies have revealed that these domains are important for the various molecular functions and for protein interactions among the family members. The BH1 and BH2 domains are necessary for the death-repression function of the anti-apoptotic molecules, whereas the BH3 domain is required for the death-promotion function of molecules. In addition, the BH4 domain, which is present mainly in the anti-apoptotic molecules, is also important for death-inhibition functions. Thus, anti-apoptotic Bcl-2 proteins contain BH domains 1–4, while the pro-death molecules can be further divided into those with only the BH3 domain and those with BH1-3 domains. It seems that the so-called "BH3-only" molecules, such as Bid, Bim, and Bad, are sensors for the peripheral death signals and are able to activate the "multidomain" executioner molecules Bax, or Bak. This process in some ways resembles the caspase cascade, in which the initiator caspases activate the effector caspases.

In mammals, at least 12 core Bcl-2 family proteins have been identified, based on three-dimensional structural similarities or predicted secondary structure. In addition, other than Bid, the predicted overall structures of the BH3-only proteins seem to be unrelated and appear to lack a close evolutionary relationship to the core members of the Bcl-2 family (Aouacheria et al. 2005). Yet, all BH3-only proteins interact with and regulate the core Bcl-2 family proteins to promote apoptosis.

The antiapoptotic members of the Bcl-2 family proteins are: Bcl-2, Bcl-xL, Bcl-w, (Bcl-2 related gene, long isoform), A1 and Mcl-1 (Myeloid cell leukaemia 1). The proapoptotic members are the effector molecules Bak (Bcl-2 antagonist killer 1) and Bax (Bcl-2 associated x protein). The BH3-only proteins are: Bad (Bcl-2 antagonist of cell death), Bid (Bcl-2 interacting domain death agonist), Bik (Bcl-2 interacting killer), Bim (Bcl-2 interacting mediator of cell death), Bmf (BCL-2 modifying factor), bNIP3 (BCL-2/adenovirus E1B 19-KD protein interacting protein 3), HRK (Harakiri), Noxa and Puma (p53-upregulated modulator of apoptosis). Several of the above proteins have been knocked out in mice to reveal their physiological roles, redundancy and interactions *in vivo* (Youle and Strasser 2008). Thorough phylogenetic analyses of the Bcl-2 family have generated important insights into the origins of the core Bcl-2 family members and the BH3-only proteins, and suggest that many of these proteins might have biological activities beyond regulation of cell death (Aouacheria et al. 2005). In fact, there is growing interest in the newly identified roles that both pro- and anti-death family proteins play in healthy cells. For example, in membrane fission and fusion mechanisms, mitochondrial autophagy, neuronal activity, and cellular energetic (Hetz and Glimcher 2008; Youle and Strasser 2008).

It is quite intriguing that the 3D structures of the seven core Bcl-2 family proteins (e.g. Bcl-2, Bcl-xL, Bcl-W, Mcl1, Bax, Bak and Bid) have yet to reveal any distinguishing difference between anti-apoptotic members (such as and Mcl1) and pro-apoptotic members (such as Bax and Bid). All seven proteins are helical bundles with a hydrophobic helix-turn-helix hairpin that is flanked on both sides by pairs of amphipathic helices. They appear to have C-terminal membrane-anchoring domains. In addition, pro-apoptotic Bid appears to be myristoylated to mediate membrane anchorage (Zha et al. 2000), and in three proteins, Bax, Bcl-W and Mcl1, the C-terminal anchor fits into a hydrophobic pocket formed by the BH1, 2 and 3 regions. The same pocket that sequesters the C terminal membrane anchor can also bind to peptides of the BH3-domain sequences of Bak, Bad and Bim (Liu et al. 2003), which suggests that it also functions in dimerization with BH3-only proteins and/or multi-BH-domain-containing Bcl-2 family members.

Activation of BH3-only proteins occurs in response to different stimuli specific for each member of the family, and hence their regulators serve as primary sensors for cellular stress (Youle and Strasser 2008). BH3-only protein expression can be induced by transcription factors. For example, Noxa and Puma are induced by the tumour suppressor p53 in response to DNA damage, and Bim is induced in response to growth-factor deprivation and by several transcription factors in response to endoplasmic reticulum (ER) stress. BH3-only proteins can also be activated post-translationally; for example, Bad is activated by loss of phosphorylation in response to growth-factor deprivation. Bid is activated upon cleavage by caspase 8, granzyme B and more weakly by caspase 2 and 3, and the resulting N-terminal truncated fragment (tBid) is engaged in response to death receptor stimulation, cytotoxic T-lymphocyte killing and heat shock respectively. Bim is held inactive in the cell through binding to dynein motor complex and is activated upon release form the cytoskeleton or by loss of extracellular signal-regulated kinase (ERK)-mediated phosphorylation (which targets it for ubiquitilation and proteasomal degradation in healthy cells). Bmf is activated by release from actin-myosin motor complexes; and Bik is activated by an unknown mechanism in response to inhibition of protein synthesis.

Regulation of the expression levels of antiapoptotic Bcl-2 family proteins is another way in which cells can regulate apoptosis. For example, Bcl-xL can be transcriptionally induced by growth factors to promote cell survival (Youle and Strasser 2008). Mcl1 is rapidly degraded by the ubiquitin-proteasome pathway in response to cytokine deprivation or other death stimuli (such as ultraviolet (UV) radiation) and can be upregulated post-transcriptionally to prevent apoptosis by inhibiting the rate of degradation.

Regulation of the expression levels of the pro-apoptotic proteins Bax and Bak is less apparent and the proteins appear to be constitutively expressed at more or less constant levels. Bax and Bak are primarily post-translationally regulated by other members of the Bcl-2 family.

2.5.3.2 Mechanism of Action

Protein interactions among the Bcl-2 family members are important steps for their activation and mechanism of action. Based on *in vitro* and some *in vivo* studies, several interaction types can be defined. The most common one is the interaction between the anti-death and pro-death members, such as Bcl-2 *vs* Bax or Bid. This interaction can result in antagonistic action of the two types of molecules and thus could set a rheostat control of the death program. Interestingly, not all anti-death molecules can interact with all pro-death molecules. It seems that some members of one group will preferentially bind to some members of the other group. For example, the pro-death Bok binds to Mcl-1 but not to Bcl-2, Bcl-xL, or Bcl-w. Correspondingly, these molecules may only antagonize the function of those molecules to which they bind. This type of selectivity suggests that specific amino acids required for particular interactions may only exist in some but not all of the family members. In addition, it may also suggest that in certain tissues and for certain death stimuli, a specific set of Bcl-2 family proteins is critically involved.

The second type of interaction occurs between two pro-death members, usually one BH3-only molecule and one multidomain molecule, such as Bid to Bax or Bak; and Bim to Bax. Such interactions are important for the activation of the multidomain executioner molecules, Bax or Bak, by the BH3-only sensor molecules.

The third type of interactions is multimerization of the same molecule. This has been observed in both anti-death molecules such as Bcl-2 or Bcl-xL, and pro-death molecules such as Bax and Bak. The ability of Bax of Bak to oligomerize has been considered and important factor in their role as a mitochondrial channel for releasing the intermembrane space mitochondrial apoptotic factors.

Though it has long been recognized that the ability of different Bcl-2 family members to engage in selective protein–protein interactions with other family members is integral to their anti- and pro-death functions, the exact nature of these interactions remains highly controversial in the apoptosis field. The basis of this debate have crystallized is several models relating to the molecular mechanism of these proteins (Chipuk and Green 2008). The first hypothesis to explain the mechanism of action of Bcl-2 proteins in apoptosis was conceived around the idea that the cellular decision to die relays on tipping the balance of total Bcl-2 family expression from and anti-apoptotic (or pro-survival) to a pro-apoptotic position. This hypothesis was widely accepted because at the time it was supported by various observations (Chipuk and Green 2008). Essentially, this rheostat model described that the stoichiometry between the pro- and anti-apoptotic Bcl-2 proteins dictated cellular commitment to apoptosis. The rheostat model became the foundation for our understanding of mammalian Bcl-2 protein family function and in some aspects, it continues to be time-honoured. Nevertheless, it does not explain the more recently discovered complexities within the Bcl-2 family. For example, how does a cell tolerate relatively high levels of Bax and/or Bak expression without these proteins being constitutively bound to anti-apoptotic members? How do certain BH3-only proteins engage Bax and/or Bak activation and mitochondrial outer membrane permeabilization, whereas others do not? These issues continue to fuel controversies and they have been the core of other competing hypotheses regarding the activation of Bax and Bak.

One of these models is known as the anti-apoptotic protein neutralization model, also referred as to the "indirect" model. It proposes that anti-apoptotic proteins must continually inhibit the function of Bax and Bak to ensure mitochondrial integrity and survival. The signal for mitochondrial outer membrane permeability is the moment that all anti-apoptotic proteins are functionally neutralized (i.e. the hydrophobic binding groove of the anti-apoptotic protein is occupied by the BH3 domain of a BH3-only protein) by activated (either transcriptional or post-translational) BH3-only proteins. This indirectly promotes Bax and/or Bak liberation, homo-oligomerization into proteolipid pores, and cytochrome *c* release. Due to the restricted ability of each BH3-only protein to bind only certain anti-apoptotic proteins, by determining the affinities of BH3 domain peptides for the anti-apoptotic Bcl-2 proteins it was hypothesized what combination of BH3-only proteins are required for complete neutralization of anti-apoptotic proteins and mitochondrial permeabilization (Willis et al. 2007). In all cases, it is assumed that the anti-apoptotic proteins equally inhibit Bax and Bak, and that neutralizing a subset of anti-apoptotic proteins causes the newly liberated effectors to quickly be inhibited by any available anti-apoptotic proteins.

The anti-apoptotic protein neutralization scenario is a modern interpretation of the rheostat model, because it is based on the hypothesis that the pro-apoptotic protein function overcomes inhibition by the anti-apoptotic proteins. One caveat of this hypothesis is that not all endogenous effector molecules appear to be actively sequestered by anti-apoptotic proteins (Chipuk and Green 2008). In addition, the majority of data concerning the interactions between BH3-only protein and anti-apoptotic members is based on synthetic BH3 domain peptides and their interaction with immobilized, recombinant anti-apoptotic proteins. Furthermore, several lines of evidence suggest that BH3-only proteins rarely bind effector proteins in the absence of a lipid environment and that unique binding surfaces are created by anti-apoptotic and pro-apoptotic protein interactions. In fact, many examples hint that the calculated *in vitro* affinity between two Bcl-2 proteins does not directly translate into activity, because both environment (e.g. a cytoplasmic versus a hydrophobic membrane) and binding partners dictate the cellular response.

In the direct activation model, the key event necessary to engage mitochondrial outer membrane permeabilization is the interaction between direct activator BH3-only proteins and the effector molecules (e.g. Bid or Bim interacting with Bax or Bak) at the outer mitochondrial membrane. Preventing this interaction is crucial for mitochondrial integrity and cellular survival, because recovery after mitochondrial permeabilization is unusual. To ensure that Bax and/or Bak activation only occurs when appropriate, the cell expresses a variety of anti-apoptotic Bcl-2 proteins that sequester the occasionally activated Bid or Bim molecules. Under conditions of cellular stress and transformation, direct activator proteins are induced to engage Bax and/or Bak activation (Certo et al. 2006). Yet a cell can overcome a death signal by sequestering the direct activator BH3-only proteins onto its repertoire of anti-apoptotic Bcl-2 proteins. Although this temporarily preserves cellular survival, it presents the cell with the duty of increasing anti-apoptotic proteins to counteract future stress-induced direct activator BH3-only protein expression – this challenge is also referred to as "Bcl-2 addiction" and is thought to occur during tumour initiation (Certo et al. 2006).

Notably, the cooperation between effector molecules and the direct activator BH3-only proteins has been described, but the affinity appears to be weak and is not readily detectable (Kuwana et al. 2002). However, the weak association between BH3-only activators and effectors has led to speculation that the interaction is not essential for apoptosis to proceed and that the crucial step towards mitochondrial outer membrane permeabilization is in fact the robust interaction between BH3-only proteins and anti-apoptotic proteins. Taken together, there are several points for important consideration. First, although the interaction between effector molecules and direct activators can be weak in vitro, the conditions in which this occurs in the cell are not well understood and their affinity might be higher *in vivo*. This leads one to question what structural and cellular components are essential for effector-direct activator BH3-only functional association; for example, perhaps a component(s) of the outer mitochondrial membrane.

It is clear that the two models differ primarily in whether Bax/Bak is constitutively active or requires activation and which interactions are crucial to the anti apoptotic function of for instance. However, there is also considerable overlap between these two competing models: both recognize that anti-apoptotic members such as directly binds to Bcl-xL and inhibits a pro-apoptotic Bcl-2 family protein that is directly involved in membrane permeabilization, while other pro-apoptotic Bcl-2 family proteins indirectly induce apoptosis by binding to and preventing this function. It has been difficult to sort out which interaction is important in cells, where multiple Bcl-2 members are present simultaneously. Moreover, some authors postulate that both models may be relevant in different cell types or in the same cell type under different circumstances (e.g., normal versus cancerous) (Deng et al. 2007).

On the basis of recent data demonstrating the importance of dynamic interactions of Bcl-2 family members with membranes, a third model termed "embedded together" has been proposed (Billen et al. 2008). In this model, interactions of Bcl-2 proteins with each other changes after binding to the membrane as this causes conformational changes that alter and/or expose new binding surfaces (Youle 2007). In this model neither antiapoptotic (Bcl-xL) nor proapoptotic (Bax) bind to membranes alone. However, the addition of tBid recruits molar excesses of either protein to membranes, indicating that tBid activates both pro- and anti-apoptotic members of the Bcl-2 family. Bcl-xL competes with Bax for the activation of soluble, monomeric Bax through interaction with membranes, tBid, or tBid-activated Bax, thereby inhibiting Bax binding to membranes, oligomerization, and membrane permeabilization. Experiments in which individual interactions were abolished by mutagenesis indicate that both Bcl-xL-tBid and Bcl-xL-Bax binding contribute to the anti-apoptotic function of Bcl-xL. By out-competing Bax for the interactions leading to membrane permeabilization, Bcl-xL ties up both tBid and Bax in non-productive interactions and inhibits Bax binding to membranes. Accordingly, because Bcl-xL does not oligomerize and forms pores, it inhibits apoptosis by acting as if it is a dominant-negative version of Bax in the membrane permeabilization process (Billen et al. 2008). Thus, in contrast to the other linear models, this one provides the strongest evidence to date that uses multiple mechanisms to neutralize functional Bax activation. Most certainly, an exciting avenue for further research will be the development of assays in reconstituted systems that allow detection and monitoring of dynamic (rather than static) interactions between specific anti- and pro-apoptotic Bcl-2 family members plus additional cellular proteins in membranes.

Nevertheless, the precise biochemical mechanism that leads to the activation of Bax and Bak remains a mystery and constitutes the "holy grail" of apoptosis research. Besides the obvious scientific interest, the importance of unravelling the mechanism(s) of action of the Bcl-2 family proteins relays in part on their possible clinical applications. Considering the prospects for the use of BH3 mimetic drugs in the treatment of neoplasia, the distinction might have life or death consequences (Chipuk and Green 2008). Thus, if apoptosis versus cellular survival depends on the extent to which anti-apoptotic Bcl-2 family members are neutralized, then tumour cells should be more resistant to such drugs than their normal counterparts are, because tumours tend to have increased expression of anti-apoptotic Bcl-2 proteins. By contrast, if the direct activator model is correct, then there might exists a therapeutic window within which some tumours are more sensitive to apoptosis induction than are healthy cells, owing to a tumour cell's elevated levels of direct activators.

2.5.3.3 Mitochondrial Permeabilization

Bcl-2 family members interact with mitochondria either constitutively or on induction of apoptosis and, although they might have activities in other cellular compartments, it is clear that they regulate apoptosis by their impact on the mitochondrial outer membrane. The anti-apoptotic Bcl-2 protein is embedded in the ER, the nuclear envelope and the mitochondrial outer membrane by a hydrophobic C-terminal membrane-spanning domain, with most of its amino acids in the cytosol. Although Bcl-2 in any of these subcellular locations can block apoptosis, the functions of Bcl-2 at the ER and the nuclear envelope are less clear than those on mitochondria. In contrast to Bcl-2, Bax is mostly cytosolic and sequesters its hydrophobic C-terminal membrane anchor in its BH3-binding pocket, with a minor fraction lightly bound to the mitochondrial outer membrane (Hsu et al. 1997). Bax appears to exist as a monomer in the cytosol of cells. Both pro- and anti-apoptotic Bcl-2 family members undergo dramatic conformational changes during apoptosis to unfold and insert deeply into the lipid bilayer. During apoptosis induction two steps in the activation process of Bax can be discerned: an initial translocation to mitochondria, and then the N-terminal conformational change that is likely to be coupled to membrane insertion and oligomerization (Heath-Engel and Shore 2006) to form large channels (Antonsson et al. 2000). The translocation step of Bax, although reversible in certain situations, usually correlates closely with the irreversible commitment of cells to die and to the cytochrome *c* release step. Bak also changes conformation and oligomerizes during apoptosis (Mikhailov et al. 2003;

Heath-Engel and Shore 2006). Although the mitochondrial outer membrane channel protein VDAC2 (voltage-dependent anion channel-2) also reportedly binds to Bak (Cheng et al. 2003), and is important for Bak import into the outer membrane (Setoguchi et al. 2006), *Vdac1/Vdac2/Vdac3* triple knockout mice display normal apoptosis (Baines et al. 2007), indicating there is little or no role for VDACs in Bcl-2 family protein regulation (Mannella and Kinnally 2008). How many units of Bax or Bak form these oligomers is still unclear, though it appears that they are distributed over a wide range of molecular weights.

The mechanisms responsible for the release of mitochondrial mediators of cell death are still a subject of lively discussion (Peixoto et al. 2004). There are two candidate pathways. Both result in permeabilization of the outer membrane but do so by activation of two distinct channels. These channels are the permeability transition pore PTP in the inner membrane (Kroemer et al. 2007) and the mitochondrial apoptosis-induced channel MAC in the outer membrane (Kinnally and Antonsson 2007).

The Mitochondrial Apoptotic Channel: The Knife

Consistent with their common structure, both the anti and the pro-apoptotic Bcl-2 proteins present channel-forming activity *in vitro*, although the channel forming conditions differ. In particular, oligomeric Bax forms slightly cation-selective, voltage-independent channels when reconstituted into liposomes, with a variable conductance from a few pS of up to several nS (Roucou et al. 2002). The diameter estimated for the Bax channels from the peak conductance of 1–5.4 nS is 2.7–5.4 nm, indicating many Bax channels could easily allow the passage of the approximately 3 nm cytochrome *c* (Dejean et al. 2005). Even more, Bax channels as well as the release of various compounds from liposomes containing Bax channels, are blocked by the anti-apoptotic proteins Bcl-2 and Bcl-xL (Antonsson et al. 1997).

The apoptosis-induced channel MAC was first detected in patch-clamp experiments on mitochondria isolated from apoptotic FL5.12 cells 12 h after withdrawal of the growth factor interleukin-3 (IL-3) (Pavlov et al. 2001) and has since been studied in several systems (Guihard et al. 2004; Guo et al. 2004; Dejean et al. 2005). MAC is a heterogeneous high-conductance channel. The mean conductance of MAC of apoptotic HeLa and FL5.12 cells is 3.3 and 4.5 nS, respectively (Dejean et al. 2005), and presents multiple sub-conductance levels. The pore diameter calculated with the polymer exclusion method indicates that channels with a conductance between 1.5 and 5 nS have pore sizes in the range of 3.3–6.0 nm. Hence, MAC activity is remarkably similar to that of the recombinant Bax, but differs from other well-characterized large channels, like the TOM and TIM channels, found on both mitochondrial membranes (Campo et al. 1997; Kinnally et al. 2000; Peixoto et al. 2007). Importantly, MAC has only been characterized early in apoptosis at the time of cytochrome *c* release. This "gigantic" outer membrane channel is exquisitely regulated by Bcl-2 family proteins. Over-expression of Bcl-2 blocks formation of MAC and release of cytochrome *c* (Pavlov et al. 2001). Many experimental approaches indicate that both, Bax and Bak, are putative components of this channel (Kinnally and

Antonsson 2007). It is likely that the mean pore size of MAC increases as apoptosis progresses possibly by incorporation of additional Bax and/or Bak molecules into the MAC channel complex, or perhaps by other, as yet unidentified, components. In addition, the relationship between MAC and the multi-domain proapoptotic proteins Bax and Bak has also been examined using molecular approaches. Previous studies using single and double knock out cell lines for Bax and Bak showed these two proteins are functionally redundant with respect to their roles in apoptosis (Danial and Korsmeyer 2004). That is, cytochrome *c* release occurs in Bax and Bak single knockout cells but not in the double Bax/Bak knockout cells during staurosporine treatment (Wei et al. 2001; Danial and Korsmeyer 2004; Dejean et al. 2005). Similarly, MAC is detected in single knockout, but not the double knockout cell lines during apoptosis. These data indicate that Bak can replace Bax as a structural component of MAC in Bax deficient cells. That is, Bax and Bak are functionally redundant with respect to MAC. In this scenario, it has been proposed that MAC forms in the mitochondrial outer membrane early in apoptosis and directly provides a pathway for the release of cytochrome *c* from the intermembrane space to the cytosol, which initiates activation of executioner caspases and cell death. Hence, it has been hypothesized that MAC is the "knife" that cuts cytochrome *c* from mitochondria (Dejean et al. 2006).

The Mitochondrial Permeability Transition

The so-called mitochondrial permeability transition was originally described over 25 years ago as a sudden increase of the permeability of mitochondrial inner membrane to solutes with molecular mass up to 1.5 kDa. The mitochondrial structure to which the permeability transition has been attributed resides in the inner mitochondrial membrane and is known as the permeability transition pore (PTP). Interestingly, despite the time elapsed since its discovery, the exact nature of the pore forming components of PTP has not been clearly established yet. However, a consensus has started to emerge about some of the proteins that may be involved (Peixoto et al. 2004). In addition, it is alleged that PTP builds up at the contact sites (i.e. sites at which the outer and inner mitochondrial membranes juxtapose). Traditionally the voltage-dependent anion-selective channel or VDAC, the adenine nucleotide translocator (ANT), and cyclophilin D, among others, have been linked with PTP (Crompton et al. 2002). PTP can be activated by a myriad of effectors including Ca^{2+} plus phosphate and ROS. Often PTP can be reversibly closed by removal of Ca^{2+} with EGTA or by the addition of cyclosporine A, magnesium or ADP.

Before the discovery of MAC, it was hypothesized that the opening of the PTP of the inner membrane would cause swelling of the matrix space, which would rupture the outer membrane, and spill cytochrome *c* and other proapoptotic proteins into the cytosol. However, at this time, the role of the PTP in apoptosis is quite controversial. Indeed, some studies implicate PTP as an initiating event in the mitochondrial pathway. In this sense, compounds that induce the permeability transition in isolated mitochondria often cause apoptosis and PTP is blocked or delayed by the antiapoptotic drug cyclosporine A (Gunter and Pfeiffer 1990). Also, suppression of the Ca^{2+} activation of the PTP by overexpression of Bcl-2 supports a role for this channel in apoptosis (Belzacq et al. 2003). On the contrary, recent data has shown that sustained PTP opening is primarily involved in necrosis and ischemia-reperfusion (Nakagawa et al. 2005). Cyclophilin-D is a regulator of the PTP. Remarkably, cyclophilin-D deficient cells died normally in response to apoptotic stimuli known to activate both the extrinsic and intrinsic pathways, but showed resistance to necrotic cell death induced by reactive oxygen species and Ca^{2+} overload (Nakagawa et al. 2005). Furthermore, cytochrome *c* release can occur in the absence of mitochondrial depolarization and without loss of outer membrane integrity. These observations indicate that, instead of rupturing, a more selective mechanism of permeabilization is operating, like the formation of a pore in the outer membrane (Dejean et al. 2006). Nonetheless, it would be premature and an oversimplification to conclude that PTP opening is irrelevant to apoptosis. The opening of PTP, although subsequent to MAC, may result in a more complete release of cytochrome *c* and other pro-apoptotic factors from the cristae, which would synergistically amplify the apoptotic cascade initiated by MAC. Furthermore, the delay in the progression of apoptosis caused by PTP inhibitors like cyclosporine A may be related to their ability to delay depolarization and energy production as well as any synergistic effects of further cytochrome *c* release from the cristae (Kinnally and Antonsson 2007).

Mitochondrial Cristae Remodelling

Besides the matrix and the intermembrane space of mitochondria, recent morphological studies performed with electron microscopic tomography have lead to the identification of an additional mitochondrial compartment: the intracristae space, which communicates with the intermembrane space through bottleneck-like junctions so tight that they create a diffusion barrier. Most cytochrome *c* (85%) is contained in this space. Cristae remodelling is the process by which the narrow junctions that delimitate the intracristae space get enlarged. This is thought to facilitates the mobilization of the cytochrome *c* to the intermembrane space (Scorrano et al. 2002). Several proteins that are involved in mitochondrial dynamics (fusion and fission) may play a major role in the proapoptotic reorganization of cristae. For instance, Drp1 (dynamin-related protein 1) participates in mitochondrial fission and is required for the optimal release of cytochrome *c*, presumably through its contribution to cristae remodelling (Germain et al. 2005). Conversely, inhibition of mitochondrial fusion by loss of a different mitochondrial dynamin family member, Opa1 (optic atrophia 1), induces spontaneous apoptosis (Olichon et al. 2003). Moreover, recent work shows that Opa 1 controls mitochondrial cristae formation, and that tight cristae junctions can inhibit cytochrome *c* during apoptosis. Together these data suggest that mediators of mitochondrial fission and fusion might have a role in cytochrome *c* release and apoptosis (Cipolat et al. 2006). Also unexpectedly, healthy cells that lack both Bax and Bak present altered mitochondrial morphology and slower mitochondrial fusion rates, which indicates that Bax and Bak affect mitochondrial morphogenesis machineries even in the absence of apoptotic stimuli

(Karbowski et al. 2006). In contrast to cytochrome *c* and Opa1, the release of other mitochondrial intermembrane space proteins (such as DIABLO) is not inhibited by *Drp1* knockdown (Parone et al. 2006), which underscores the suggestion that the role of the mitochondrial fission machinery in apoptosis might be indirectly linked to the cytochrome *c* release step. Given the multiple relationships that have been already characterized between mitochondrial dynamics and apoptosis (Parone and Martinou 2006), it will be interesting to learn in the future also the more specific molecular liaisons between mitochondrial fission (which often occurs during apoptosis) and cristae remodelling. Moreover, it will be a challenge for further investigation to determine how different pro-apoptotic effectors interact with the molecular machinery that mediates cristae remodelling (Kroemer et al. 2007). In any case, cristae remodelling has elucidated a novel checkpoint for apoptosis, which may determine sensitivity to apoptosis *in vivo*, during adult animal life.

2.6 Concluding Remarks: Apoptosis, a Battle with Treaty

Apoptosis is a battle cells fight against themselves. The battlefield is thus the cell itself. The endeavour of this particular battle is to sacrifice the individual on behalf of the whole. Hence, apoptosis shows us death's second face, which looks remarkably like that of life. For this reason it can not be considered that apoptosis is a battle cells always lose. On the contrary, complex organisms do relay so much on apoptosis that we should only wish this process keeps continuously looking after each one of us, until we finally get defeated by the ultimate battle: that without treaty. There are many variations, footnotes and uncertainties to the basic picture exposed above. This picture gets more and more complex by the day, and so does the relevant literature, which literally has exploded during recent years. It is then right to apologize to all the researches whose work has not been appropriately cited and to encourage them to keep unveiling all the intriguing check and balance molecular mechanisms deeply involved in apoptosis. At last, understanding in detail the workings of the roads to cell death is the best way to design the many strategies for protection against intracellular pathogens and cellular transformation and to understand how viruses and tumours evade immune destruction. Inevitably, research in this area will also help us comprehend important mechanisms in normal cellular metabolism and some of the strategies cells have at their disposal to deal with outside stress.

Acknowledgments This work was supported by grants PRI07A072 and BFU2008-00475/BMC funded by Junta de Extremadura and the Spanish Ministry of Science and Innovation, respectively.

References

Adrain C, Murphy BM, Martin SJ (2005) Molecular ordering of the caspase activation cascade initiated by the cytotoxic T lymphocyte/natural killer (CTL/NK) protease granzyme B. J Biol Chem 280:4663–4673

Alnemri ES, Livingston DJ, Nicholson DW et al. (1996) Human ICE/CED-3 protease nomenclature. Cell 87:171
Antonsson B, Conti F, Ciavatta A et al. (1997) Inhibition of Bax channel-forming activity by Bcl-2. Science 277:370–372
Antonsson B, Montessuit S, Lauper S et al. (2000) Bax oligomerization is required for channel-forming activity in liposomes and to trigger cytochrome *c* release from mitochondria. Biochem J 345:271–278
Aouacheria A, Brunet F, Gouy M (2005) Phylogenomics of life-or-death switches in multicellular animals: Bcl-2, BH3-only, and BNip families of apoptotic regulators. Mol Biol Evol 22:2395–2416
Ashkenazi A, Dixit VM (1998) Death receptors: signaling and modulation. Science 281:1305–1308
Augstein P, Dunger A, Salzsieder C et al. (2002) Cell surface trafficking of Fas in NIT-1 cells and dissection of surface and total Fas expression. Biochem Biophys Res Com 290:443–451
Baines CP, Kaiser RA, Sheiko T et al. (2007) Voltage-dependent anion channels are dispensable for mitochondrial-dependent cell death. Nature Cell Biol 9:550–555
Belzacq A-S, Vieira HLA, Verrier F et al. (2003) Bcl-2 and Bax modulate adenine nucleotide translocase activity. Cancer Res 63:541–546
Bergeron L, Perez GI, Macdonald G et al. (1998) Defects in regulation of apoptosis in caspase-2-deficient mice. Genes Dev 12:1304–1314
Billen LP, Kokoski CL, Lovell JF et al. (2008) Bcl-XL inhibits membrane permeabilization by competing with Bax. PLoS Biol 6:1268–1280
Blink EJ, Jiansheng Z, Hu W et al. (2005) Interaction of the nuclear localizing cytolytic granule serine protease granzyme B with importin α or β: Modulation by the serpin inhibitor PI-9. J Cell Biochem 95:598–610
Blomgren K, Zhu C, Wang X et al. (2001) Synergistic activation of caspase 3 by m-calpain after neonatal hypoxia-ischemia: a mechanism of "pathological apoptosis"? J Biol Chem 276:10191–10198
Boatright KM, Renatus M, Scott FL et al. (2003) A unified model for apical caspase activation. Mol Cell 11:529–541
Bossi G, Griffiths GM (2005) CTL secretory lysosomes: biogenesis and secretion of a harmful organelle. Semin Immunol 17:87–94
Boya P, Gonzalez-Polo R-A, Poncet D et al. (2003) Mitochondrial membrane permeabilization is a critical step of lysosome-initiated apoptosis induced by hydroxychloroquine. Oncogene 22:3927–3936
Campo ML, Tedeschi H, Muro C et al. (1997) Effects of carbonyl cyanide phenylhydrazone on two mitochondrial ion channel activities. J Bioenerg Biomem 29:225–233
Cascino I, Fiucci G, Papoff G et al. (1995) Three functional soluble forms of the human apoptosis-inducing Fas molecule are produced by alternative splicing. J Immunol 154:2706–2713
Casciola-Rosen L, Garcia-Calvo M, Bull HG et al. (2007) Mouse and human granzyme B have distinct tetrapeptide specificities and abilities to recruit the bid pathway. J Biol Chem 282:4545–4552
Certo M, Moore VDG, Nishino M et al. (2006) Mitochondria primed by death signals determine cellular addiction to antiapoptotic BCL-2 family members. Cancer Cell 9:351–365
Chen C-J, Kono H, Golenbock D et al. (2007) Identification of a key pathway required for the sterile inflammatory response triggered by dying cells. Nature Med 13:851–856
Chen W, Li N, Chen T et al. (2005) The lysosome-associated apoptosis-inducing protein containing the pleckstrin homology (PH) and FYVE domains (LAPF), representative of a novel family of PH and FYVE domain-containing proteins, induces caspase-independent apoptosis via the lysosomal-mitochondrial pathway. J Biol Chem 280:40985–40995
Cheng EHY, Sheiko TV, Fisher JK et al. (2003) VDAC2 inhibits BAK activation and mitochondrial apoptosis. Science 301:513–517

Chipuk JE, Green DR (2005) Do inducers of apoptosis trigger caspase-independent cell death? Nat Rev Mol Cell Biol 6:268–275

Chipuk JE, Green DR (2008) How do BCL-2 proteins induce mitochondrial outer membrane permeabilization? Trends Cell Biol 18:157–164

Chowdhury D, Lieberman J (2008) Death by a thousand cuts: granzyme pathways of programmed cell death. Annual Rev Immunol 26:389–420

Chowdhury I, Tharakan B, Bhat GK (2008) Caspases – An update. Comp Biochem Physiol B Biochem Mol Biol 151:10–27

Cipolat S, Rudka T, Hartmann D et al. (2006) Mitochondrial rhomboid PARL regulates cytochrome c release during apoptosis via OPA1-dependent cristae remodeling. Cell 126:163–175

Creagh EM, Martin SJ (2001) Caspases: cellular demolition experts. Biochem Soc Trans 29:696–702

Crompton M, Barksby E, Johnson N et al. (2002) Mitochondrial intermembrane junctional complexes and their involvement in cell death. Biochimie 84:143–152

Cuervo AM (2004) Autophagy: in sickness and in health. Trends Cell Biol 14:70–77

Cullen SP, Adrain C, Lüthi AU et al. (2007) Human and murine granzyme B exhibit divergent substrate preferences. J Cell Biol 176:435–444

Danial NN, Korsmeyer SJ (2004) Cell death: critical control points. Cell 116:205–219

Debatin K-M, Krammer PH (2004) Death receptors in chemotherapy and cancer. Oncogene 23:2950–2966

Degterev A, Boyce M, Yuan J (2003) A decade of caspases. Oncogene 22:8543–8567

Dejean LM, Martinez-Caballero S, Guo L et al. (2005) Oligomeric Bax is a component of the putative cytochrome c release channel MAC, mitochondrial apoptosis-induced channel. Mol Biol Cell 16:2424–2432

Dejean LM, Martinez-Caballero S, Kinnally KW (2006) Is MAC the knife that cuts cytochrome *c* from mitochondria during apoptosis? Cell Death Differ 13:1387–1395

Deng J, Carlson N, Takeyama K et al. (2007) BH3 Profiling identifies three distinct classes of apoptotic blocks to predict response to ABT-737 and conventional chemotherapeutic agents. Cancer Cell 12:171–185

Deveraux QL, Reed JC (1999) IAP family proteins – Suppressors of apoptosis. Genes Dev 13:239–252

Deveraux QL, Roy N, Stennicke HR et al. (1998) IAPs block apoptotic events induced by caspase-8 and cytochrome c by direct inhibition of distinct caspases. EMBO J 17:2215–2223

Duan H, Orth K, Chinnaiyan AM et al. (1996) ICE-LAP6, a novel member of the ICE/Ced-3 gene family, is activated by the cytotoxic T cell protease granzyme B. J Biol Chem 271:16720–16724

Earnshaw WC, Martins LM, Kaufmann SH (1999) Mammalian caspases: structure, activation, substrates, and functions during apoptosis. Annu Rev Biochem 68:383–424

Eckhart L, Ballaun C, Uthman A et al. (2005) Identification and characterization of a novel mammalian caspase with proapoptotic activity. J Biol Chem 280:35077–35080

Eckhart L, Kittel C, Ballaun C et al. (2006) Caspase-15 is autoprocessed at two sites that contain an aspartate residue in the P1 position. Biochem Biophys Res Com 350:955–959

Espinosa JM (2008) Mechanisms of regulatory diversity within the p53 transcriptional network. Oncogene 27:4013–4023

Fernandes-Alnemri T, Armstrong RC, Krebs J et al. (1996) In vitro activation of CPP32 and Mch3 by Mch4, a novel human apoptotic cysteine protease containing two FADD-like domains. PNAS, USA 93:7464–7469

Fischer U, Jänicke RU, Schulze-Osthoff K (2003) Many cuts to ruin: a comprehensive update of caspase substrates. Cell Death Differ 10:76–100

Frank S, Gaume B, Bergmann-Leitner ES et al. (2001) The role of dynamin-related protein 1, a mediator of mitochondrial fission, in apoptosis. Dev Cell 1:515–525

Fuentes-Prior P, Salvesen GS (2004) The protein structures that shape caspase activity, specificity, activation and inhibition. Biochem J 384:201–232

Fuster JJ, Sanz-Gonzalez SM, Moll UM et al. (2007) Classic and novel roles of p53: Prospects for anticancer therapy. Trends Mol Med 13:192–199
Galluzzi L, Joza N, Tasdemir E et al. (2008) No death without life: vital functions of apoptotic effectors. Cell Death Differ 15:1113–1123
Galluzzi L, Maiuri MC, Vitale I et al. (2007) Cell death modalities: classification and pathophysiological implications. Cell Death Differ 14:1237–1243
Germain M, Mathai JP, McBride HM et al. (2005) Endoplasmic reticulum BIK initiates DRP1-regulated remodelling of mitochondrial cristae during apoptosis. EMBO J 24:1546–1556
Golks A, Brenner D, Fritsch C et al. (2005) c-FLIPR, a new regulator of death receptor-induced apoptosis. J Biol Chem 280:14507–14513
Guihard G, Bellot G, Moreau C et al. (2004) The mitochondrial apoptosis-induced channel (MAC) corresponds to a late apoptotic event. J Biol Chem 279:46542–46550
Gunter TE, Pfeiffer DR (1990) Mechanisms by which mitochondria transport calcium. Am J Physiol 258:C755–C786
Guo L, Pietkiewicz D, Pavlov EV et al. (2004) Effects of cytochrome c on the mitochondrial apoptosis-induced channel MAC. Am J Physiol Cell Physiol 286:C1109–1117
Guo Y, Srinivasula SM, Druilhe A et al. (2002) Caspase-2 induces apoptosis by releasing proapoptotic proteins from mitochondria. J Biol Chem 277:13430–13437
Gupta S (2003) Molecular signaling in death receptor and mitochondrial pathways of apoptosis. Int J Oncol 22:15–20
Hail N Jr, Carter BZ, Konopleva M et al. (2006) Apoptosis effector mechanisms: a requiem performed in different keys. Apoptosis 11:889–904
Hara T, Nakamura K, Matsui M et al. (2006) Suppression of basal autophagy in neural cells causes neurodegenerative disease in mice. Nature 441:885–889
Heath-Engel HM, Shore GC (2006) Regulated targeting of Bax and Bak to intracellular membranes during apoptosis. Cell Death Differ 13:1277–1280
Hetz C, Glimcher L (2008) The daily job of night killers: alternative roles of the BCL-2 family in organelle physiology. Trends Cell Biol 18:38–44
Hofmann K (1999) The modular nature of apoptotic signaling proteins. Cell Mol Life Sci 55:1113–1128
Hsu Y-T, Wolter KG, Youle RJ (1997) Cytosol-to-membrane redistribution of Bax and Bcl-XL during apoptosis. Proc Natl Acad Sci USA 94:3668–3672
Jans DA, Briggs LJ, Jans P et al. (1998) Nuclear targeting of the serine protease granzyme A (fragmentin-1). J Cell Sci 111:2645–2654
Johnson CR, Jarvis WD (2004) Caspase-9 regulation: an update. Apoptosis 9:423–427
Johnson DE (2000) Noncaspase proteases in apoptosis. Leukemia 14:1695–1703
Juo P, Woo MSA, Kuo CJ et al. (1999) FADD is required for multiple signaling events downstream of the receptor Fas. Cell Growth Differ 10:797–804
Karantza-Wadsworth V, Patel S, Kravchuk O et al. (2007) Autophagy mitigates metabolic stress and genome damage in mammary tumorigenesis. Genes Dev 21:1621–1635
Karbowski M, Norris KL, Cleland MM et al. (2006) Role of Bax and Bak in mitochondrial morphogenesis. Nature 443:658–662
Keefe D, Shi L, Feske S et al. (2005) Perforin triggers a plasma membrane-repair response that facilitates CTL induction of apoptosis. Immunity 23:249–262
Kerr JF (2002) History of the events leading to the formulation of the apoptosis concept. Toxicology 181–182:471–474
Kerr JF, Wyllie AH, Currie AR (1972) Apoptosis: a basic biological phenomenon with wide-ranging implications in tissue kinetics. Br J Cancer 26:239–257
Kinnally KW, Antonsson B (2007) A tale of two mitochondrial channels, MAC and PTP, in apoptosis. Apoptosis 12:857–868
Kinnally KW, Muro C, Campo ML (2000) MCC and PSC, the putative protein import channels of mitochondria. J Bioenerg Biomem 32:47–54

Klionsky DJ (2007) Autophagy: from phenomenology to molecular understanding in less than a decade. Nat Rev Mol Cell Biol 8:931–937

Krammer PH (2000) CD95's deadly mission in the immune system. Nature 407:789–795

Krammer PH, Kaminski M, Kießling M et al. (2007) No life without death. Adv Cancer Res 97C:111–138

Kroemer G, Galluzzi L, Brenner C (2007) Mitochondrial membrane permeabilization in cell death. Physiol Rev 87:99–163

Kroemer G, Martin SJ (2005) Caspase-independent cell death. Nat Med 11:725–730

Krueger A, Schmitz I, Baumann S et al. (2001) Cellular FLICE-inhibitory protein splice variants inhibit different steps of caspase-8 activation at the CD95 death-inducing signaling complex. J Biol Chem 276:20633–20640

Kuribayashi K, El-Deiry WS (2008) Regulation of programmed cell death by the p53 pathway. Adv Exp Med Biol 615:201–221

Kurschus FC, Bruno R, Fellows E et al. (2005) Membrane receptors are not required to deliver granzyme B during killer cell attack. Blood 105:2049–2058

Kurschus FC, Fellows E, Stegmann E et al. (2008) Granzyme B delivery via perforin is restricted by size, but not by heparan sulfate-dependent endocytosis. Proc Natl Acad Sci USA 105:13799–13804

Kuwana T, Mackey MR, Perkins G et al. (2002) Bid, Bax, and lipids cooperate to form supramolecular openings in the outer mitochondrial membran. Cell 111:331–342

Lane JD, Lucocq J, Pryde J et al. (2002) Caspase-mediated cleavage of the stacking protein GRASP65 is required for Golgi fragmentation during apoptosis. J Cell Biol 156:495–509

Lavrik IN, Golks A, Baumann S et al. (2006) Caspase-2 is activated at the CD95 death-inducing signaling complex in the course of CD95-induced apoptosis. Blood 108:559–565

Levine AJ, Hu W, Feng Z (2006) The P53 pathway: what questions remain to be explored? Cell Death Differ 13:1027–1036

Levine B, Sinha S, Kroemer G (2008) Bcl-2 family members: dual regulators of apoptosis and autophagy. Autophagy 4:600–606

Li N, Zheng Y, Chen W et al. (2007) Adaptor protein LAPF recruits phosphorylated p53 to lysosomes and triggers lysosomal destabilization in apoptosis. Cancer Res 67:11176–11185

Liu X, Dai S, Zhu Y et al. (2003) The structure of a Bcl-xL/Bim fragment complex: implications for Bim function. Immunity 19:341–352

Lum JJ, Bauer DE, Kong M et al. (2005) Growth factor regulation of autophagy and cell survival in the absence of apoptosis. Cell 120:237–248

Lüthi AU, Martin SJ (2007) The CASBAH: A searchable database of caspase substrates. Cell Death Differ 14:641–650

Mandic A, Hansson J, Linder S et al. (2003) Cisplatin induces endoplasmic reticulum stress and nucleus-independent apoptotic signaling. J Biol Chem 278:9100–9106

Mannella CA, Kinnally KW (2008) Reflections on VDAC as a voltage-gated channel and a mitochondrial regulator. J Bioenerg Biomembr 40:149–155

Marchenko ND, Moll UM (2007) The role of ubiquitination in the direct mitochondrial death program of p53. Cell Cycle 6:1718–1723

Marchetti P, Castedo M, Susin SA et al. (1996) Mitochondrial permeability transition is a central coordinating event of apoptosis. J Exp Med 184:1155–1160

Martinvalet D, Dykxhoorn DM, Ferrini R et al. (2008) Granzyme a cleaves a mitochondrial complex I protein to initiate caspase-independent cell death. Cell 133:681–692

Mathiasen IS, Jäättelä M (2002) Triggering caspase-independent cell death to combat cancer. Trends Mol Med 8:212–220

Mehlen P, Bredesen DE (2004) The dependence receptor hypothesis. Apoptosis 9:37–49

Meulmeester E, Jochemsen AG (2008) p53: A guide to apoptosis. Curr Cancer Drug Targets 8:87–97

Micheau O, Thome M, Schneider P et al. (2002) The long form of FLIP is an activator of caspase-8 at the Fas death-inducing signaling complex. J Biol Chem 277:45162–45171

Mikhailov V, Mikhailova M, Degenhardt K et al. (2003) Association of Bax and Bak homo-oligomers in mitochondria: Bax requirement for Bak reorganization and cytochrome c release. J Biol Chem 278:5367–5376
Moll UM, Wolff S, Speidel D et al. (2005) Transcription-independent pro-apoptotic functions of p53. Curr Opinion Cell Biol 17:631–636
Motyka B, Korbutt G, Pinkoski MJ et al. (2000) Mannose 6-phosphate/insulin-like growth factor II receptor is a death receptor for granzyme B during cytotoxic T cell-induced apoptosis. Cell 103:491–500
Nakagawa T, Shimizu S, Watanabe T et al. (2005) Cyclophilin D-dependent mitochondrial permeability transition regulates some necrotic but not apoptotic cell death. Nature 434:652–658
Nakagawa T, Yuan J (2000) Cross-talk between two cysteine protease families: activation of caspase-12 by calpain in apoptosis. J Cell Biol 150:887–894
Nicholson DW (1999) Caspase structure, proteolytic substrates, and function during apoptotic cell death. Cell Death Differ 6:1028–1042
O'Reilly LA, Ekert P, Harvey N et al. (2002) Caspase-2 is not required for thymocyte or neuronal apoptosis even though cleavage of caspase-2 is dependent on both Apaf-1 and caspase-9. Cell Death Differ 9:832–841
Oldenhuis C, Stegehuis JH, Walenkamp A et al. (2008) Targeting TRAIL death receptors. Curr. Opinion Pharmacol. 8:433–439
Olichon A, Baricault L, Gas N et al. (2003) Loss of OPA1 perturbates the mitochondrial inner membrane structure and integrity, leading to cytochrome c release and apoptosis. J Biol Chem 278:7743–7746
Oppenheim JJ, Yang D (2005) Alarmins: chemotactic activators of immune responses. Host-pathogen interactions/immunological techniques. Curr Opinion Immunol 17:359–365
Palacios G, Crawford HC, Vaseva A et al. (2008) Mitochondrially targeted wild-type p53 induces apoptosis in a solid human tumor xenograft model. Cell Cycle 7:2584–2590
Pan G, Bauer JH, Haridas V et al. (1998) Identification and functional characterization of DR6, a novel death domain-containing TNF receptor. FEBS Lett 431:351–356
Parone PA, James DI, Cruz SD et al. (2006) Inhibiting the mitochondrial fission machinery does not prevent Bax/Bak-dependent apoptosis. Mol Cell Biol 26:7397–7408
Parone PA, Martinou J-C (2006) Mitochondrial fission and apoptosis: an ongoing trial. Nat Rev Mol Cell Biol 1763:522–530
Pavlov EV, Priault M, Pietkiewicz D et al. (2001) A novel, high conductance channel of mitochondria linked to apoptosis in mammalian cells and Bax expression in yeast. J Cell Biol 155:725–732
Peixoto PMV, Graña F, Roy T et al. (2007) Awaking TIM22, a dynamic ligand-gated channel for protein insertion in the mitochondrial inner membrane. J Biol Chem 282:18694–18701
Peixoto PMV, Martinez-Caballero S, Grigoriev SM et al. (2004) The ins and outs of mitochondrial protein import from an electrophysiological point of view. In: Pandalai SG (ed) Recent research developments in biophysics, vol 3. Transworld Research Network, Kerala, pp 413–474
Penaloza C, Lin L, Lockshin R et al. (2006) Cell death in development: shaping the embryo. Histochem Cell Biol 126:149–158
Peter ME (2000) The trail discussion: it is FADD and caspase-8! Cell Death Differ 7:759–760
Peter ME, Hellbardt S, Schwartz-Albiez R et al. (1995) Cell surface sialylation plays a role in modulating sensitivity towards APO-1-mediated apoptotic cell death. Cell Death Differ 2:163–171
Pipkin ME, Lieberman J (2007) Delivering the kiss of death: progress on understanding how perforin works. Curr Opinion Immunol 19:301–308
Pluquet O, Hainaut P (2001) Genotoxic and non-genotoxic pathways of p53 induction. Cancer Lett 174:1–15
Renatus M, Zhou Q, Stennicke HR et al. (2000) Crystal structure of the apoptotic suppressor CrmA in its cleaved form. Structure 8:789–797

Riedl SJ, Shi Y (2004) Molecular mechanisms of caspase regulation during apoptosis. Nat Rev Mol Cell Biol 5:897–907

Robertson JD, Enoksson M, Suomela M et al. (2002) Caspase-2 acts upstream of mitochondria to promote cytochrome c release during etoposide-induced apoptosis. J Biol Chem 277:29803–29809

Roucou X, Rostovtseva T, Montessuit S et al. (2002) Bid induces cytochrome c-impermeable Bax channels in liposomes. Biochem J 363:547–552

Rupinder SK, Gurpreet AK, Manjeet S (2007) Cell suicide and caspases. Vascul Pharmacol 46:383–393

Salvesen GS, Duckett CS (2002) IAP proteins: blocking the road to death's door. Nat Rev Mol Cell Biol 3:401–410

Scorrano L, Ashiya M, Buttle K et al. (2002) A distinct pathway remodels mitochondrial cristae and mobilizes cytochrome c during apoptosis. Dev Cell 2:55–67

Setoguchi K, Otera H, Mihara K (2006) Cytosolic factor- and TOM-independent import of C-tail-anchored mitochondrial outer membrane proteins. EMBO J 25:5635–5647

Shi Y (2002) Mechanisms of caspase activation and inhibition during apoptosis. Mol Cell 9:459–470

Shimizu S, Kanaseki T, Mizushima N et al. (2004) Role of Bcl-2 family proteins in a non-apoptopic programmed cell death dependent on autophagy genes. Nat Cell Biol 6:1221–1228

Sodeman T, Bronk SF, Roberts PJ et al. (2000) Bile salts mediate hepatocyte apoptosis by increasing cell surface trafficking of Fas. Am J Physiol Gastrointest Liver Physiol 278:G992–G999

Soussi T (2005) The p53 pathway and human cancer. Br J Surg 92:1331–1332

Sprick MR, Rieser E, Stahl H et al. (2002) Caspase-10 is recruited to and activated at the native TRAIL and CD95 death-inducing signalling complexes in a FADD-dependent manner but can not functionally substitute caspase-8. EMBO J 21:4520–4530

Stennicke HR, Deveraux QL, Humke EW et al. (1999) Caspase-9 can be activated without proteolytic processing. J Biol Chem 274:8359–8362

Stennicke HR, Ryan CA, Salvesen GS (2002) Reprieval from execution: the molecular basis of caspase inhibition. Trends Biochem Sci 27:94–101

Susin SA, Lorenzo HK, Zamzami N et al. (1999) Molecular characterization of mitochodrial apoptosis-inducing factor. Nature 397:441–446

Suzuki Y, Takahashi-Niki K, Akagi T et al. (2004) Mitochondrial protease Omi/HtrA2 enhances caspase activation through multiple pathways. Cell Death Differ 11:208–216

Takano J, Tomioka M, Tsubuki S et al. (2005) Calpain mediates excitotoxic DNA fragmentation via mitochondrial pathways in adult brains: evidence from calpastatin mutant mice. J Biol Chem 280:16175–16184

Takle H and Andersen O (2007) Caspases and apoptosis in fish. J Fish Biol 71:326–349

Thorburn A (2008) Apoptosis and autophagy: regulatory connections between two supposedly different processes. Apoptosis 13:1–9

Thornberry NA, Lazebnik Y (1998) Caspases: enemies within. Science 281:1312–1316

Thornberry NA, Rano TA, Peterson EP et al. (1997) A combinatorial approach defines specificities of members of the caspase family and granzyme B: Functional relationships established for key mediators of apoptosis. J Biol Chem 272:17907–17911

Trinchieri G, Sher A (2007). Cooperation of Toll-like receptor signals in innate immune defence. Nat Rev Immunol 7:179–190

Truneh A, Sharma S, Silverman C et al. (2000) Temperature-sensitive differential affinity of TRAIL for its receptors: DR5 is the highest affinity receptor. J Biol Chem 275:23319–23325

Tsujimoto Y, Shimizu S (2005) Another way to die: autophagic programmed cell death. Cell Death Differ 12:1528–1534

Tu S, McStay GP, Boucher L-M et al. (2006) In situ trapping of activated initiator caspases reveals a role for caspase-2 in heat shock-induced apoptosis. Nat Cell Biol 8:72–77

Turk V, Turk B, Turk D (2001) Lysosomal cysteine proteases: facts and opportunities. EMBO J 20:4629–4633

Vaux DL, Cory S, Adams JM (1988) Bcl-2 gene promotes haemopoietic cell survival and cooperates with c-myc to immortalize pre-B cells. Nature 335:440–442
Veugelers K, Motyka B, Frantz C et al. (2004) The granzyme B-serglycin complex from cytotoxic granules requires dynamin for endocytosis. Blood 103:3845–3853
Wagner KW, Engels IH, Deveraux QL (2004) Caspase-2 can function upstream of bid cleavage in the TRAIL apoptosis pathway. J Biol Chem 279:35047–35052
Wang ECY, Thern A, Denzel A et al. (2001) DR3 regulates negative selection during thymocyte development. Mol Cell Biol 21:3451–3461
Wang KKW (2000) Calpain and caspase: can you tell the difference? Trends Neurosci 23:20–26
Wang L, Yu C, Lu Y et al. (2007) TMEM166, a novel transmembrane protein, regulates cell autophagy and apoptosis. Apoptosis 12:1489–1502
Wang Y, Singh R, Massey AC et al. (2008) Loss of macroautophagy promotes or prevents fibroblast apoptosis depending on the death stimulus. J Biol Chem 283:4766–4777
Wei MC, Zong WX, Cheng EH et al. (2001) Proapoptotic BAX and BAK: A requisite gateway to mitochondrial dysfunction and death. Science 292:727–730
Williams JR, Little JB, Shipley WU (1974) Association of mammalian cell death with a specific endonucleolytic degradation of DNA. Nature 252:754–755
Willis SN, Fletcher JI, Kaufmann T et al. (2007) Apoptosis initiated when BH3 ligands engage multiple Bcl-2 homologs, not Bax or Bak. Science 315:856–859
Wolff S, Erster S, Palacios G et al. (2008) p53's mitochondrial translocation and MOMP action is independent of Puma and Bax and severely disrupts mitochondrial membrane integrity. Cell Res 18:733–744
Wood DE, Newcomb EW (1999) Caspase-dependent activation of calpain during drug-induced apoptosis. J Biol Chem 274:8309–8315
Xu G, Cirilli M, Huang Y et al. (2001) Covalent inhibition revealed by the crystal structure of the caspase-8/p35 complex. Nature 410:494–497
Yamashima T (2004) Ca^{2+}-dependent proteases in ischemic neuronal death. A conserved 'calpain-cathepsin cascade' from nematodes to primates. Cell Calcium 36:285–293
Youle RJ (2007) Cellular demolition and the rules of engagement. Science 315:776–777
Youle RJ, Strasser A (2008) The BCL-2 protein family: opposing activities that mediate cell death. Nat Rev Mol Cell Biol 9:47–59
Yu L, Wan F, Dutta S et al. (2006) Autophagic programmed cell death by selective catalase degradation. Proc Natl Acad Sci USA 103:4952–4957
Zakeri Z, Lockshin RA (2008) Cell death: history and future. Adv Exp Med Biol 615:1–11
Zang Y, Beard RL, Chandraratna RAS et al. (2001) Evidence of a lysosomal pathway for apoptosis is induced by the synthetic retinoid CD437 in human leukemia HL-60 cells. Cell Death Differ 8:477–485
Zha J, Weiler S, Oh KJ et al. (2000) Posttranslational N-myristoylation of BID as a molecular switch for targeting mitochondria and apoptosis. Sciences 290:1761–1765
Zitvogel L, Kroemer G (2008) Introduction: the immune response against dying cells. Curr Opinion Immunol 20:501–503

Chapter 3
Apoptosis in Exocrine Acinar Cells

J.A. Pariente

Abstract Apoptosis or programmed cell death is an essential physiological process that is required for the normal development and maintenance of the tissue homeostasis. When misregulated, apoptosis can contribute to various diseases. Different studies have revealed that apoptosis is a major mechanism of pancreatic acinar cells death in various experimental models of acute pancreatitis, suggesting a role for apoptosis in the pathophysiology of the disease. In addition, the relationship between reactive oxygen species and apoptosis in the pathophysiology of acute pancreatitis in acinar cells has been reported. A variety of studies indicate that oxidative stress may play an important role in the initiation and development of acute pancreatitis. In particular, inhibition of pancreatic secretions, an elevation in serum enzyme levels, cytoplasmic vacuolization, death of acinar cells, oedema formation, and an infiltration of inflammatory cells into the pancreas are some of the effect of pancreatitis. It has been shown that both apoptosis and necrosis occur in various models of pancreatitis and, in particular, in cerulein-induced pancreatitis, however, the mechanisms underlying theses processes remain controversial. In addition, apoptosis of salivary epithelial acinar cells plays a major role in the pathogenesis of Sjögren's syndrome. Finally, Ca^{2+} is one of the key regulators of cell survival, but Ca^{2+} can also induce apoptosis in response to a variety of pathological conditions.

Keywords Pancreas · Salivary gland · Cholecystokinin · Sjögren's syndrome · Hydrogen peroxide

3.1 Introduction

The number of published papers on apoptosis or programmed cell death in different cell types and different tissues has considerably increased over the last decade. However, most of the studies carried out to analyze apoptosis in exocrine acinar cells

J.A. Pariente (✉)
Department of Physiology, University of Extremadura, Avda Elvas s/n, 06071 Badajoz, Spain
e-mail: pariente@unex.es

G.M. Salido, J.A. Rosado (eds.), *Apoptosis: Involvement of Oxidative Stress and Intracellular Ca^{2+} Homeostasis*, DOI 10.1007/978-1-4020-9873-4_3,

have basically focused in the role that apoptosis of pancreatic acinar cells plays in acute pancreatitis. Acute pancreatitis is a severe and debilitating inflammatory disease with a steadily increasing incidence over recent years and significant mortality. Inflammation and acinar cell death are the hallmarks of both human and experimental pancreatitis. Although significant progress in understanding the inflammatory response has been achieved in the last years, the mechanisms of acinar cell death in pancreatitis remain largely unexplored. In fact, the events that regulate the severity of acute pancreatitis are, for the most part, unknown. The mechanisms through which the diverse etiologic factors can induce pancreatitis are likewise unclear, but once the disease process is initiated common inflammatory and repair pathways are invoked. This chapter attempts to discuss our current understanding of the contribution of exocrine acinar cells injury (especially, exocrine acinar cell apoptosis) to severity of acute pancreatitis.

3.2 General Morphology of Exocrine Cells

The majority of an exocrine gland is composed of acinar cells. The acinus is bound by a connective tissue matrix, including the basal lamina, which does not course between the lateral areas of contiguous acinar cells. Gap junctions connect the cytoplasm of the acinar cells and allows for electric coupling within the acini. However, there are also tight junction structures that connect the acinar cells with each other (Iwatsuki and Petersen 1979).

The ultrastructural appearance of acinar cells reveals that they are specialised for enzyme secretion. By light microscopy, these acinar cells are pyramidally shaped and polarised, with the apex facing the lumen of the acinus (Fig. 3.1). Areas of specialisation are easily appreciated within acinar cells (Gorelick and Jamieson 1987). Secretory granules are restricted to the apical cytoplasm of the cells and vary in number, depending on the stage of development and the state of stimulation by agonists (Burgoyne and Morgan 2003). An area between the basally located nucleus and the apex is often paler staining by light microscopy. This represents the Golgi complex, which consists of two elements: stacks of flattened cisternae of varying size and numerous small, smooth-surface vesicles whose dimensions are variable. These vesicles are particularly prominent on the *cis* face (or forming face) and lateral margins of the cisternae. They are orientated towards elements of the rough endoplasmic reticulum and are partly devoid of ribosomes. On the *trans* face (or maturing face) of the Golgi complex and in some species in the torus of the Golgi cisternae, forming secretory granules and lysosomes are found. The initial stage of secretory granule formation (condensing vacuoles) is distinguished from lysosomes on the basis of their size, configuration and histochemical staining patterns. Condensing vacuoles derived from the trans-Golgi cisternae mature secretory granules by a process involving aggregate formation among the secretory products. Mature secretory granules are stored in the apical portion of the acinar cells and accumulate beneath the luminal plasma membrane, which projects numerous microvilli into the luminal space (Reddy et al. 1979).

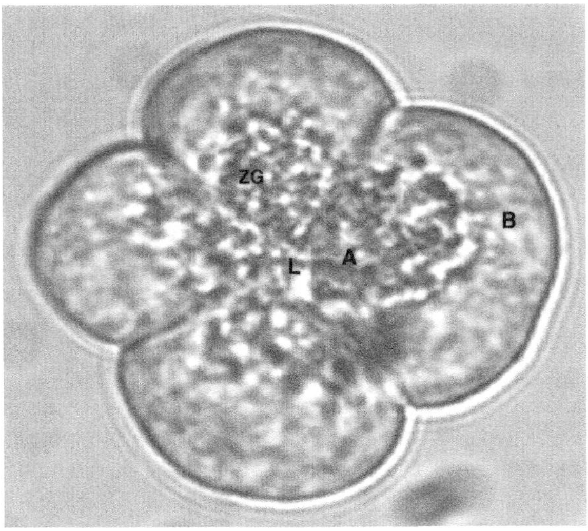

Fig. 3.1 Transmitted light photomicrograph of typical exocrine acinar cells, such as the mouse pancreatic acinus. A: Apical cytoplasm rich in zymogen (secretory) granules (ZG). B: Basal, basophilic staining regions representing rough endoplasmic reticulum. L: Lumen of the acinus (modified from Granados et al. 2005)

A third zone characterized by intense basophilia is located in the basal region of the cell and represents endoplasmic reticulum. By conventional thin section electron microscopy, the rough endoplasmic reticulum consists of a highly convoluted tridimensional array of cisternae bounded by a bilayer 60–70 Å in thickness. Continuity among the rough endoplasmic reticulum elements is especially evident when their cisternae are distended with secretory products. Immunocytochemical studies (Bendayan et al. 1980) suggest that all rough endoplasmic reticulum cisternae are probably equivalent in their capacity to synthesize individual secretory proteins, a finding consistent with the proposal that the rough endoplasmic reticulum is a continuous space. Finally, the mitochondria are large, abundant, and tend to be elongated. Occasionally, small, smooth-surface vesicles are seen in the cytoplasm. Microtubules and microfilaments are found between secretory granules and under the apical cell membrane. These filamentous structures are involved in transport and discharge of secretory granules (Williams 1977).

3.3 Apoptosis and Necrosis in Exocrine Acinar Cells

Programmed cell death leading to apoptosis is essential for normal tissue development and homeostasis, and it contributes to certain forms of pathological cell loss. Although apoptosis was originally defined by characteristic morphological features, including cell shrinkage, nuclear chromatin condensation and the formation of apoptotic bodies, it is now better defined by its macromolecular underpinnings (Yu et al. 2001) mediated by a family of cystein aspartases (caspases), which are expressed as inactive zymogens and are proteolytically processed to activate state following an apoptotic stimulus.

Two separable pathways leading to caspase activation have been characterised. One pathway is the so-called extrinsic pathway initiated by the binding of an extracellular death ligand, such as FasL, to its cell-surface death receptor, such as Fas (Ashkenazi and Dixit 1998). The second pathway is the intrinsic pathway, which is mediated by mitochondrial alterations. In response to apoptotic stimuli, several proteins are released from the intermembrane space of mitochondria into the cytoplasm (Green and Reed 1998). Some of well-characterized proteins include cytochrome *c*, which mediates the activation of caspase 9 (Li et al. 1997), triggering a cascade of caspase activation, including caspase 3, which promotes cellular apoptosis. The release of cytochrome *c* by mitochondria is almost a universal feature found in response to various intracellular stimuli, including DNA damage, glucocorticoids, oxidative injury, and growth factor deprivation.

Mitochondrial membrane permeabilisation is regulated by the opposing actions of pro and anti-apoptotic Bcl-2 family members. Based on both structural and functional criteria, three subgroups of Bcl-2 homologues have been identified. One of them includes pro-apoptotic effectors like Bak, Bax, and Bok, characterised by containing three of the prototypical BH domains (BH1, BH2, and BH3) (Adams and Cory 1998; Reed 1998). A functional BH3 domain has been shown to be critical for the apoptotic activity of these proteins (Chittenden et al. 1995; Simonen et al. 1997). A second subgroup contains pro-apoptotic molecules structurally related by the presence of only one family domain corresponding to BH3. Members of this subfamily are thus known as BH3-only proteins (Bouillet and Strasser 2002). The third subgroup includes anti-apoptotic homologues like Bcl-2, Bcl-w, and Bcl-XL, and a distinct structural feature is the presence of a BH4 domain in addition to BH domains 1, 2, and 3 (Adams and Cory 1998; Reed 1998). Bcl-2 family proteins have a propensity to dimerise, and the fine balance between pro- and anti-apoptotic members often defines whether a cell will survive or will commit to death in response to a particular insult (Reed 1998; Danial and Korsmeyer 2004).

There is considerable cross-talk between the extrinsic and intrinsic pathways. For example, caspase 8 can proteolytically activate Bid, which can then facilitate cytochrome *c* release (Bhatia 2004b). This apparently amplifies the apoptotic signal following death receptor activation, and different cell types may be more reliant on this amplification pathway than others. Conversely, activators of the intrinsic pathway can sensitise the cell to extrinsic death ligands (Bhatia 2004a).

In other hand, recent studies revealed that apoptosis is a major mechanism of exocrine acinar cell death in various experimental models of acute pancreatitis, suggesting a role for apoptosis in the pathophysiology of the disease (Kaiser et al. 1995; Sandoval et al. 1996). Inflammation and death of pancreatic acinar cells are the hallmark of both human and experimental pancreatitis (Balkwill and Burke 1996; Cohen and Cphen 1996). Death can occur by either necrosis or apoptosis. Some evidence has indicated that mild acute pancreatitis was characterized by very little necrosis but a high degree of apoptosis, while severe acute pancreatitis was found to involve extensive acinar cell necrosis (Bhatia 2004a; Gukovskaya et al. 1996; Kaiser et al. 1995). Histological observation of pancreatic acinar cells treated with cerulein showed predominantly apoptotic cells with condensed chromatin, shrunken nuclei with convoluted shapes, and compact cytoplasm but fewer necrotic cells with swelling and disrupted plasma membrane (Yu et al. 2007). Similar phenomena were observed on pancreatic acinar cells exposed to oxidative stress, showing apoptotic cells with characteristic nuclear condensation and shrinkage and large vacuole (Song et al. 2003). Sandoval et al. (1996) showed that cerulein stimulates pancreatic production of platelet-activating factor (PAF). PAF mediates both apoptosis and neutrophil chemotaxis in the pancreas. Neutrophils, in turn, may convert acinar cells undergoing apoptosis into necrotic cells. Therefore, an early stage of apoptosis might be controlled in order to prevent pancreatic necrosis by inflammatory cells. Others demonstrate that the induction of apoptosis may be beneficial because it prevents cellular necrosis and subsequent inflammation (Bhatia et al. 1998; Mareninova et al. 2006).

Finally, apoptosis causing the loss of acinar cells has been noted in several studies in animals (Su et al. 2001; Hashimoto et al. 2000). Parameters of apoptosis have been investigated in human chronic pancreatitis, and the Fas and FasL pathway was shown to be activated (Kornmann et al. 2000).

3.4 Caspase Activation and Apoptosis in Exocrine Acinar Cells

Caspase activation may be one of the mechanisms of apoptosis in pancreatitis. There are several studies showing that different pathogenic factors in acute pancreatitis lead to an activation of caspase in acinar cells. In fact, apoptotic cell death is attributed to the activation of caspase 3 and the accumulation of p53, a known signalling molecule that acts upstream of caspase 3 (Kim et al. 2002). The signalling pathways leading to caspase activation during apoptosis cause the release of cytochrome *c* and other apoptogenic factors from injured mitochondria. Apoptosis inducing factor (AIF) is a conserved mitochondrial protein that is released into the cytoplasm and nucleus during apoptotic stimuli, inducing chromatin condensation and DNA fragmentation in human and rodent cells. AIF induces apoptotic changes in purified nuclei, even in the presence of caspase inhibitors (Cande et al. 2002; Susin et al. 1999) suggesting that the mediator function of AIF for apoptosis acts in a caspase-independent fashion. Caspase 3 activation and AIF expression were induced by cerulein in pancreatic acinar AR42J cells (Yu et al. 2007), suggesting that cerulein causes mitochondria to activate caspase 3 and transcriptional regulation and thus induces the expression of apoptotic genes such as AIF. In addition, stimulation of isolated rat pancreatic acini with cholecystokinin (CCK), which serves as a model for human acute oedematous pancreatitis, leads to a rapid redistribution and activation of caspase 8 (Beil et al. 2002).

In addition to apoptosis, caspases also regulate other physiological processes required for some cellular functions. In particular, we have recently demonstrated an early activation of caspase 3 by cholecystokinin (CCK) in mouse pancreatic acinar cells, which is required for pancreatic secretion. Treatment of mouse pancreatic acinar cells with a caspase 3 inhibitor clearly reduced amylase release stimulated by

CCK (Fig. 3.2), suggesting that caspase 3 activation is involved in the physiological activation of amylase secretion in pancreatic acinar cells (Rosado et al. 2006). Additionally, treatment with CCK was able to induce a concentration-dependent increase in phosphatidylserine externalisation (Fig. 3.2B) (Rosado et al. 2006).

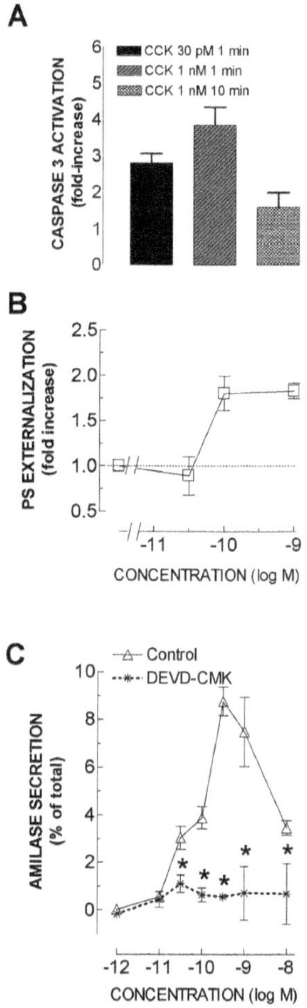

Fig. 3.2 CCK induces rapid activation of caspase 3 that is required for amylase secretion. (**A**) Freshly isolated pancreatic acinar cells were stimulated with CCK and caspase 3 was analysed. Values are presented as mean±SEM ($n = 6$) expressed as the percentage of resting values. (**B**) Pancreatic acinar cells were stimulated with different concentrations of CCK-8 and phosphatidylserine externalisation was analysed. Values are presented as mean ± SEM ($n = 6$) expressed as fold-increase over the pre-treatment level (experimental/control). (**C**) Pancreatic acinar cells were incubated with DEVD-CMK (an inhibitor of caspase 3) or the vehicle (DMSO) as control and then were stimulated with different concentrations of CCK-8. Amylase activity was expressed as the percentage of cellular amylase released into the extracellular medium and presented as the increase over basal. Results are mean ± SEM of six separate experiments in which each point was determined in duplicate. $^*P < 0.05$ compared with amylase release in cells stimulated with CCK-8 in the absence of the inhibitor. (For experimental details see Rosado et al. 2006)

3.5 Genes Expression, Transcription Factors and Acinar Cells Apoptosis

In addition to caspases, the apoptotic regulators Bcl-2 family is another one the functional components comprised in the apoptosis pathway (Bhatia 2004b). Different studies have shown that acinar cells apoptosis observed in acute pancreatitis is also at least partly attributable to the greatly increases proapoptotic *bax* gene expression. It has been reported that acute pancreatitis induced a increase in pancreatic *bax* mRNA levels closely followed by an increase in Bax protein levels in the cells and it suggested that acinar cells apoptosis observed during caerulein-induced acute pancreatitis is attributable, at least in part, to the greatly increased *bax* expression (Gomez et al. 2001). Besides Bcl-2 family, SIP gene has also been identified. SIP expression is localised in acinar cells, and it is overexpressed during pancreatitis and promoted cell apoptosis, thus suggesting that SIP is a stress-inducible gene (Tomasini et al. 2001).

Moreover, transcription factors may be also involved in apoptosis. The transcription factor p53 can initiated the apoptosis by inducing *bax* expression; however, the role of this factor remains unclear in acinar cells apoptosis. It has been reported that p53 mRNA levels are temporally coordinated with those of *bax* during acute pancreatitis (Gomez et al. 2001). However, others have reported that the mechanisms of pancreatic acinar cells apoptosis correlated with the expression of apoptosis-regulated gene *bax*, but had no relationship with the expression of p53 (Yuan et al. 2001).

Others transcription factors, such as nuclear factor kappa B (NF-κB) and activator protein-1 (AP-1), may also be involved in the signal transduction pathways leading to apoptosis. Early studies have suggested a possible link between apoptosis and activation of NF-κB in pancreatic acinar cells (Gukovsky et al. 1998). The potential role of NF-κB in cells death was suggested by activation of TNF-α transcription (Gukovsky et al. 1998).

The low-affinity neurotrophin receptor p75 (p75NTR) mediates apoptosis in many cell types *in vivo* and *in vitro*. In chronic pancreatitis samples, cells with elevated p75NTR mRNA and protein were present in the outer part of acinar lobules, most ducts, and in ductular structures, which suggest that p75NTR is involved in the apoptotic process of the exocrine pancreas (Zhu et al. 2003).

3.6 Neutrophils, Citokines and Apoptosis in Exocrine Acinar Cells

Neutrophils activation is an important determinant of severity of acute pancreatitis. Some reports have recently indicated that depletion of neutrophils results in a significant increase in acinar cells undergoing apoptosis. It has been reported a dramatic increase in apoptotic acinar cells by administration of antineutrophil serum, which, comparable to aICAM-1 (an ICAM-1 neutrophilising antibody), significantly reduced the extent of necrosis and the inflammatory infiltrate (Fujimoto et al. 1997; Bhatia, 2004a). Yet another study, using taurocholate-induced severe acute pancreatitis as the experimental model, showed that neutrophils, via a tumor necrosis factor (TNF)-α-dependent mechanism, may be involved in the development of apoptotic as well as necrotic forms of acinar cell death (Rau et al. 2001).

In addition to their role in inducing inflammation, cytokines, such as TNF-α or interleukin (IL)-1β, which are released by neutrophils (and to a lesser extent, by acinar cells) have been shown to induce apoptosis in pancreas (Gukovskaya et al. 1997). Pancreatic acinar cells have been reported produce and release TNF-α (Gukovskaya et al. 1997). These authors have found that TNF-α was involved in the development of pancreatitis, and it mediated apoptosis in acinar cell suspension in vitro and also in vivo in the cerulein of acute pancreatic.

3.7 Induction of Pancreatic Acinar Cell Apoptosis by Physiological Agents

High-dose CCK, a major agonist of pancreatic acinar cells, through its G-protein-coupled receptor, stimulates death signalling pathways in rat pancreatic acinar cells, including caspase activation, cytochrome *c* release, and mitochondrial depolarization, leading to apoptosis (Gukovskaya et al. 2002). The mitochondrial dysfunction is mediated by upstream caspase(s). CCK causes mitochondrial alterations through both permeability transition pore (PTP)-dependent and PTP-independent mechanisms. In addition to apoptosis, caspases also regulate other processes in the pancreatic acinar cell that play key roles in pancreatitis; in particular, caspases negatively regulate necrosis and intra-acinar cell activation of trypsin. Caspase-mediated protection against necrosis and trypsin activation can explain the inverse correlation between the extent of apoptosis, on the one hand, and necrosis and the severity of the disease, on the other hand, observed in experimental models of pancreatitis. These signalling mechanisms may play an important role in acinar injury and death in pancreatitis. CCK also increases the number of cells with apoptotic nuclear morphology and stimulates internucleosomal DNA fragmentation (Gukovskaya et al. 2002). In addition, we have recently reported that stimulation of mice pancreatic acinar cells with CCK induces rapid activation of caspase 3 that was detected within 1 minute of stimulation and a concentration-dependent increase in phosphatidylserine exposure (Rosado et al. 2006). These results, taken together, demonstrate that CCK stimulates apoptosis in pancreatic acinar cells.

Animal models of acute pancreatic insufficiency have been created by inducing pancreatic acinar cells apoptosis, either by administration of the synthetic cholecystokinin (CCK-8) analogue cerulein (Reid and Walker 1999) or by physical blockage of the pancreas by ductal ligation (Hamamato et al. 2002). High doses of cerulein result in experimental pancreatitis, which is characterized by a deregulation of the production and secretion of digestive enzymes, particularly the inhibition of pancreatic secretion, and an elevation in their serum levels (hyperamylasemia), cytoplasmic vacuolization, the death of acinar cells, oedema formation similar to human pancreatitis (Lerch and Adler 1994) and high production of lipid peroxide (an index of oxidative cell damage in the pancreas of the rats (Choi and Kim 1998)). In addition, supraphysiological concentrations of cerulein induce reactive oxygen species (ROS) production (Yu et al. 2005) and apoptosis in rat pancreas. The role of ROS on altering the cytoskeleton function in the pancreas has been shown (Jungerman et al. 1995). Cytoskeleton disruption may induce disturbances in the intracellular transport of digestive enzymes, leading to their premature intracellular activation (Dabrowski et al. 1999). Finally, CCK-8 has also the ability to activate apoptosis-related caspase 3 by ROS generation (Trulsson et al. 2001).

Several apoptotic regulatory peptides and pathways have been reported (Borner and Monney 1999) but very little is known about apoptotic regulation on the exocrine pancreas with regard to nitric oxide (NO). In hepatocytes, endogenous NO has a capacity to suppress apoptosis by interrupting the caspase activation (Li et al. 1999), and in pancreatic acinar cells endogenous NO generation, which is inhibited by Nω-nitro-L-arginine (L-NNA), reduces apoptosis both during basal condition and during simultaneous stimulation with CCK-8 (Trulsson et al. 2002).

ROS produced during pancreatitis are also the important mediators of apoptosis (Kaiser et al. 1995; Sandoval et al. 1996). In cerulein-induced pancreatitis, a high degree of ROS production and apoptosis were observed in pancreatic acinar cells (Gukovskaya et al. 1997; Kimura et al. 1998). In pancreatic acinar AR42J cells, cerulein-induced increases in apoptotic indices including apoptosis inducing factor (AIF) expression, DNA fragmentation, TUNEL staining, and caspase 3 activity (Yu et al. 2007). Additionally, we have recently shown that H_2O_2 stimulates apoptosis in rat pancreatic acinar AR42J cells, which was mediated by mitochondrial depolarization and cytochrome *c* release, and caspase 3 activation (Morgado et al. 2008).

3.8 Induction of Pancreatic Acinar Cell Apoptosis by Agents That Do Not Cause Acute Pancreatitis

CCK is the major agonist of pancreatic acinar cells and it can, however, also induce both pancreatic acinar cell apoptosis and acute pancreatitis (Bhatia 2004a). In addition to CCK, other inducers of pancreatic acinar cell apoptosis have been used to investigate cell death in relation to acute pancreatitis. Examples of these inducers are menadione and crambene (1-cyano-2-hydroxy-3-butene-CHB). Menadione is a quinone that is metabolised by flavoprotein reductase to semiquinone, which can be oxidised back to quinine in the presence of molecular oxygen. In this redox cycle the superoxide anion radical, hydrogen peroxide and other ROS are generated (Monks et al. 1992).

Menadione can cause elevations in the cytosolic free Ca^{2+} concentration ($[Ca^{2+}]_c$) contributing to cell death (Nicotera et al. 1992). In mouse pancreatic acinar cells it has been shown that menadione induces repetitive cytosolic Ca^{2+} spikes, partial mitochondrial depolarisation through induction of the permeability transition pore, efflux of cytochrome *c* into the cytoplasm and activation of caspase 9 and 3 (Gerasimenko et al. 2002). Crambene is a stable plant nitrile found in many cruciferous vegetables, and it is also a potent inducer of "phase II" detoxification enzymes, including certain glutathione S-transferases and quinone reductase, which are important enzymes associated with conjugation and elimination of reactive chemical intermediates and carcinogens (Wallig et al. 1998; Niedoborski et al. 2001). The mechanism of induction of pancreatic acinar cells apoptosis by crambene is not yet clear. Induction of pancreatic acinar cells apoptosis with crambene protects mice against acute pancreatitis induced by cerulein (Bhatia et al. 1998; Bhatia and Wallig 2004). This is despite the fact that crambene does not alter the interaction of CCK to its receptor on pancreatic acinar cells (Bhatia et al. 1998). The maximal protective effect of crambene against acute pancreatitis is observed when the pancreatic acinar cells have become committed to apoptosis (Bhatia et al. 1998; Bhatia and Wallig 2004).

3.9 Apoptosis in Salivary Acinar Cells

Apoptosis of salivary epithelial acinar cells plays a major role in the pathogenesis of Sjögren's syndrome, an autoimmune disease where the exocrine glands are targeted, characterised by lymphocyte infiltration into salivary and lachrymal glands with concomitant tissue destruction (Nakamura et al., 1998). Patients typically have dry mouth (xerostomia) and dry eyes (kerato-conjunctivitis sicca) (Talal et al. 1987). It has been reported that acinar epithelial cells in Sjögren's syndrome express FAS and FAS ligand (FasL) and that those cells died by apoptosis (Kong et al. 1997), which suggest that the Fas death pathway may be an important mechanism leading to the glandular destruction found in Sjögren's syndrome. In addition, it has been shown that caspase 1 plays an important role in Fas-mediated apoptosis of SMG-C6, an immortalized salivary acinar cell line (Aiba-Masago et al. 2001). Salivary epithelial cells are also sensitive to radiation-induced apoptosis, a major concern to patients with undergoing radiotherapy in the head/neck region (Stephens et al. 1991).

3.10 Apoptosis in Gastric Parietal Cells

The gastric mucosa is frequently infected with the gram-negative bacterium *Helicobacter pylori*, colonising the antral part of the gastric epithelium in which the pH is elevated to 3–4, which allows optimal growth (Blaser 1998). Previous studies have revealed that urease activity of *Helicobacte pylori* depends on urea uptake through a specific channel, which is open at pH values between 3 and 6 but closes at levels >7 (Scott et al. 1998; Athmann et al. 2000). During long-term infection as well as long-term therapy with proton pump inhibitors, the infection may also proceed to the gastric corpus, resulting in atrophy of gastric glands, hypochlorhydria, and possibly the development of gastric adenocarcinoma (Gillen et al. 1998; Stolte and Meining 1998).

One mechanism by which *Helicobacter pylori* generates mucosal atrophy may be the induction of apoptosis of the epithelial mucosal barrier followed by destruction of the remaining mucosa by exposure to luminal acid. Indeed, *in vivo* and *in vitro* studies demonstrate that infection with *Helicobacter pylori* is associated with apoptosis of gastric epithelial cells (Shirin and Moss 1998; Moss et al. 2001; Neu et al. 2002). Coculture of the human gastric cancer cell line AGS with *Helicobacter pylori in vitro* resulted in growth inhibition predominantly at the G(0)-G(1) checkpoint, possibly mediated by changes of the regulatory proteins p53, p21, and cyclin E (Ahmed et al. 2000). These studies have suggested that adherence of *Helicobacter pylori* to the gastric mucosa triggers the production of proinflammatory cytokines such as interleukin (IL)-8, TNF-α, and IL-1β (Crabtree 1996). The release of these cytokines in turn induces the accumulation of inflammatory cells and also leads to a sustained functional impairment of mucosal cells. For example, incubation of gastric parietal cells and enterochromafin-like cells with IL-1β induced apoptosis of these cells that are crucial for acid secretion (Crabtree 1996; Prinz et al. 1997; Schepp et al. 1998; Mahr et al. 2000).

3.11 Ca^{2+} Signalling, Oxidant Agents and Apoptosis

Cytosolic Ca^{2+} concentration may play an important role in the apoptosis process. Elevations of $[Ca^{2+}]_c$ are used as general signalling mechanisms, which activate different processes even in the same cell. Exocrine acinar cells, in particular pancreatic acinar cells, have proved useful models for the study of Ca^{2+} signalling. Physiological stimulation, with the neurotransmitter acetylcholine or the circulating peptide hormone CCK, evokes increases in $[Ca^{2+}]_c$. Supramaximal agonist stimulation, eliciting a sustained global cytosolic Ca^{2+} elevation, has been shown to induce intracellular enzyme activation and vacuole formation in the apical granular pole (Parekh 2000; Sandoval et al. 2000), which can induce apoptosis. Thapsigargin, a specific inhibitor of sarcoplasmic reticulum Ca^{2+}-ATPase, can induce apoptosis in a wide variety of epithelial and non-epithelial cell lineages (Baffy et al. 1993; Qi et al. 1997).

ROS and oxidant agents are known to be mediators in alteration of normal Ca^{2+} homeostasis, which precede other morphological and functional alterations of cell injury induced by oxygen free radicals. In fact, the production of ROS is associated with many forms of apoptosis (Suzuki et al. 1997; Lopez et al. 2007), as well as the death that occurs in stroke, ischemia, and many neurodegenerative diseases (Tan et al. 1998). In pancreatic acinar cells, it has been shown that ROS and other oxidant agents, such as menadione or H_2O_2, at concentrations inducing Ca^{2+} signal, evoke apoptosis as defined by mitochondrial depolarisation, release of cytochrome *c* from mitochondria, activation of caspases, phosphatidylserine externalisation and, at later time points, condensation of nuclear chromatin (Gerasimenko et al. 2002; Lopez et al. 2007). The blockade of Ca^{2+} signal by loading the cytosol with the Ca^{2+} chelator BAPTA prevent the induction of apoptosis evoked by ROS (Bejarano et al. 2008). Furthermore, Ca^{2+} released from intracellular stores may trigger apoptosis owing to their proximity to mitochondria (mitochondria Ca^{2+} uptake and consequent overload may result in transition pore opening and cytochrome *c* release) (Hajnoczky et al. 2000). In addition, mitochondrial ROS production has been reported to induce Ca^{2+} release from intracellular stores, which in turn, stimulated mitochondrial Ca^{2+} loading. The amplification of oxidative stress and Ca^{2+} loading culminated in mitochondrial membrane permeabilisation and cell death (Jacobson and Duchen 2002).

Acknowledgments This work was supported by MEC-DGI grant BFU2007-60091.

References

Adams JM, Cory S (1998) The Bcl-2 protein family: arbiters of cell survival. Science 281:1322–1326

Ahmed A, Smoot D, Littleton G et al. (2000) *Helicobacter pylori* inhibits gastric cell cycle progression. Microbes Infect 2:1159–1169

Aiba-Masago S, Masago R, Vela-Roch N et al. (2001) Fas-mediated apoptosis in rat acinar cell line is dependent on caspase-1 activity. Cell Signal 13:617–624

Ashkenazi A, Dixit VM (1998) Death receptors: signaling and modulation. Science 281:1305–1308

Athmann C, Zeng N, Kang T et al. (2000) Local elevation mediated by the intrabacterial urease of Helicobacter pylori cocultured with gastric cells. J Clin Invest 106:339–347

Baffy G, Miyashita T, Williamson JR et al. (1993) Apoptosis induced by withdrawal of interleukin-3 (IL-3) from an IL-3-dependent haematopoietic-cell line is associated with repartioning of intracellular calcium and is blocked by enforce bcl-2 oncoprotein production. J Biol Chem 268:6511–6519

Balkwill FR, Burke F (1996) The cytokine network. Immunol Today 10:299–303

Beil M, Leser J, Lutz MP et al. (2002) Caspase 8-mediated cleavage of plectin precedes F-actin breakdown in acinar cells during pancreatitis. Am J Physiol Gastrointest Liver Physiol 282:G450–G460

Bejarano I, Lozano GM, Ortiz A et al. (2008) Caspase 3 activation in human spermatozoa in response to hydrogen peroxide and progesterone. Fertil Steril 90:1340–1347

Bendayan M, Roth J, Perrelet A et al. (1980) Immunocytochemical localization of secretory proteins in pancreas. J Histochem Cytochem 28:149–160

Bhatia M (2004a) Apoptosis of pancreatic acinar cells in acute pancreatitis. Is it good or bad? J Cell Mol Med 8:402–409

Bhatia M (2004b) Apoptosis versus necrosis in acute pancreatitis. Am J Physiol 286:G189–G196

Bhatia M, Wallig MA (2004) 1-Cyano-2-hydroxy-3-butene: a plant nitrile that induces apoptosis in pancreatic acinar cells and reduces the severity of acute pancreatitis. In: Tan BK-H, Bay B-H, Zhu Y-Z (eds) Novel compounds from natural products in the new millenium: potential and challenges. World Scientific Publishing Co, New Jersey, pp 92–99

Bhatia M, Wallig MA, Hofbauer B et al. (1998) Induction of apoptosis in pancreatic acinar cells reduces the severity of acute pancreatitis. Biochem Biophys Res Commun 246:476–483

Blaser MJ (1998) *Helicobacter pylori* and gastric diseases. BMJ 316:1507–1510

Borner C, Monney L (1999) Apoptosis without caspases: an inefficient molecular guillotine? Cell Death Differ 6:497–507
Bouillet P, Strasser A (2002) BH3-only proteins – evolutionarily conserved proapoptotic Bcl-2 family members essential for initiating programmed cell death. J Cell Sci 115:1567–1574
Burgoyne RD, Morgan A (2003) Secretory granule exocytosis. Physiol Rev 83:581–632
Cande C, Cecconi F, Dessen P et al. (2002) Apoptosis-inducing factor (AIF): Key to the conserved caspase-independent pathways of cell death? J Cell Sci 115:4727–4734
Chittenden T, Flemmington C, Houghton AB et al. (1995) A conserved domain in Bak, distinct from BH1 and BH2, mediates cell death and protein binding functions. EMBO J 14:5589–5596
Choi JY, Kim KH (1998) Effects of small molecular antioxidants on cerulein-induced acute pancreatitis in rat. Korean J Physiol Pharmacol 2:629–635
Cohen MC, Cphen S (1996) Cytokine function: a study in biologic diversity. Am J Clin Pathol 105:589–598
Crabtree JE (1996) Gastric mucosal inflammatory responses to *Helicobacter pylori*. Aliment Pharmacol Ther 10:29–37
Dabrowski A, Konturek SJ, Konturek JW et al. (1999) Role of oxidative stress in the pathogenesis of caerulein-induced acute pancreatitis. Eur J Pharmacol 377:1–11
Danial NN, Korsmeyer SJ (2004) Cell death: critical control points. Cell 116:205–219
Fujimoto K, Hosotani R, Doi R et al. (1997) Role of neutrophils in cerulein-induced pancreatitis in rats: possible involvement of apoptosis. Digestion 58:421–430
Gerasimenko JV, Gerasimenko OV, Palejwala A et al. (2002) Menadiaone-induced apoptosis: roles of cytosolic Ca^{2+} elevations and the mitochondrial permeability transition pore. J Cell Sci 115:485–497
Gillen D, El-Omar EM, Wirz AA et al. (1998) The acid response to gastrin distinguishes duodenal ulcer patients from *Helicobacter pylori*-infected healthy subjects. Gastroenterology 114:50–57
Gomez G, Lee HM, He Q et al. (2001) Acute pancreatitis signals activation of apoptosis-associated and survival genes in mice. Exp Biol Med 226:692–700
Gorelick FS, Jamieson JD (1987) Structure-function relationship of the pancreas. In: Johnson LR (ed) Physiology and the gastrointestinal tract, 2nd edn. Raven Press, New York, pp 1089–1108
Granados MP, Salido GM, Pariente JA et al. (2005) Effect of H2O2 on CCK-8-evoked changes in mitochondrial activity in isolated mouse pancreatic acinar cells. Biol Cell 97:847–856
Green DR, Reed JC (1998) Mitochondria and apoptosis. Science 281:1309–1312
Gukovskaya AS, Gukovsky I, Jung Y et al. (2002) Cholecystokinin induces caspase activation and mitochondrial dysfunction in pancreatic acinar cells. Roles in cell injury processes of pancreatitis. J Biol Chem 277:22595–22604
Gukovskaya AS, Gukovsky I, Zaninovic V et al. (1997) Pancreatic acinar cells procedure, release, and respond to tumor necrosis factor-alpha. Role in regulating cell death and pancreatitis. J Clin Invest 100:1853–1862
Gukovskaya AS, Perkins P, Zaninovic V et al. (1996) Mechanisms of the cell death after pancreatic duct obstruction in the opossum and the rat. Gastroenterology 110:875–884
Gukovsky I, Gukovskaya AS, Blinman TA et al. (1998) Early NF-kappaB activation is associated with hormone-induced pancreatitis. Am J Physiol 275:G1402–G1414
Hajnoczky G, Csordas G, Madesh G et al. (2000) Control of apoptosis by IP3 and ryanodine receptor driven calcium signals. Cell Calcium 28:349–363
Hamamato N, Ashizawa N, Niigaki M et al. (2002) Morphological changes in the rat exocrine pancreas after pancreatic duct ligation. Histol Histopathol 17:1033–1041
Hashimoto T, Yamada T, Yokoi T et al. (2000) Apoptosis of acinar cells is involved in chronic pancreatitis in Wbn/Kob rat: role of glucocorticoides. Pancreas 21:296–304
Iwatsuki N, Petersen OH (1979) Electrical coupling and uncoupling of the exocrine acainar cells. J Cell Biol 79:533–545
Jacobson J, Duchen MR (2002) Mitochondrial oxidative stress and cell death in astrocytes-requirement for stored Ca^{2+} and sustained opening of the permeability transition pore. J Cell Sci 115:1175–1188

Jungermann J, Lerch MM, Weidenbach H et al. (1995) Disassembly of rat pancreatic acinar cell cytoskeleton during supramaximal secretagogue stimulation. Am J Physiol 268:G328–G338

Kaiser AM, Saluja AK, Sengupta A et al. (1995) Relationship between severity necrosis and apoptosis in five models of experimental pancreatitis. Am J Physiol 269:G1295–G1304

Kim SJ, Ju JW, Oh CD et al. (2002) ERK-1/2 and p38 kinase oppositely regulate nitric oxide-induced apoptosis of chondrocytes in association with p53, caspase-3, and differentiation status. J Biol Chem 277:1332–1339

Kimura K, Shimosegawa T, Sasano H et al. (1998) Endogenous glucocorticoids decrease the acinar cell sensibility to apoptosis during cerulein pancreatitis in rats. Gastroenterology 114:372–381

Kong L, Ogawa N, Nakabayashi T et al. (1997) Fas and Fas ligand expression in the salivary glands of patients with primary Sjögren's syndrome. Arthritis Rheum 40:87–97

Kornmann M, Ishiwata T, Maruyama H et al. (2000) Coexpression of FAS and FAS-ligand in chronic pancreatitis: correlation with apoptosis. Pancreas 20:123–128

Lerch MM, Adler G (1994) Experimental animal models of acute pancreatitis. Int J Pancreatol 15:159–170

Li J, Bombeck CA, Yang S et al. (1999) Nitric oxide suppresses apoptosis via interrupting caspase activation and mitochondrial dysfunction in cultured hepatocytes. J Biol Chem 274:17325–17333

Li P, Nijhawan D, Budihardjo I et al. (1997) Cytochrome *c* and dATP-dependent formation of Apaf-1/caspase-9 complex initiates an apoptotic protease cascade. Cell 91:479–489

Lopez JJ, Salido GM, Gómez-Arteta E et al. (2007) Thrombin induces apoptotic events through the generation of reactive oxygen species in human platelets. J Thromb Haemost 5:1283–1291

Mahr S, Neumayer N, Gerhard M et al. (2000) IL-1β-induced apoptosis in rat gastric enterochromaffin-like cells is mediated by iNOS, NF-κB, and Bax protein. Gastroenterology 118:515–524

Mareninova OA, Sung KF, Hong P et al. (2006) Cell death in pancreatitis: caspase protect from necrotizing pancreatitis. J Biol Chem 281:3370–3381

Monks TJ, Hanzlik RP, Cohen GM et al. (1992) Quinone chemistry and toxicity. Toxicol Appl Pharmacol 112:2–16

Morgado S, Granados MP, Bejarano I et al. (2008) Role of intracellular calcium on H_2O_2-induced mitochondrial apoptosis in rat pancreatic acinar AR42J cells. J Appl Biomed 6:215–228

Moss SF, Sordillo EM, Abdalla AM et al. (2001) Increased gastric epithelial cell apoptosis associated with colonization with cagA *Helicobacter pylori* strains. Cancer Res 61:1406–1411

Nakamura H, Koji T, Tominaga M et al. (1998) Apoptosis in labial salivary glands from Sjögren's syndrome (SS) patients: comparison with human T lymphotropic virus-I (HTLV-I)-seronegative and – seropositive SS patients. Clin Exp Immunol 114:106–112

Neu B, Randlkofer P, Neuhofer M et al. (2002) Helicobacter pylori induces apoptosis of rat gastric parietal cells. Am J Physiol Gastrointest Liver Physiol 283:G309–G318

Nicotera P, Bellomo G, Orrenius S (1992) Calcium-mediated mechanisms in chemically induced cell death. Annu Rev Pharmacol Toxicol 32:449–470

Niedoborski TE, Klein BP, Wallig MA (2001) Rapid isolation and purification of 1-cyano-2-hydroxy-3-butene (crambene) from Crambe abyssinica seed meal using immiscible solvent extration and high-performance liquid chromatography. J Agric Food Chem 49:3594–3599

Parekh AB (2000) Calcium signalling and acute pancreatitis: specific response to a promiscuous messenger. Proc Natl Acad Sci USA 97:12933–12934

Prinz C, Neumayer N, Mahr S et al. (1997) Functional impairment of rat enterochromaffin-like cells by interleukin 1β. Gastroenterology 112:364–375

Qi XM, He H, Zhonh H et al. (1997) Baculovirus p35 and Z-VAD-fmk inhibit thapsigargin-induced apoptosis of breast cancer cells. Oncogene 15:1207–1212

Rau B, Paszkowski A, Esber S et al. (2001) Anti-ICAM-1 antibody modulates late onset of acinar cell apoptosis and early necrosis in taurocholate-induced experimental acute pancreatitis. Pancreas 23:80–88

Reddy JK, Rao MS, Warren JR et al. (1979) Concanavalin A agglutinability and surface microvilli of dissociated normal and neoplastic pancreatic acinar cells. Exp Cell Res 120:55–61

Reed JC (1998) Bcl-2 family proteins. Oncogene 17:3225–3236

Reid LE, Walker NI (1999) Acinar cell apoptosis and the origin of tubular complexes in caerulein-induced pancreatitis. Int J Exp Pathol 80:205–215

Rosado JA, López JA, Gómez-Arteta E et al. (2006) Early caspase-3 activation independent of apoptosis is required for cellular function. J Cell Physiol 209:142–152

Sandoval D, Gukovkaya A, Reavey P et al. (2000) Calcium-dependent enzyme activation and vacuole formation in the apical granular region of pancreatic acinar cells. Proc Natl Acad Sci USA 97:13126–13131

Schepp W, Dehne K, Herrmuth H et al. (1998) Identification and functional importance of IL-1 receptors on rat parietal cells. Am J Physiol Gastrointest Liver Physiol 275:G1094–G1105

Scott DR, Weeks D, Hong C et al. (1998) The role of internal urease in acid resistance of Helicobacter pylori. Gastroenterology 114:58–70

Shirin H, Moss SF (1998) Helicobacter pylori induced apoptosis. Gut 43:592–594

Simonen M, Keller H, Heim J (1997) The BH3 domain of Bax is sufficient for interaction of Bax with itself and with other family members and it is required for induction of apoptosis. Eur J Biochem 249:85–91

Song JY, Lim JW, Kim H et al. (2003) Oxidative stress induces nuclear loss of DNA repair proteins Ku70 and Ku80 and apoptosis in pancreatic acinar AR42J cells. J Biol Chem 278:36676–36687

Stephens LC, Schultheiss TE, Price RE et al. (1991) Radiation apoptosis of serous acinar cells of salivary and lacrimal glands. Cancer 67:1539–1543

Stolte M, Meining A (1998) Helicobacter pylori and gastric cancer. Oncology 3:124–128

Su SB, Motoo Y, Xie MJ et al. (2001) Apoptosis in rat spontaneous chronic pancreatis: role of the Fas and Fas ligand system. Dig Dis Sci 46:175–2001

Susin SA, Lorenzo HK, Zamzami N et al. (1999) Molecular characterization of mitochondrial apoptosis-inducing factor. Nature 397:441–446

Suzuki YS, Forman HJ, Sevanian A (1997) Oxidants as stimulators of signal transduction. Free Radical Biol Med 22:269–285

Talal S, Moutsopoulos HM, Kassa SS (1987) Sjögren's Syndrome-Clinical and Immunological Aspects. Springer-Verlag, Berlin.

Tan S, Sagara Y, Liu Y et al. (1998) The regulation of reactive oxygen species production during programmed cell death. J Cell Biol 141:1423–1432

Tomasini R, Samir AA, Vaccaro MI et al. (2001) Molecular and functional characterization of the stress-induced protein (SIP) gene and its two transcripts generated by alternative splicing. SIP induced by stress and promotes cell death. J Biol Chem 276:44185–44192

Trulsson LM, Gasslander T, Sundqvist T et al. (2002) The influence of nitric oxide on basal and cholecystokinin-8-induced proliferation and apoptosis in the rat pancreas. Regulatory Peptides 106:97–104

Trulsson LM, Svanvik J, Permert J et al. (2001) Cholecystokinin octapeptide induces both proliferation and apoptosis in the rat pancreas. Regulatory Peptides 98:41–48

Wallig MA, Kingston S, Staack R et al. (1998) Induction of rat pancreatic glutathione S-transferase and quinona reductase activities by a mixture of glucosinolate breakdown derivatives found in Brussels sprouts. Food Chem Toxicol 36:365–373

Williams JA (1977) Effects of cytochalasin B on pancreatic acinar cell structure and secretion. Cell Tissue Res 179:453–466

Yu JH, Kim KH, Kim DG et al. (2007) Diphenyleneiodonium suppresses apoptosis in cerulein-stimulated pancreatic acinar cells. Int J Biochem Cell Biol 39:2063–2075

Yu JH, Lim JW, Kim H et al. (2005) NADPH oxidase mediates interleukin-6 expression in cerulein-stimulated pancreatic acinar cells. Int J Biochem Cell Biol 37:1458–1469
Yu SP, Canzoniero LMT, Choi DW (2001) Ion homeostasis and apoptosis. Curr Opinion Cell Biol 13:405–411
Yuan Y, Gong Z, Lou K et al. (2001) Effects and mechanisms of somatostatin analogs on apoptosis of pancreatic acinar cells in acute pancreatitis in mice. J Gastroenterol Hepatol 16:683–688
Zhu Z, Friess H, Shi X et al. (2003) Up-regulation of p75 neutrophin receptor (p75NTR) is associated with apoptosis in chronic pancreatitis. Dig Dis Sci 48:717–725

Chapter 4
Apoptosis in the Liver

J. González-Gallego and M.J. Tuñón

Abstract The importance of apoptosis is evident in the liver during development and homeostasis of the biliary tree, and apoptotic cell death, induced by reactive oxygen species via the intrinsic signalling pathway, is also increased in the aging liver. Apoptosis plays an important role in liver pathogenesis, and disarrangement of death receptor pathways has been identified as a major contributor to the initiation and aggravation of acute and chronic liver injury. Dysregulation of the apoptotic process, resulting in too much or too little cell death, has potentially devastating effects. Over activation involving mediators such as Fas, TNF-α, TGF-β1, and members of the Bcl-2 family, can lead to significant acute injuries, such as fulminant hepatic failure or even chronic sustained hepatocellular damage, as occurs with toxic liver injury, viral hepatitis, alcoholic and non alcoholic liver disease or cholestatic disorders. On the contrary, inhibition of apoptosis can promote the proliferation and transformation of cells and possibly hepatocellular carcinoma, protecting malignant hepatocytes from cellular suicide and avoiding removal of cells carrying mutated genes. Agents that modulate apoptosis may be of therapeutic benefit in a number of liver diseases, and research related to cell type-specific activation or inhibition of apoptotic signaling pathways will provide new strategies for treatment.

Keywords Liver · Hepatocytes · Hepatitis · Alcohol · Hepatocarcinoma · Cholestasis · Fulminant hepatic failure · Steatohepatitis · Fibrosis · Biliary tree · Aging

J. González-Gallego (✉)
Institute of Biomedicine, University of León, Campus Universitario, 24071 León, Spain
e-mail: jgonga@unileon.es

G.M. Salido, J.A. Rosado (eds.), *Apoptosis: Involvement of Oxidative Stress and Intracellular Ca^{2+} Homeostasis*, DOI 10.1007/978-1-4020-9873-4_4,

4.1 Introduction

4.1.1 Structure of the Liver

The liver is one of the largest and most important organs in the body. It is involved with almost all the biochemical pathways that allow growth, fight against disease, supply nutrients or provide energy. A large number of vital functions of the liver have been identified. Some of the more well-known include:

- Production of bile which breaks down fats in the small intestine during digestion and carries away toxic molecules.
- Synthesis of proteins for blood plasma.
- Metabolism of lipoproteins to help carry fats through the body.
- Conversion of excess glucose into glycogen for storage.
- Excretion of potentially toxic substances, including drugs.
- Regulation of haemostasis.
- Resisting infections by producing immune factors.

The liver receives blood from two systems. The portal vein collects blood from the gastrointestinal tract, while the hepatic artery supplies blood from the aorta. The liver is made up of parenchymal cells (hepatocytes) and non-parenchymal cells. Basically, the liver parenchyma is organized in hexagonal lobes that contain a central vein surrounded by six portal spaces. The portal space consists of a vein, a hepatic artery, and a biliary duct that drains spaces between the hepatocytes and is surrounded by capillary vessels termed sinusoids.

Hepatocytes are the main functional cells of the liver, responsible for delivering most of the hepatic functions important for body homeostasis. These cells function as a biochemical defence against toxic chemicals entering through the food and as a re-processor of absorbed food ingredients. Nutrients are transformed into secreted proteins (albumin, most coagulation factors, several plasma carrier proteins, etc), lipids sent as lipoproteins to other tissues, and carbohydrates stored in the liver as glycogen. Synthesis of bile is essential for absorption of fat and lipophilic nutrients. Hepatocytes are the first cells of the liver to enter into the cell cycle and undergo proliferation, and they produce mitogenic signals for other hepatic cell types (Michalopoulos 2007).

The non-parenchymal cells line the normal sinusoid and bile ducts and consist of five different cell types: Kupffer cells, endothelial cells, pit cells, stellate cells, and cholangiocytes. These cells differ in origin, population, kinetic, phenotypic and functional characteristics.

Kupffer cells represent a large population of macrophages. They have an strategic position to control the presence of a variety of substances in the blood, mostly located in the sinusoidal lumen and anchored on or between the endothelial cells of the sinusoidal lining. Their ability to eliminate and detoxify microorganisms, endotoxins, degenerated cells, immune complexes and toxic agents is an important physiological function. Due to their key location, they also function as antigen-presenting cells and participate in tumour surveillance and the regeneration process in the liver. A role of Kupffer cells in innate immune response and host defence through the expression of soluble inflammatory mediators has been described. Finally, interaction with lippopolysaccharide may be the intiating event leading to hepatotoxicity in various types of liver injury (Koilos et al. 2006).

Stellate cells, fat-storing cells or Ito cells are cells of mesenchymal origin situated in the space of Disse, between the endothelial lining and the parenchymal cells. Stellate cells take part together with the parenchymal cells in the storage and metabolism of retinoids in the liver. Quiescent cells contribute to the secretion of extracellular matrix in normal liver, mainly expressing collagen type III and IV, and laminin. When activated, are responsible for a more diverse spectrum of connective tissue molecules, containing sizeable amounts of collagen type I, III, and IV, fibronectin, laminin and tenascin. Quiescent stellate cells also express hepatocyte growth factor, which stimulates parenchymal cell proliferation. In activated cells there is an increased expression of the fibrogenic cytokine TGF-β (Sato et al. 2003).

Sinusoidal endothelial cells are characterized by the fenestrations, which are organized in sieve plates. They have no basal lamina and are responsible for the major part of the transport between the blood and the hepatocytes. Pit cells represent a special, liver-associated population of large granular lymphocytes. They are responsible for NK activity, i.e. spontaneous cytotoxicity to some tumour cells which is not restricted by the presence of major histocompatibility antigens on the target cell surface (Nakatani et al. 2004). Cholangiocytes line the biliary tree and are of considerable intrinsic biologic interest, particularly with regard to their roles in the transport of water, ions, and solutes, and to their heterogeneity and proliferative capacity. They are also involved in angiogenesis, being an important source of vascular endothelial growth factor (Gaudio et al. 2006).

4.1.2 Apoptotic Signalling in the Liver

Apoptotic signalling within the cells is traduced mainly via two molecular pathways: the extrinsic or death receptor pathway and the intrinsic or mitochondrial pathway. Apoptotic events in the hepatocytes can be regulated by different stimuli that bind to death receptors in the cell membranes, such as Fas ligand (FasL), tumour necrosis factor (TNF-α) o TNF-related apoptosis-induced ligand (TRAIL), which activate the extrinsic pathway. Other factors, particularly the transforming growth factor-beta (TGF-β), do not bind the death receptors, but its intracellular signals couple to the apoptotic machinery through activation of the intrinsic pathway. Binding of FasL or TNF-α to their corresponding death receptors induce the recruitment of procaspase 8–10 to form the death-inducing signaling complex, leading to cell death. TRAIL, which is regulated by two death receptors (TRAIL-R1 and TRAIL-R2), induces apoptosis in transformed cell lines but not in normal tissue (Yoon and Gores 2002) (Fig. 4.1).

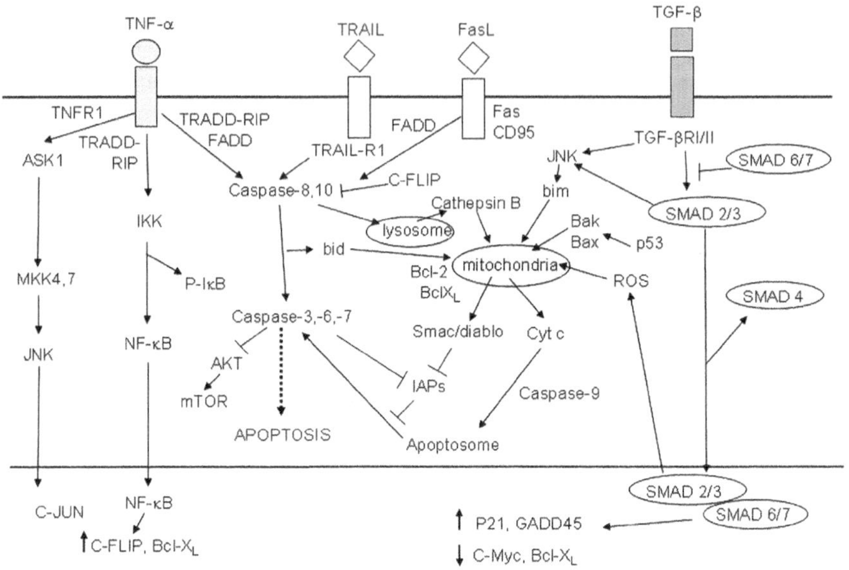

Fig. 4.1 Signal transduction pathway in hepatocytes

The intrinsic pathway is triggered by different intracellular or extracellular signals that induce hepatocytic mitochondrial dysfunction, resulting in the cyosolic release of mitochondrial proteins such as cytochrome c, Smac/DIABLO, apoptosis-inducing factors or endonuclease G among others. Several intracellular proteins, in particular of Bcl-2 family, are important regulators of the intrinsic pathway, integrating death and survival signals (Fabregat et al. 2007). This family includes both pro-apoptotic (t-Bid or Bax) and anti-apoptotic (Bcl-2 or Bcl-Xl) proteins (Fig. 4.1).

There is less information on the apoptotic signalling pathways in other liver cells. Cholangiocytes constitutively express Fas, but the apoptotic potential role of TNF-α or TRAIL is less clear (Tanaka et al. 2000). Activated hepatic stellate cells become sensitive to Fas-mediated apoptosis, with increased TRAIL-R protein expression (Yoon and Gores 2002). Kuppfer cells appear to regulate the apoptotic machinery with potential relevance for immunoregulation. Finally, endothelial cells express Fas, and Fas activation can induce apoptosis of endothelial cells from liver sinusoids (Vekemans et al. 2004).

4.2 Mechanisms of Apoptosis in the Normal Liver

Apoptosis is required together with cell proliferation for a proper development of the liver parenchyma and the biliary tree. The biliary tree initially develops from a double-layered cylinder of cells termed the ductal plate. Increasing levels of

Bcl-2 have been reported in later stages of ductal plate development, resulting in a decrease of the number of apoptotic cells, while Fas remained constant (Terada and Nakanuma 1995). The importance of apoptosis for duct development is confirmed by the fact that in foetuses with an autosomal recessive syndrome (Meckel syndrome) in which the ductal plate does not disappear, increased levels of Bcl-2 and decreased number of apoptotic cells are present (Bai and Odin 2003).

Little is known on the rate of cell turnover and apoptosis in the adult liver parenchyma. In fact, physiologically, apoptosis is almost undetectable in the liver, with only 1–5 apoptotic cells/10,000 cells detected (Eichhorst 2005). This apoptotic cell turnover shows slight variations by dietary factors (Albright et al. 1997). In Fas-knockout mouse strains there is a massive production of lymphocytes and a substantial liver hyperplasia that is accompanied by the enlargement of nuclei in the hepatocytes (Ghavami et al. 2005). Apoptotic alterations caused by Fas mutations are also present in a rare autoimmune lymphoproliferative syndrome (ALPS). Children with ALPS present massive non-malignant lymphadenopathy, hepatosplenomegaly, autoimmunity and increased number of circulating and tissue TCR-CD3+ CD4-CD8- T cells (Fisher et al. 1995).

An interesting aspect of the liver function in relation to apoptosis comes from its role in immunity and its acceptance of grafts. It has been demonstrated that the liver constitutes the main site for apoptosis of blood cells and a large body of evidence demonstrated that the liver is accountable for the induction of immune tolerance. Thus, portal vein transfusions of donor leukocytes induce tolerance to allografts and even xenografts, and liver grafts can actually inhibit the rejection of other solid organs and the graft-versus-host response (Sun and Shi 2001). Available data suggest that the number of apoptotic cells that specifically accumulate in the liver are partially responsible for the distinct immunological function of the liver. The best established examples of apoptotic cell accumulating in the liver are activated apoptotic T-cells. As Kuppfer cells account for 80% of all macrophages in the organism, it is reasonable to assume that the liver is a specialized organ for inducing or clearing apoptotic cells, and it may be postulated that Kupffer cells are responsible for the establishment of immune tolerance in the liver (Selmi et al. 2007).

Numerous reports suggest that aging is accompanied by alterations in the apoptotic behaviour of a variety of cell types and tissues, with modification on caspase activities. The presence of apoptotic ultrastructural alterations and increases in the cytochrome *c* mitochondrial release, the Bax to Bcl-2 relative expression and the activity of caspase 3, with no significant changes in Fas ligand expression and caspase 8 activity, has been recently reported in the liver of senescent rats (Molpeceres et al. 2007). Melatonin administration was able to abrogate these changes, supporting that liver apoptotic cell death is induced by reactive oxygen species (ROS), via the intrinsic signalling pathway. It remains, however, to be established if enhanced levels of apoptosis serve as a self-protective mechanisms to remove increased number of dysfunctional cells as a result of aging, or play a destructive role, causing excessive cell death and the decline of organ function attendant with aging (Zhang et al. 2002).

4.3 Apoptosis in Liver Diseases

In the context of liver disease, over activation of the apoptotic process may lead to hepatocellular damage, while inhibition of apoptosis may promote cell proliferation and transformation. This section gives an overview of the importance of changes in the apoptosis signalling pathways as a mechanism of liver disease.

4.3.1 Fulminant Liver Failure

Fulminant hepatic failure (FHF), induced by drugs (e.g. paracetamol overdose), toxins (e.g. mushroom-derived amanitin) or viral hepatitis, is characterized by a severe hepatocellular dysfunction with a massive death of hepatocytes, in which apoptosis may play a role in addition to necrosis (Pretet et al. 2003). It has been indicated that serum cytochrome c is a possible new marker for acute liver failure in humans and correlates to serum levels of AST and ALT (Sakaida et al. 2005). Because FHF is associated with a high risk of lethal results and orthotopic liver transplantation is not always available in a timely fashion, the intervention on apoptotic pathways could be molecular targets for potential therapeutical approaches to FHF, contributing to temporary support while awaiting a liver transplant.

Different studies have tried to block the apoptotic cascade in FHF. Expression of antiapoptotic molecules such as Bcl-2, Bcl-X_L or a dominant negative FADD/MORT1 molecule, as well as treatment with the caspase inhibitor Ac.Y.VAD-cmk have been reported to rescue mice from anti-CD95 or TNF-induced failure (Schuchmann and Galle 2001). Pretreatment with colchicine protects mice from a lethal dose of anti-CD95 antibody Jo2, probably by down-regulating the surface expression of CD95 on hepatocytes (Feng and Kaplowitz 2000). CD95 and TNF-induced liver damage may also be prevented by the neurokinin 1-receptor antagonists and the polysulfonated derivative of urea suramin (Eichhorst 2005).

It has been shown by data on animal survival, clinical feature, histological data, changes in blood chemistry and intracranial pressure monitoring that the rabbit hemorrhagic disease (RHD) fulfils many of the requirements of an animal model of FHF, being useful to improve our insight into the metabolic and physiological derangements of FHF (Tuñón et al. 2003). In the RHD model of FHF, apoptotic cell death is induced via both the intrinsic and the extrinsic signalling pathways, and N-acetylcysteine exerts an antiapoptotic effect related to the modulation of *bcl-2* and *bax* genes. The protective effect provided by the antioxidant N-acetylcysteine suggests that oxidative stress is a primary pathway for apoptosis in this model of FHF (San Miguel et al. 2006). In this same line, intraperitoneal injection of an antioxidant lignan compound from *Schisandra fructus*, has been reported to reduce DNA fragmentation and to attenuate the elevation of serum TNF-α and activation of caspase 3 in D-galactosamine and lipopolysaccharide-induced liver failure (Kim et al. 2008).

Suppressor of cytokine signaling-1 (SOCS1), which is a negative-feedback molecule for cytokine signalling, has been demonstrated to be rapidly induced during liver injury. Very recently it has been shown, using liver-specific SOCS1-conditional-knockout mice, that SOCS1 deletion in hepatocytes enhanced concanavalin A-induced hepatitis, and enhances proapoptotic signals, including signal transducers and activators of transcription 1 (STAT1) and Jun-terminal kinase (JNK) activation, and higher expression of Fas antigen. These findings indicate that SOCS1 plays important negative roles in FHF and his forced expression could be of therapeutical use (Torisu et al. 2008).

RNA interference (siRNA) is a relatively new and powerful approach with encouraging success in experimental models of apoptotic liver damage. Studies in mice have demonstrated the silencing effect of siRNA duplexes targeting the gene Fas in models of autoimmune hepatitis induced by concanavalin A (Song et al. 2003). Injection of caspase 8 siRNA during AdFasL- and adenovirus wild type-mediated liver failure prevents apoptosis and significantly attenuates acute liver damage (Zender et al. 2003).

4.3.2 Cholestatic Liver Disease

Cholestasis is present in a variety of clinical syndromes and is characterized by an accumulation in the liver of hydrophobic bile acids which are thought to play a key role in liver injury. Although some studies have suggested that the predominant mode of cell death during obstructive cholestasis *in vivo* is oncotic necrosis (Gujral et al. 2004; Fickert et al. 2005), they have focused on cell death occurring in bile infarcts, and this is a late occurrence in which identification of cell killing mechanism by morphologic criteria may be difficult (Malhi et al. 2006). Moreover, pan caspase inhibitors prevent bile infarcts in bile duct-ligated mice (Canbay et al. 2004b), although mechanisms other than suppression of apoptosis, such as blocking of cytokine signalling, might be involved (Martinon and Tschopp 2004). In any case, widespread necrosis is not a prominent feature in most cholestatic liver diseases, and a higher rate of apoptosis compared with healthy controls occurs in patients with primary biliary cirrhosis (Koga et al. 1997).

Hepatic retention of hydrophobic bile acids has long been implicated as a major cause of liver damage, and it is well-known that hepatic levels of toxic bile acids correlate with the degree of liver damage (Rust and Gores 2000). The mechanisms of apoptosis by bile acids have been elucidated in recent years. Although there is evidence of a Fas-dependent apoptotic signalling (Faubion et al. 1999), bile acids may also induce apoptosis in Fas deficient lpr mice (Canbay et al. 2002) through a mechanism that involves transcriptional induction and oligomerization of TRAIL-R2, with recruitment of FADD and cleavage of caspase 8 (Higuchi et al. 2001). In this model of apoptosis, signals converge to produce mitochondrial permeabilization, release of cytochrome *c* and activation of downstream caspases and cathepsin B (Roberts et al. 1997). This cysteine protease also participates as an important effector mechanism, because apoptosis and subsequent fibrosis are absent in mice genetically deficient in cathepsin B after bile duct ligation (Canbay et al. 2002).

Mitochondrial dysfunction also appears to contribute to hepatocyte apoptosis in cholestatic liver disease. *De novo* expression of Bcl-2 has been observed in hepatocytes of bile duct-ligated rats, suggesting an adaptative mechanism to resist apoptosis by toxic bile acids. In fact, cholangiocytes, which are in direct contact with bile continuously, express Bcl-2 constitutively (Rust and Gores 2000). Induction and translocation of Bax to mitochondria, probably mediated by *t*Bid, have been reported after bile duct ligation in lpr mice (Higuchi et al. 2001).

The non toxic hydrophilic bile acid ursodeoxycholate (UDC) reduces apoptosis induced by toxic bile acids through a mechanism that seems to be associated with prevention of the mitochondrial permeability transition (Botla et al. 1995). It has been demonstrated that the taurine conjugate of UDDC protects against bile acid-induced apoptosis via the activation of survival pathways such as mitogen-activated protein quinases (Schoemaker and Moshage 2003). More recently, a cross-talk of UDC with nuclear steroids receptors has been suggested to contribute to the modulation of hepatocyte apoptosis (Solá et al. 2006).

4.3.3 Viral Hepatitis

Viral hepatitis is the most common cause of liver disease world wide. Deregulated apoptosis appears both in acute and chronic viral hepatitis. Apoptotic bodies, previously recognized as Councilmen bodies, are present in the liver of patients with viral hepatitis (Yoon and Gores 2002). The degree of hepatic inflammation in hepatitis C infection correlates with the degree of caspase activation (Bai and Odin 2003), and patients with active hepatitis C virus (HCV) have elevated levels of caspase-generated cytokeratin 18 cleavage fragments compared with healthy controls (Bantel et al. 2004).

Some data suggest that the apoptotic events may be mediated by the host immune system. Studies on human liver biopsy material have shown increased number of TUNEL-positive liver cells expressing Fas receptor closely associated with Fas ligand expressing cytotoxic T lymphocytes (Jaeschke et al. 2004). The activation of immune-mediated apoptotic pathways is corroborated by the fact that the transcripts for perforin/granzyme B are elevated in patients with HCV-related cirrhosis compared to normal liver or other no inflammatory causes of liver cirrhosis (Kountouras et al. 2003).

Hepatitis B virus (HBV) is one of the leading causes of chronic liver disease and is often associates with hepatocarcinogenesis. The virus consists of a nucleocapsid and an outer envelope composed mainly of three surface antigens. The nucleocapsid contains the core antigen, the viral genome and cellular proteins. The X protein of HBV (HBx) is a potent transactivator essential for virus replication and shows oncogenic properties in animal models (Ghavami et al. 2005). HBx forms complexes with p53 in the cytoplasm and retains p53 from entering the nucleus

(Elmore et al. 1997). HBx may also directly inhibit caspase 3 and activate PKB survival pathway (Eichhorst 2005). Induction of nuclear factor kappaB (NF-kappaB) or JNK pathway may also contribute to the protection of HBx against anti-Fas-mediated apoptosis in human liver cells (Schattenberg et al. 2006). However, HBx may also sensitize cells to apoptotic killing by TNF-α without up-regulation of TNF-R1 expression, through activation of MAPK kinase 1 and n-Myc (Su 1997). TNFα activation, in turn, may lead to inactivation of the inhibitor protein of caspase 8, c-FLIP (Kim and Seong 2003). HBV viral variants containing two core promoter mutations associated with fulminant hepatitis have been reported to induce apoptosis in primary Tupaia hepatocytes. The core mutations resulted in two amino acid changes of the HBx protein, which suggests that HBx may be a potential candidate mediating this effect (Baumert et al. 2007).

HCV may frequently result in liver cirrhosis and hepatocellular carcinoma. HCV belongs to the flaviviridae and is a RNA virus whose RNA encodes for a polyprotein that is cleaved into different structural (E1, E2, core, p7) and non-structural proteins (NS2, NS·, NS4, NS4B, NS5, NS5B). Six different genotypes which differ genetically from one another by at least 30% have been identified. This genetic heterogeneity makes difficult to compare apoptotic pathways (Fischer et al. 2007). Multiple HCV proteins expressed in transgenic mice have been shown to inhibit Fas-mediated apoptosis by preventing the release of cytochrome *c* from mitochondria and activation of caspase 9, 3 and 7 (Guicciardi and Gores 2005). The HCV non-structural NS2 protein has been reported to bind and protect from cell death-inducing DFF45-like effector (CIDE). Interactions block cytochrome *c* release from mitochondria and cell death triggering, and double staining of NS2 and CIDE reveals partial overlapping signals in the perinuclear region (Erdtmann et al. 2003). NS5A protein has sequence homologies to Bcl-2 and inhibits apoptosis (Chung et al. 2003). In addition, NS5A may sequester p53 in the cytoplasm and thereby inhibit apoptosis, activate the P13-kinase-Akt/PKB survival pathway (Street et al. 2004), or activate STAT3 with enhanced expression of Bcl-X_L and p21 (Sarcar et al. 2004). By contrast, a direct inhibition by proapoptotic Bin 1, a tumour suppressor protein with a SH3 domain has been shown in hepatoma cells (Nanda et al. 2006), and there are reports of a direct NS5A-induced apoptosis (McDonald and Harris 2004).

It has been shown that core protein inhibits c-myc mediated apoptosis in Chinese hamster ovarian cells and ciplastin-mediated apoptosis in human cervical epithelial cells (Kanzler and Galle 2000). Direct binding to the downstream death domain of FADD and c-FLIP has been reported to result in anti-apoptotic effects (Saito et al. 2006). Moreover, inhibition of the TGF-β pathway by direct interaction with the DNA-binding domain of Smad3, important apoptosis mediators of TGF-β receptor I/II, has been demonstrated (Fischer et al. 2007). However, over expression of HCV core protein protects against chemotherapeutic drug-induced apoptosis, but not against CD95-induced apoptosis in hepatoma cells (Ruggieri et al. 1997). Moreover, core protein can bind to the cytoplasmic domain of TNF-R1 and this interaction promotes apoptosis in mouse and human cell lines (Yoon and Gores 2002). Other studies have shown core-induced apoptosis through

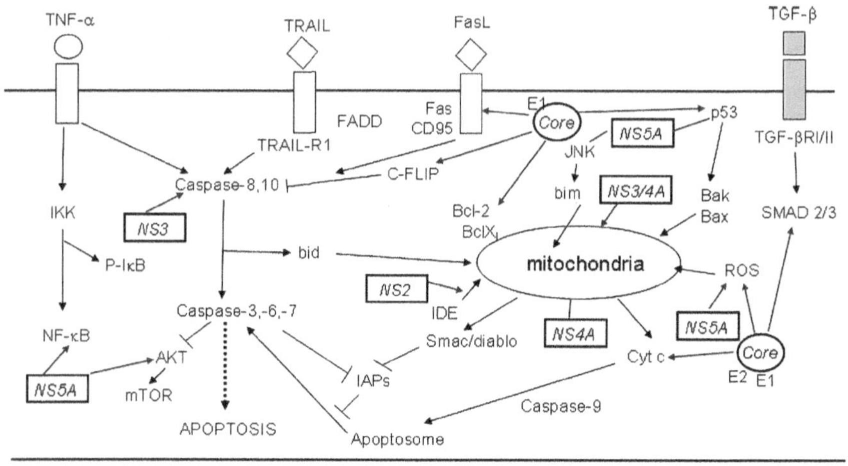

Fig. 4.2 Interference of hepatitis C virus (HCV) with the apoptotic pathway

mitochondrial cytochrome *c* release and indirect activation of Bax and TRAIL-induced apoptosis in hepatoma cells (Fischer et al. 2007). Moreover, HCV triggers production of ROS and lowering of the mitochondrial transmembrane potential, leading to consecutive caspase-independent cell death (Machida et al. 2006). Envelope proteins E1 and E2 inhibit CD95L-mediated apoptosis in a transgenic mouse model expressing HCV proteins (Fig. 4.2). Inhibition of TRAIL-induced apoptosis in hepatoma cells by E2 has been demonstrated, with no effect of E1. By contrast, E2 induces mitochondria-related and caspase-dependent apoptosis in the same hepatoma cell lines (Fischer et al. 2007).

Thus, the role of virus proteins in apoptosis is not totally clarified, may sensitize hepatocytes to apoptosis or inhibit apoptosis, and requires to be investigated in more detail before therapeutical goals can be achieved. The now available infectious tissue culture systems as well as future *in vivo* systems may give answer to these questions, may better reflect the *in vivo* situation and may help to clarify the role of the apoptotic pathways in the pathogenesis of HCV infection (Fischer et al. 2007). In spite of those limitations, some cell death-targeted approaches to viral elimination have been reported. Bid engineered to contain a specific cleavage site for the NS3/NS4A protease and delivered into mice with chimeric human liver is specifically activated and causes a considerably decline in HCV serum titres (Hsu et al. 2001). Effective reduction of liver damage and improvement of survival in mice injected with small interfering RNA (siRNA) against caspase 8 has been shown in models of acute hepatitis, and the same approach prevents the development of fibrosis in a model of chronic hepatitis (Guicciardi and Gores 2005). Differential sensitivity of HBV-infected hepatocytes to TRAIL might be therapeutically useful for removing hepatocytes bearing CCC DNA which permit HBV to resume replication even during or after antiviral therapy (Yoon and Gores 2002). In murine models of autoimmune hepatitis induced by injection of concavalin A or a Fas agonistic antibody, a nitric oxide derivative of ursodeoxycholic acid (NCX-1000) inhibits caspase activity (Fiorucci et al. 2001). Although longer studies are merited, the potent inhibitor of caspases IDN-6556 has been recently shown, in a multicenter, double-blind, placebo-controlled study with a 14-day dosing period, to significantly lower aminotransferase activity in HCV patients (Pockros et al. 2007).

4.3.4 Non-Alcoholic Steatohepatitis

Non-alcoholic steatohepatitis (NASH) is characterized by an accumulation of fat in the liver along with inflammation in patients with no history of alcohol consumption or drug abuse. Many aspects of the pathophysiology of NASH are still unclear, but it is known that hepatocyte apoptosis is increased in patients with NASH and correlates with the disease severity and stage of fibrosis. Death receptor expression, especially Fas and TNF-R1, is also significantly enhanced (Guicciardi and Gores 2005).

Induction of the prooxidant cytochrome P450 2E1 (CYP2E1) is a general finding, and over expression of CYP2E1 causes in hepatocytes activation of ERK1/2 and sensitizes cells to TNF-induced apoptosis through activation of JNK (Schattenberg et al. 2004). The fact that antioxidants reduce both MAPK activation and TNF-induced cell death in CYP2E1 expressing hepatocytes suggests a contribution of ROS to increased apoptosis and hepatocellular injury in this disease (Schattenberg et al. 2006).

Liver injury is associated with subclinical insulin resistance and diabetes mellitus in patients with NASH and evidence has been provided of a possible crosslink between metabolic and inflammatory pathways that leads to increased activation of apoptotic pathways. Thus, JNK activation causes insulin resistance *in vitro* and *in vivo*, and application of adiponectin, whose levels are decreased in patients with non-alcoholic fatty liver disease, counteracts TNF-induced liver injury in mice (Masaki et al. 2004).

4.3.5 Alcoholic Liver Disease

Although the pathogenesis of alcoholic liver disease is still poorly understood, different studies have shown the importance of apoptosis in this type of liver damage. The severity of the disease can be correlated with the amount of apoptotic cells in liver biopsies from patients (Rust and Gores 2000). Increased expression of cytochrome CYP2E1, ROS and lipoperoxides may contribute to hepatocyte apoptosis (Lieber 2004). Production of ROS, which is driven by increased availability of the reduced form of nicotinamide adenine dinucleotide as a consequence of acetaldehyde metabolism, may cause mitochondrial dysfunction and release of cytochrome c into the cytosol (Guicciardi and Gores, 2005). Antioxidants reduce acute hepatocellular injury in rats receiving ethanol through prevention of the mitochondrial release of apoptotic factors (Schattenberg et al. 2006).

In patients with alcoholic liver disease high Fas ligand expression has been found in hepatocytes. This may be induced by ROS or by TNF-α-induced activation of NF-kappa B, which may upregulate the transcription of Fas and FasL genes (Guicciardi and Gores 2005). In fact, TNF-α levels are also increased during alcoholic hepatitis, and chronic ethanol administration has been reported to increase TNF-R1 expression on hepatocytes (Rust and Gores 2000). The important role of TNF-α in the pathophysiology of alcoholic liver disease is reinforced by the demonstration that reduced TNF expression with antibodies anti-TNF, adiponectin therapy or impaired TNF signaling through TNF-R1 have reduce fatty acid infiltration and inflammation in experimental models of alcoholic liver disease (García-Ruiz and Fernández-Checa 2007).

A role for TGF-β signalling appears to be also important. TGF-β has been reported to be produced by hepatic stellate cells in alcoholic liver disease and might have a double impact on the progression of the disease by promoting fibrogenesis and killing hepatocytes by apoptosis (Rust and Gores 2000). Moreover, studies on HepG2 cells indicate that potentiation of TGF-β-induced cell death by CYP2E1 may contribute to mechanisms of alcohol-induced liver disease (Zhuge and Cerdebaum 2006).

T-cells and NK T-cells have been implicated in the development of alcoholic liver disease by the increased numbers found in human liver following ethanol injury. Cytotoxic T cells can be generated in response to acetaldehyde-modified cells, and it has been reported that aldehyde-modified proteins at high levels can cause cell death and apoptosis (Willis et al. 2002). Therefore, the build-up of these adducts in the liver could potentially increase the level of apoptosis (Duryee et al. 2007).

4.3.6 Hepatocellular Carcinoma

Insufficient apoptosis has been associated with development and progression of tumours of the liver and the biliary tree (Guicciardi and Gores 2005). Hepatocellular carcinoma (HCC) is the most common primary malignant hepatic tumour, with a vast incidence through the world. The available evidence indicates that disruption of apoptosis in several steps of HCC development, especially in its promotion stage. Fas is partially or completely loss in HCC (Strand et al. 1996) and Fas expression negatively correlates to the degree of HCC differentiation and patient survival (Ito et al. 1998; Fukuzawa et al. 2001). On the contrary, p53-mediated up-regulation of Fas expression and increased apoptosis have been reported in human hepatoma cells following treatment with different chemotherapeutic agents (Müller et al. 1997). Nevertheless, down-regulation of Fas alone may not be sufficient to escape the immune response and many HCC also over express Bcl-X_L, which confers resistance to mitochondrial-induced apoptosis. As the death receptor-mediated pathway of apoptosis is linked to the mitochondrial pathway in hepatocytes, over expression of Bcl-X_L may contribute to Fas resistance in HCC (Takehara et al. 2001).

Several chemotherapeutic agents have been shown to up regulate Fas expression and increase sensitivity to Fas mediated apoptosis in hepatoma cells (Guicciardi and Gores 2005).

Disruption of TGF-β signalling appears to be another important factor in HCC. In HCC a significant reduction in the mRNAs for both TGF-β type 1 (TβR-I) and TGF-β type 2 (TβRII) receptor has been reported and it has also been shown that type 2 receptor, although expressed in tumoral hepatocytes, is detected in the cytoplasm but not in the plasma membrane (Sue et al. 1995). However, changes at the receptor levels do not appear to be as important as they are in other types of gastrointestinal tumours (Rossmanith and Schulte-Hermann 2001), and other factors disrupting TGF-β signalling may exist. Dysregulation of the EGF pathway has been observed in HCC (Breuhahn et al. 2006) and EGF receptor ligands might protect tumoral cells from TGF-β-induced apoptosis, perhaps through over expression of TAC/ADAM17 (Fabregat et al. 2007).

Mutations in the p53 gene are also common alterations in HCC. A dysfunctional p53 allows the tumour cells to escape apoptosis and results in cancer development (Guicciardi and Gores 2005). Moreover, because several chemotherapeutic drugs induce apoptosis of tumour cells by activation of p53, tumours with disrupt p53 are generally resistant to chemotherapy and associated with an unfavorable prognosis (Anderson et al. 1998). It has been reported that decreased p53 function contributes to the loss of CD95 expression and reduces sensitivity of HCC cells towards this apoptosis pathway (Müller et al. 1997). Microinjections of wild-type p53 and treatment with belomycin restore sensitivity towards CD95-induced apoptosis (Müller et al. 1997). Nevertheless, delivery of p53 recombinant DNA into rodent models of HCC does not suppress tumour growth (Bao et al. 1996) and it has been hypothesized that the effect of p53 loss is probably associated to the presence of intact or dysfunctional telomeres (Farazi et al. 2006). Only preliminary and non conclusive results are available to date from p53 gene therapy trials for HCC (Guicciardi and Gores 2005).

Inhibitors of apoptosis proteins (IAPS), which inhibit caspase activation, are also required for apoptosis inhibition in HCC. One member of the IAP family, survivin, may play an important role in progression of HCC by promoting cell proliferation, and is positively correlated with high risk of disease recurrence and poor prognosis in HCC. Its expression may therefore serve as a prognostic factor for patients with HCC after hepatectomy (Ye et al. 2007). Transfection of liver tumour cells HepG2 with antisense oligonucleotide (ASO) against survivin results in significant cells growth inhibition and reduction expression of survivin. Furthermore, systemic treatment with ASO significantly inhibits tumour growth in an orthotopic transplant model of HCC in nude mice, suggesting that ASO could potentially be a promising gene therapy approach to treatment of HCC (Sun et al. 2006).

Levels of c-FLIP are increased in HCC, contributing to apoptosis induced by CD95, TRAIL-R1 and TRAIL-R2. Down regulation of c-FLIP by pre-treatment with cycloheximide and actinomycin D results in partial resensitization of tumour cells to apoptotic stimuli (Eichhorst 2005). Increased expression of TRAIL-R1 may be a mechanism of cisplatin-induced sensitization to TRAIL-induced apoptosis in some HCC cells (Shin et al. 2002). Different studies have also shown that various agents, such as selective COX-2 inhibitors, interferon-gamma and Pk111195, a mitochondrial benzodiazepine receptor antagonist, increase the sensitivity of HCC to chemotherapeutic agents and radiation therapy. Enhancement of tumour sensitivity to apoptotic signals may improve the prognosis of patients with HCC (Bai and Odin 2003).

New approaches to HCC treatment could benefit from siRNA. Myeloid cell leukemia-1 (Mcl-1) is an anti-apoptotic member of the Bcl-2 protein family which interferes with mitochondrial activation. This molecule is highly expressed in tissues of human HCC, and its specific down-regulation by RNA interference sensitizes HCC cells to different chemotherapeutic agents (Schulze-Begkamen et al. 2006). Inhibition of glioma-associated antigen-1 (Gli-1) mRNA in Huh7 cells through Gli-1 siRNA also induces apoptosis through down-regulation of Bcl-2 (Chen et al. 2008).

To avoid the deleterious effects of potential therapies from injury to healthy liver tissue tools allowing type-specific targeting may be important. Thus, adenoviral, tumour cell-specific delivery of TRAIL or caspase 8 has been reported to induce increased rates of apoptosis in human HCC cells (Jacob et al. 2005; Yamaguchi et al. 2006). However, some conflicting results, suggesting that intrarterial injection of an adenovirus vector with IOEs can result in cancer-selective but not effective gene therapy for HCC, have been reported (Shiba et al. 2008).

4.3.7 Apoptosis and Fibrosis

Repair mechanisms induced by acute and chronic liver diseases may cause excessive disposition of scar matrix (liver fibrosis) leading to liver cirrhosis. Although the relationship between apoptosis and liver fibrosis has not been fully explored it is generally assumed that pro-apoptotic stimuli induce hepatocytes apoptosis. Engulfment of the hepatocyte apoptotic bodies by Kupffer and stellate cells enhances their expression of profibrogenic genes and death ligands, and persistent activation of these cells promotes further hepatocyte apoptosis, which culminates in generation of chemokines and sustained stellate cell activation (Canbay et al. 2004). Activation of stellate cells plays an important role in this process by transformation into myofibroblastic cells that synthesize scar tissue, and causes most of the pathological changes in cirrhosis. The activation of stellate cells involves the transdifferentiation from a quiescent state into myofibroblast-like cells, which are distinguished by accelerated proliferation and enhanced production of extracellular matrix components (Prosser et al. 2006).

It therefore appears that both inhibition of apoptosis and selective targeting of hepatic stellate cells with apoptotic stimuli could be adequate antifibrotic strategies. From the first perspective it has been reported that Fas-specific sRNA attenuates hepatic fibrosis following repeated concavalin A administration in mice (Song et al. 2003). More recently it has been reported that cytosolic caspase 3 activity and cytosolic fractions of Bax, Bcl-2, cytochrome c, and calpain-mu protein expressions are decreased in rats receiving carbon tetrachloride plus *S. miltiorrhiza* extract, with no change in caspase 8 which suggests that the antiapoptotic effect is related to the antioxidant properties of *S. miltiorrhiza* (Lee et al. 2006).

Several studies have confirmed that apoptosis of activated stellate cells may prevent fibrogenesis (Malhi and Gores 2008). Thus, the fungal metabolite gliotoxin reduces hepatic fibrosis, an effect accompanied by reduction by apoptosis of the numbers of activated hepatic stellate cells in the liver (Dekel et al. 2003). Because the use of apoptosis-inducing drugs may be limited due to lack of cell specificity, with a risk of severe adverse effects, very recently gliotoxin has been coupled to mannose-6-phosphate-modified human serum albumin (M6P-HSA), which selectively accumulates in liver fibrogenic cells. Administration of gliotoxin-M6P-HSA to bile duct-ligated revealed a significant decrease in alpha-smooth muscle actin mRNA levels and a reduced staining for this marker of hepatic stellate cells in fibrotic livers (Hagens et al. 2008).

Proteasome inhibitors induce apoptosis in transformed cells, especially those cells dependent upon NF-kappaB activation. Stimulated stellate cells also trigger NFkB activation, and it has been reported that when the immortalized human stellate cell line, LX-2 or primary rat stellate cells are treated with the proteasome inhibitors bortezomib and MG132, there is an apoptotic response that is not related to proteasome inhibition-induced alterations in TRAIL, death receptor 5, and Bim, but to a blocked activation of NF-kappaB dependent upon the NF-kappaB target gene A1 (Anan et al. 2006). Other studies have also demonstrated the possibility of acting on additional apoptotic mediators. Thus, atorvastatin induces apoptosis in activated rat stellate cells through an ERK-dependent cleavage of Bid and a highly increased protease activity of caspase 9 and 3 (Aprigliano et al. 2008).

4.4 Summary and Conclusions

Apoptosis in the liver has become the focus of research since there is a large body of evidence that dysregulation of the apoptotic programme is involved in the pathophysiology of liver diseases. Cell death by apoptosis can increase in toxic, viral, alcoholic and non alcoholic liver injury. A reduction of apoptosis could be beneficial in these diseases and the goal for the development of new drugs. Since multiple apoptotic pathways are activated in any of these conditions, combination therapies targeting different pathways may be useful. On the contrary, hepatocellular carcinomas seem to escape immune surveillance and apoptosis. Induction of apoptosis might therefore be a new option for the prevention and treatment of hepatocarcinoma. Finally, liver fibrosis may require targeted approaches aimed to limit apoptosis to parenchymal cells while accelerating apoptosis of hepatic stellate cells.

Acknowledgments CIBEREHD is founded by Instituto de Salud Carlos III (Spain).

References

Albright CD, Liu R, Mar MH et al. (1997) Diet, apoptosis, and carcinogenesis. Adv Exp Med Biol 422:97–107

Anan A, Baskin-Bey ES, Bronk SF et al. (2006) Proteasome inhibition induces hepatic stellate cell apoptosis. Hepatology 43:335–344

Anderson SC, Johnson DE, Harris MP et al. (1998) p53 gene therapy in a rat model of hepatocellular hepatocarcinoma: intra-arterial delivery of a recombinant adenovirus. Clin Cancer Res 4:1649–1659

Aprigliano I, Dudas J, Ramadori G et al. (2008) Atorvastatin induces apoptosis by a caspase-9-dependent pathway: an *in vitro* study on activated rat hepatic stellate cells. Liver Int 28:546–557

Bai J, Odin JA (2003) Apoptosis and the liver: relation to autoimmunity and related conditions. Autoimmun Rev 2:36–42

Bantel H, Lügering A, Heidemann J et al. (2004) Detection of apoptotic caspase activation in sera from patients with chronic HCV infection is associated with fibrotic liver injury. Hepatology 40:1078–1087

Bao JL, Zhang W, Kuo MT (1996) Adenoviral delivery of recombinant DNA into transgenic mice bearing hepatocellular carcinomas. Human Gen Ther 7:355–365

Baumert TF, Thimme R, von Weizsacker F (2007) Pathogenesis of hepatitis B virus infection. World J Gastroenterol 7:82–90

Botla R, Spivey JR, Aguilar H et al. (1995) Ursodeoxycholate (UDCA) inhibits the mitochondrial membrane permeability transition induced by glycodeoxycholate: a mechanism of UDCA cytoprotection. J Pharmacol Exp Ther 272:930–938

Breuhahn K, Longerich T, Schirmacher P (2006) Dysregulation of growth factor signaling in human hepatocellular carcinoma. Oncogene 25:3787–3800

Canbay A, Friedmann S, Gores GJ (2004) Apoptosis: the nexus of liver injury and fibrosis. Hepatology 39:3273–3278

Canbay A, Feldstein A, Baskin-Bey E et al. (2004b) The caspase inhibitor IDN-6556 attenuates hepatic injury and fibrosis in the bile duct ligated mouse. J Pharmacol Exp Ther 308:1191–1196

Canbay A, Higuchi H, Bronk SF et al. (2002) Fas enhances fibrogenesis in the bile duct ligated mouse: a link between apoptosis and fibrosis. Gastroenterology 123:1323–1330

Chen XL, Cao LQ, She MR et al. (2008) Gli-1 siRNA induced apoptosis in Huh7 cells. World J Gastroenterol 14:582–589

Chung YL, Sheu ML, Yen SH (2003) Hepatitis C virus NS5A as a potential viral Bcl-2 homologue interacts with Bax and inhibits apoptosis in hepatocellular carcinoma. Int J Cancer 107:65–73

Dekel R, Zvibel I, Brill S et al. (2003) Gliotoxin ameliorates development of fibrosis and cirrhosis in a thioacetamide rat model. Dig Dis Sci 48:1642–1647

Duryee MJ, Klassen LW, Thiele GM (2007) Immunological response in alcoholic liver disease. Worl J Gastroenterol 13:4938–4946

Eichhorst ST (2005) Modulation of apoptosis as a target for liver disease. Expert Opin Ther Targets 9:83–89

Elmore LW, Hancock AR, Chang SF et al. (1997) Hepatitis B virus X protein and p53 tumour suppressor interactions in the modulation of apoptosis. Proc Natl Acad Sci USA, 94:14707–14712

Erdtmann L, Franck N, Lerat H et al. (2003) The hepatitis C virus NS2 protein is an inhibitor of CIDE-B-induced apoptosis. J Biol Chem 278:18256–18264

Fabregat I, Roncero C, Fernández M (2007) Survival and apoptosis: a dysregulated balance in liver cancer. Liver Int 2007:1551–1562

Farazi PA, Glickman J, Hormer J et al. (2006) Cooperative interactions of p53 mutation, telomere dysfunction, and chronic liver damage in hepatocellular carcinoma progression. Cancer Res 66:4766–4763

Faubion WA, Guicciardi ME, Miyoshi H et al. (1999) Toxic bile salts induce rodent hepatocyte apoptosis via direct activation of Fas. J Clin Invest 103:137–145

Feng G, Kaplowitz N (2000) Colchicine protects mice from the lethal effect of an agonistic anti-Fas antibody. J Clin Invest 105:329–339

Fickert P, Trauner M, Fuchsbichler A et al. (2005) Oncosis represents the main type of cell death in mouse models of cholestasis. J Hepatol 42:378–385
Fisher GH, Fisher GH, Rosenberg FJ et al. (1995) Dominant interfering Fas gene mutations impair apoptosis in a human autoimmune lymphoproliferative syndrome. Cell 81:935–946
Fiorucci S, Mencarelli A, Palazzetti B et al. (2001) An NO derivative of ursodeoxycholic acid protects against Fas-mediated liver injury by inhibiting caspase activity. Proc Natl Acad Sci USA 98:2652–2657
Fischer R, Baumert T, Blum HE (2007) Hepatic C virus infection and apoptosis. World J Gastroenterol 28:4865–4872
Fukuzawa K, Takahashi K, Furuta K et al. (2001) Expression of FAS/Fas ligand and its involvement in infiltrating lynmphocytes in hepatocellular carcinoma (HCC). J Gastroenterol 36:681–686
García-Ruiz C, Fernández-Checa JC (2007) Redox regulation of hepatocyte apoptosis. J Gastroenterol Hepatol 22:S38–S42
Gaudio E, Franchitto A, Pannarale L et al. (2006) Cholangiocytes and blood supply. World J Gastroenterol 14:3546–3552
Ghavami S, Hashemi M, Kadkhoda K et al. (2005) Apoptosis in liver diseases – detection and therapeutics applications. Med Sci Monit 11:RA337–RA345
Guicciardi ME, Gores GJ (2005) Apoptosis: a mechanisms of acute and chronic liver injury. Gut 54:1024–1033
Gujral JS, Liu J, Farohood, Jaeschke H (2004) Reduced oncotic necrosis in Fas receptor-deficient C57BL/6J-Jpr mice after bile duct ligation. Hepatology 40:998–1007
Hagens WI, Beljaars L, Mann DA et al. (2008) Cellular targeting of the apoptosis-inducing compound gliotoxin to fibrotic rat livers. J Pharmacol Exp Ther 324:902–910
Higuchi H, Bronk SF, Takikawa Y et al. (2001) The bile acid glycochenodeoxycholate induces trail receptor 2/DR5 expression and apoptosis. J Biol Chem 276:38610–38618
Hsu FC, His B, Hirota-Tsuchiara M et al. (2001) Modified apoptotic molecule (BID) reduces hepatitis C viral infection in mice with chimeric human livers. Nat Biotechnol 21:519–525
Ito Y, Takeda T, Umeshita K et al. (1998) Fas antigen expression in hepatocellular carcinoma tissues. Oncol Rep 5:41–44
Jacob D, Schumacher G, Bahra M et al. (2005) Fiber-modified adenoviral vector expressing the tumour necrosis factor-related apoptosis-inducing ligand gene from the human telomerase reverse transcriptase promoter induces apoptosis in human hepatocellular carcinoma cells. World J Gastroenterol 11:2552–2556
Jaeschke H, Gujral JS, Bajt ML (2004) Apoptosis and necrosis in liver disease. Liver Int 24:85–89
Kanzler S, Galle PR (2000) Apoptosis and the liver. Sem Cancer Biol 10:173–184
Kim SH, Kim YS, Kang SS et al. (2008) Anti-apoptotic and hepatoprotective effects of gomisin A on fulminant hepatic failure induced by D-galactosamine and lipopolysaccharide in mice. J Pharmacol Sci 106:225–233
Kim KH, Seong BI (2003) Pro-apoptotic function of HBV X protein is mediated by interaction with c-FLIP and enhancement of death-inducing signal. EMBO J 22:2104–2116
Koga H, Sakisaka S, Sata M et al. (1997) Nuclear DNA fragmentation and expression of Bcl-2 in primary biliary cirrhosis. Hepatology 25:1077–1084
Koilos G, Valatas V, Kouroumalis E (2006) Role of Kupffer cells in the pathogenesis of liver disease. Worl J Gastroenterol 12:7413–7420
Kountouras J, Zavos C, Chatzpoulos D (2003) Apoptosis in hepatitis C. J Viral Hepat 10:335–342
Lee TY, Chang HH, Wang GJ et al. (2006) Water-soluble extract of *Salvia miltiorrhiza* ameliorates carbon tetrachloride-mediated hepatic apoptosis in rats. J Pharm Pharmacol 58:659–665
Lieber, CS (2004) From ASH to NASH. Hepatol Res 28:1–11
Machida K, Cheng KT, Lai CK et al. (2006) Hepatitis C virus triggers mitochondrial permeability transition with production of reactive oxygen species, leading to DNA damage and STAT3 activation. J Virol 80:7199–7207
Malhi H, Gores G (2008) Cellular and molecular mechanisms of liver injury. Gastroenterology 134:1641–1654
Malhi H, Gores GJ, Lemasters JJ (2006) Apoptosis and necrosis in the liver: a tale of two deaths? Hepatology 43:S31–S44

Martinon F, Tschopp J (2004) Inflammatory caspases: linking an intracellular innate immune system to autoinflammatory diseases. Cell 117:561–554

Masaki T, Chiba S, Tatsukawa, H, et al. (2004) Adiponectin protects LPS-induced liver injury through modulation of TNF-alpha in KK-AY obese mice. Hepatology 40:177–184

McDonald A, Harris M (2004) Hepatitis C virus NS5A: tales of a promiscuous protein. J Gen Virol 85:2485–2502

Michalopoulos GK (2007) Liver regeneration. J Cell Physiol 213:286–300

Molpeceres V, Mauriz JL, García-Mediavilla MV et al. (2007) Melatonin is able to reduce the apoptotic liver changes induced by aging via inhibition of the intrinsic pathway of apoptosis. J Gerontol Biol Sci 62:687–695

Müller M, Strand S, Hug H et al. (1997) Drug-induced apoptosis in hepatoma cells is mediated by the CD95 (APO-1/Fas) receptor/ligand system and involves activation of wild-type p53. J Clin Invest 99:403–413

Nakatani K, Kaneda K, Seki S et al. (2004) Pit cells as liver-associated natural killer cells: morphology and function. Metd Electron Microsc 37:29–36

Nanda SK, Herion D, Liang TJ (2006) The SH3 binding motif of HCV (corrected) NS5A protein interacts with Bin 1 and is important for apoptosis and infectivity. Gastroenterology 130:794–809

Pockros PJ, Schiff ER, Shiffman ML et al. (2007) Oral IDN-6556, an antiapoptotic caspase inhibitor, may lower aminotransferase activity in patients with chronic hepatitis C. Hepatology 46:342–349

Pretet JL, Pelletier L, Bernard B et al. (2003) Apoptosis participates to liver damage in HSV-induced fulminant hepatitis. Apoptosis 8:655–663

Prosser CC, Yen RD, Wu J (2006) Molecular therapy for hepatic injury and fibrosis: where are we? World J Gastroenterol 12:509–515

Roberts LR, Kurosawa H, Bronk SF et al. (1997) Cathepsin contributes to bile salt-induced apoptosis of rat hepatocytes. Gastroenterology 113:1714–1726

Rossmanith W, Schulte-Hermann R (2001) Biology of transforming growth factor beta in hepatocarcinogenesis. Microsc Res Technol 52:430–436

Ruggieri A, Harada T, Matsuura Y et al. (1997) Sensitization to Fas-mediated apoptosis by hepatitis C virus core protein. Virology 229:68–76

Rust C, Gores GJ (2000) Apoptosis and liver disease. Am J Med 108:567–574

Saito K, Meyer K, Warner R et al. (2006) Hepatitis C virus core protein inhibits tumour necrosis factor alpha mediated apoptosis by a protective effect involving cellular FLICE inhibitory protein. J Virol 80:4372–4379

Sakaida I, Kimura T, Yamasaki T (2005) Cytochrome c is a possible new marker for fulminant hepatitis in humans. J Gastroenterol 40:179–185

San Miguel B, Alvarez M, Culebras JM et al. (2006) N-acetyl-cysteine protects liver from apoptotic death in an animal model of fulminant hepatic failure. Apoptosis 11:1945–1957

Sarcar B, Ghosh AK, Steele R (2004) Hepatitis C virus NS5A mediated STAT3 activation requires co-operation of Jak1 kinase. Virology 322:51–60

Sato M, Suzuki S, Senoo H (2003) Hepatic stellate cells: unique characteristics in cell biology and phenotype. Cell Struct Funct 2:105–112

Schattenberg JM, Wang Y, Rigoli RM et al. (2004) CYP2E1 overexpression alters hepatocyte death from menadione and fatty acids by activation of ERK1/2 signaling. Hepatology 39:444–455

Schattenberg JM, Galle PR, Schuchmann M (2006) Apoptosis in liver disease. Liver Int 26:904–911

Schoemaker MH, Moshage H (2003) Defying death: the hepatocytes survival kit. Clin Sci 107:13–25

Schuchmann M, Galle PR (2001) Apoptosis in liver disease. Eur J Gastroenterol Hepatol 13:785–790

Schulze-Bergkamen H, Fleischer B, Schuchmann M et al. (2006) Suppression of Mcl-1 via RNA interference sensitizes human hepatocellular carcinoma cells towards apoptosis induction. BMC Cancer 2:232

Selmi C, Mackay IR, Gershwin ME (2007) The immunological milieu of the liver. Semin Liver Dis 27:129–319
Shiba H, Misawa T, Iida T et al. (2008) Adenovirus vector-mediated gene therapy using iodized oil esters for hepatocellular carcinoma in rats. Anticancer Res 28:51–53
Shin EC, Seong YR, Kim CH et al. (2002) Human hepatocellular carcinoma cells resist to TRAIL-induced apoptosis, and the resistance is abolished by cisplatin. Exp Mol Med 34:114–122
Solá S, Amaral JD, Aranha MM et al. (2006) Modulation of hepatocyte apoptosis: cross-talk between bile acids and nucelar steroid receptors. Curr Med Chem 13:3039–3051
Song E, Lee SK, Wang J et al. (2003) RNA interference targeting Fas protects mice from fulminant hepatitis. Nat Med 9:347–351
Strand S, Hofmann WJ, Hug H et al. (1996) Lymphocyte apoptosis induced by CD95 (APO-1/Fas) ligand expressing tumour cells – a mechanism of immune invasion? Nature Med 2:1361–1366
Street A, McDonald A, Crowder K et al. (2004) The hepatitis C virus NS5A activates phosphoinositide 3-kinase-dependent survival signaling cascade. J Biol Chem 279:12232–122341
Su F, Schneider RJ (1997) Hepatitis B virus HBx protein sensitizes cells to apoptotic killing by tumour necrosis factor alpha. Proc Natl Acad Sci USA 94:8744–8749
Sue SR, Chari RS, Kong FM et al. (1995) Transforming growth factor-beta receptors and mannose 6-phosphate/insulin-like growth factor-II receptor expression in human hepatocellular carcinoma. Ann Surg 222:171–178
Sun Y, Lin R, Dai J et al. (2006) Suppression of tumour growth using antisense oligonucleotide against survivin in an orthotopic transplant model of human hepatocellular carcinoma in nude mice. Oligonucleotides 16:365–374
Sun EW, Shi YF (2001) Apoptosis: the quiet death silences the immune system. Pharmacol Ther 92:135–145
Takehara T, Liu X, Fujimoto J et al. (2001) Expression and role of Bcl-xL in human hepatocellular carcinomas. Hepatology 34:55–61
Tanaka S, Sugimachi K, Shirabe K et al. (2000) Expression and antitumour effects of TRAIL in human cholangiocarcinoma. Hepatology 32:532–537
Terada T, Nakanuma Y (1995) Detection of apoptosis and expression of apoptosis-related proteins during human intrahepatic bile duct development. Am J Pathol 146:67–74
Torisu T, Nakaya M, Watanabe S et al. (2008) Suppressor of cytokine signaling 1 protects mice against concanavalin A-induced hepatitis by inhibiting apoptosis. Hepatology 47:1644–1654
Tuñón MJ, Sánchez-Campos S, García-Ferreras J et al. (2003) Rabbit hemorrhagic viral disease: characterization of a new animal model of fulminant liver failure. J Lab Clin Med 141:272–278
Vekemans K, Rosseel L, Wisse E et al. (2004) Immuno-localization of Fas and FasL in rat hepatic endothelial cells: influence of different fixation protocols. Micron 35:303–306
Willis MS, Klassen LW, Tuma DJ et al. (2002) Malondyaldehyde-acetaldehyde-heptenated protein induces cell death by induction of necrosis and apoptosis in immune cells. Immunopharmacol 2:519–535
Yamaguchi Y, Shiraki K, Fuke H et al. (2006) Adenovirus-mediated transfection of caspase-8 sensitizes hepatocellular carcinoma to TRAIL- and chemotherapeutic agent-induced cell death. Biochim Biophys Acta 1763:844–853
Ye CP, Qiu CZ, Huang ZX et al. (2007) Relationship between survivin expression and recurrence, and prognosis in hepatocellular carcinoma. World J Gastroenterol 13:6264–6268
Yoon JH, Gores GJ (2002) Death receptor-mediated apoptosis and the liver. J Hepatol 37:400–410
Zender L, Hutker S, Liedtke C et al. (2003) Caspase 8 small interfering RNA prevents acute liver failure in mice. Proc Natl Acad Sci USA 100:7797–7802
Zhang Y, Chong E, Herman B (2002) Age-associated increases in the activity of multiple caspases in Fisher 344 rat organs. Exp. Gerontol 37:777–789
Zhuge J, Cerdebaum AI (2006) Increased toxicity by transforming growth factor-beta 1 in liver cells overexpressing CYP2E1. Free Radic Biol Med 41:110–1112

Chapter 5
Apoptosis in Nervous Cells

A. González Mateos

Abstract In the central nervous system both neurons and astrocytes play crucial roles. On a cellular level, brain activity involves continuous interactions within complex cellular circuits established between neural cells and glia. Despite it was initially considered that neurons were the major cell type in cerebral function, nowadays astrocytes are considered to contribute to cerebral function too. Astrocytes support normal neuronal activity, including synaptic function, by regulating the extracellular environment with respect to ions and neurotransmitters. In both cell types Ca^{2+} signalling plays a pivotal role. Normal Ca^{2+} homeostasis is required for cell activity, in either in neurons and astrocytes, and must be precisely regulated. On the other hand, mitochondria are the major cellular source for ATP, and are also central for Ca^{2+} homeostasis. Deregulation of cell cycle has devastating effects on the integrity of cells, and has been closely associated with the development of pathologies which can lead to dysfunction and cell death. Programmed cell death or apoptosis can be activated and/or initiated by different mechanisms involving cell membrane receptor activation, Ca^{2+} signal impairment, mitochondrial uncoupling or oxidative stress, and involves in its majority, caspase-mediated cleavage cascade. An alteration of normal neuronal/glial physiology and apoptotic processes could represent the basis of neurodegenerative processes. In this chapter we will pay attention on to the recent findings in neuronal-astrocyte connection and its relationship to apoptosis.

Keywords Astrocyte · Neuron · Oxidative stress · Cell death · Neurodegeneration

5.1 Introduction

The cerebral cortex contains a number between twelve and fifteen billion neurons and there is about one billion neurons in spinal cord, whereas there is an order of

A. González Mateos (✉)
Department of Physiology, University of Extremadura, Avda Universidad s/n, 10071 Cáceres, Spain
e-mail: agmateos@unex.es

G.M. Salido, J.A. Rosado (eds.), *Apoptosis: Involvement of Oxidative Stress and Intracellular Ca^{2+} Homeostasis*, DOI 10.1007/978-1-4020-9873-4_5,

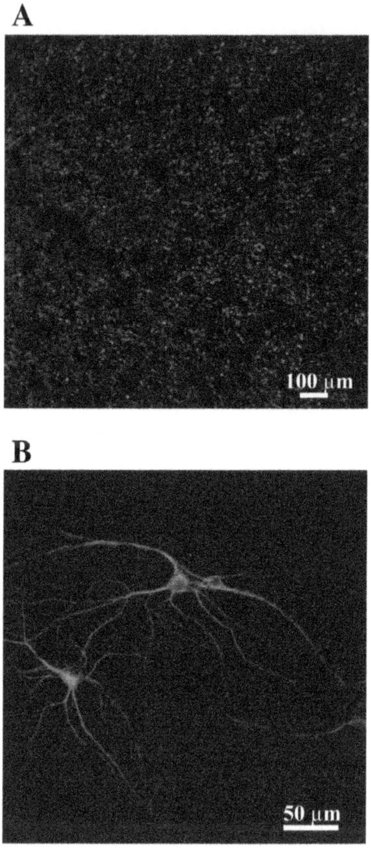

Fig. 5.1 Immunocytochemical studies of cultured hippocampal neurons. The figures show the fluorescence image of hippocampal neurons stained with FITC-labelled anti-MAP-2 antibody in brain slices (**A**) and in cell culture (**B**) (See also Plate 3 in the Color Plate Section on page 225)

ten to fifty times more glial cells than neurons in the central nervous system (CNS) (Kuffler 1984). Within this "cellular universe" that forms the CNS, neurons operate in interconnected networks where the information is received, integrated and sent as an output signal (Fig. 5.1).

Beside neurons, other cell types in the CNS are oligodendrocytes, which form myelin, astrocytes, with multiple support functions to neurons, ependymal cells, which are epithelial cells that line brain ventricles and central canal of spinal and assist in secretion and circulation of cerebral spinal fluid and microglia, which are small cells that proliferate and act as scavengers when tissue is destroyed and play an important role in defence and inflammation (Verkhratsky et al. 1998).

Among this group of cells, a cell type with an outstanding role are the astrocytes (Fig. 5.2). These glial cells act as the "glue" in the CNS, surrounding neurons in order to hold them in place (Fig. 5.3). However, glial cells play a number of other functions which are very important and represent a major support for neuronal functions: supply of nutrients and oxygen to neurons, to insulate one neuron from another, regulate extracellular pH and ion balance (for example K^+), regulate glutamate levels, produce myelin, contribute to form the blood-brain barrier, to destroy

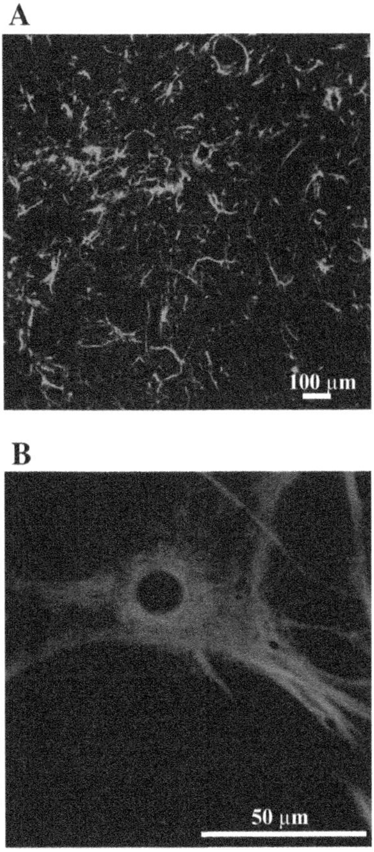

Fig. 5.2 Immunocytochemical studies of cultured hippocampal astrocytes. The figures show the fluorescence image of hippocampal astrocytes stained with FITC-labelled anti-GFAP antibody in brain slices (A) and in cell culture (B) (See also Plate 4 in the Color Plate Section on page 226)

pathogens, metabolize and remove dead neurons or for example guide migrating neurons during development (radial glia), in addition to a regulation of synaptic transmission. All these cells mentioned above, are remarkably interconnected and demonstrate considerable amount of intercellular signalling between them. Such crosstalk between these different cell types forms the basis of the physiology of the CNS.

Despite sharing several common signalling pathways, sometimes the same ligand induces opposite responses in different CNS cells, suggesting that there are cell-type-specific modulations in cellular signalling pathways. Beside their cytological differences, there is a property that clearly differentiates either type of cell: whereas neurons are electrically excitable cells, i.e. they undergo action potentials, the glia is normally considered an electrically non-excitable cell type, which will be very much related to the array of functions they can develop (Araque et al. 2001; Volterra and Meldolesi 2005). As a consequence, neurons will employ action potentials to communicate between them, whereas the glia will communicate with each other and

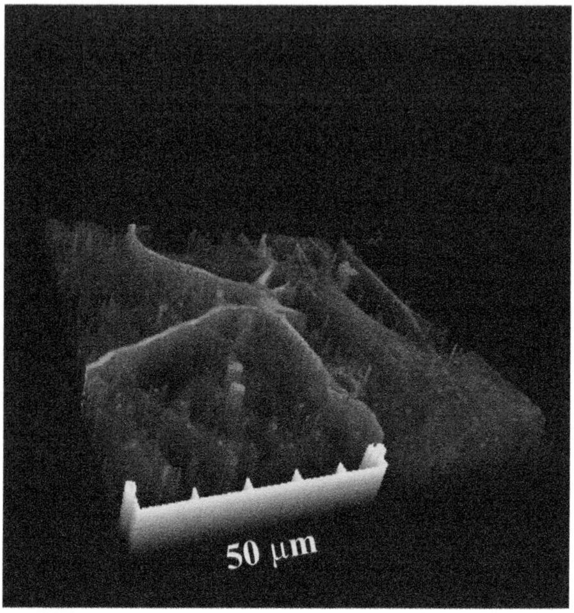

Fig. 5.3 Immunocytochemical studies of mixed cultures of hippocampal astrocytes and neurons. The image shows a three-dimensional projection of the spatial distribution of fluorescence of GFAP-positive astrocytes and MAP-2-positive neurons. The image shows the close association of both cellular types in the cultures, with astrocytic processes engulfing the neuron (See also Plate 5 in the Color Plate Section on page 227)

neurons by other means. Finally, apart from their specificity in cellular signalling, these cell types depict programmed cell death or apoptosis pathways.

Cell death presents two principal forms, apoptosis and necrosis. Physiological processes, such as the renumbering of neurons during normal development, involve apoptosis. Normally, superfluous, infected or transformed cells, i.e. unwanted cells, die by apoptosis, a controlled genetic programme for removing such cells without damaging the surrounding tissue. However, depending on the point of its initiation and when abnormally or excessively activated, apoptosis may play a role in the pathogenesis of neurodegenerative diseases (Culmsee and Landshamer 2006).

Interaction between cells in the CNS seems to rely on many communication systems and signalling molecules, which act in parallel or display regional and cellular specialization. By terms of employing this codex, there is a bidirectional signal communication system within the CNS, which might be mainly carried out by extracellular messengers released from any cell type of the CNS. It is actually assumed that the different cell types that form the cytoarchitecture of the CNS are intricately interconnected and thus, their correct and equilibrate function is of critical importance for the homeostasis in the CNS. Therefore, it is not surprising that an alteration in signalling of either cell, and hence in their function, could affect synaptic activity and plasticity, the cell itself or other cells survival and, globally, brain homeostasis. Moreover, on the basis of cell specificity, this communication interplay becomes amazingly interesting from the point of view of apoptotic processes in neurodegenerative diseases and neuroprotection.

The brain is an organ that consumes much energy. This is partially due to the character of neurons; they possess excitable plasma membrane and a large amount of ATP is indispensable for maintaining ion gradient. Once neurons experience energy failure, Ca^{2+} accumulates in the intracellular space as a result of disturbed ion homeostasis. This, in turn, activates many cellular processes, which can culminate in cell death. In this cellular catastrophic cascade, many organelles play important roles. In addition to the plasma membrane, cytosol is the "organelle" that first becomes exposed to the increased level of Ca^{2+}. Many proteases, kinases and lipases are localized here, and are activated directly or indirectly within the death process. Mitochondria, the endoplasmic reticulum (ER), the Golgi apparatus and lysosomes as well as the nucleus not only mediate energy synthesis, intracellular Ca^{2+} homeostasis, protein processing, etc., but also are organelles involved in apoptotic cell death in some situations and thus regulate cell death signal activation.

It is of general knowledge that there are three principal pathways by which the molecular events of apoptosis can be initiated: the intrinsic and extrinsic pathways and by endoplasmic reticulum stress (ER stress). In all of them Ca^{2+} ions and reactive oxygen species (ROS) play critical roles as signalling molecules (Krieger and Duchen 2002; Harukuni and Bhardwaj 2006; Shang et al. 2006; Sui et al. 2006; Yoshida et al. 2006; Hara and Snyder 2007; Keeble and Gilmore 2007; Mattson 2007). The accumulation of neurotransmitters and cytotoxic substances released from neighbouring cells, and accumulation of heavy metals and amyloid beta protein are all conditions that potentially can initiate the cascades of events underlying apoptosis and, among others, involve Ca^{2+} signalling and generation of ROS (Alexandrov et al. 2005; Han et al. 2007; Bezprozvanny and Mattson 2008; Cheng et al. 2008; Del Río et al. 2008; Hellmich et al. 2008; Prabhakaran et al. 2008; Resende et al. 2008).

The extrinsic pathway of apoptosis plays a central role in physiological programmed cell death, as well as in various neoplastic and immune system diseases. Programmed cell death is initiated in the extrinsic pathway by the binding of an extracellular messenger to any of a number of death receptors which belong to the tumor necrosis factor (TNF) superfamily (Zha and Shu 2002; Keane et al. 2006). The receptors recruit and bind other death domain-containing proteins. This interaction permits autocleavage, and activation, of initiator caspases such as caspase 8. In turn, activation of caspase 8 can mediate cell death *via* two different ways: cleavage of downstream effector caspases, such as caspase 3, which then cleave cytoskeletal and nuclear proteins leading to cell death, and/or activation of members of the Bcl-2 family, such as Bid, which will induce mitochondrial release of cytochrome *c* and other mitochondrial mediators of cell death (Schuster et al. 2003; Guthmann et al. 2005; Basu et al. 2006; Hu et al. 2006; Nagy et al. 2006; Yang et al. 2006; Wang et al. 2006).

Alternatively, another mechanism that can mediate apoptosis is the activation of the mitogen-activated protein kinase kinase kinase ASK1 (apoptosis signalling kinase) (Hsu et al. 2007; Karunakaran et al. 2007; Nadeau et al. 2007; Nagai et al. 2007; Matsuzawa and Ichijo 2008; Takenouchi et al. 2008) which, in turn, can activate c-jun N-terminal kinase (JNK).

Cell death signalling through TNF-receptor (TNFR) can also proceed by interaction with TNFR-1-associated death domain protein (TRADD), which will interact with Fas-associated death domain (FADD), resulting in the activation of caspase 8 to establish the "death-inducing signalling complex" (Dempsey et al. 2003; Micheau and Tschopp 2003; Schneider-Brachert et al. 2004; Jin and El-Deiry 2006; Zheng et al. 2006; Sharma et al. 2008).

In the intrinsic pathway, the mitochondria play a critical role. The release of protein mediators from mitochondria is the pivotal event that initiates the cell death cascades of events. Mitochondrial components can be released in response to Ca^{2+} overload and overproduction of ROS, which normally occur under conditions of excitotoxicity. Mitochondrial control this pathway and the end result is the formation of a large protein complex, the apoptosome. The most prominent pro-apoptotic factor released from mitochondria is cytochrome *c* (Hans et al. 2005; Muranyi and Li 2005; Wu et al. 2007b), but there are other mitochondrial proteins that are involved, which include apoptosis inducing factor (AIF), endonuclease G, Smac/DIABLO, and Omi/HtrA2, and also play roles in the regulation of apoptosis (Brustovetsky et al. 2005; Hans et al. 2005; Nakka et al. 2008). Release of pro-apoptotic factors may occur with or without opening of the mitochondrial permeability transition (MPT) pore. These events are controlled by members of the Bcl-2 family. Within them, there are anti- and pro-apoptotic members, and the fate of the cell is determined by which one is predominant (Ouyang and Giffard 2004; Keeble and Gilmore 2007). There are two classes of pro-apoptotic Bcl-2 family members, depending on their biochemical structures. Bax and Bak form one group, whereas the other is formed by Bid, Bim, and Bad (Nicholls and Budd 2000; Burke 2008).

In this pathway, the release of cytochrome *c* from the mitochondria results in the activation of a caspase cascade, consisting of a group of initiator caspases (2, 8, 9, and 10) and a group of effector caspases (3, 6, and 7) (Burke 2008). Following the release of cytochrome *c* from mitochondria, caspase 9 becomes activated upon association with a cytoplasmic protein, apoptosis-protease-activating factor-1, to form the apoptosome (Nicholls and Budd 2000; Cregan et al. 2002; Scorrano and Stanley 2003; Ouyang and Giffard 2004). Activated caspase 9 then leads to activation of caspase 3 and other effector caspases.

Finally, the ER can transmit molecular signals to initiate apoptosis too. ER activation of apoptosis could involve the participation of transcription factors, ATF4 and ATF6 (Williams and Lipkin 2006; Lacour et al. 2007; Penas et al. 2007), or an activation of caspase 12, that is localized to the ER (Larner et al. 2004; Chao et al. 2007; Penas et al. 2007).

A hallmark of all forms of neurodegenerative diseases is impairment of neuronal functions, and in many cases neuronal cell death. Although the etiology of neurodegenerative diseases may be distinct, different diseases display a similar pathogenesis, for example abnormal immunity within the CNS, activation of macrophage/microglia and the involvement of proinflammatory cytokines. Recent studies show that neurons in a neurodegenerative state undergo this highly regulated program of cell death called apoptosis. The activation of cellular pathways of apoptosis has been an important hypothesis underlying the origin of adult neurodegenerative diseases such as Parkinson's disease, Alzheimer's disease, amyotrophic lateral sclerosis, Huntington's disease or multiple sclerosis (Manfredi and Xu 2005;

Sofo et al. 2005; Goodall and Morrison 2006; Hsu et al. 2007; Karunakaran et al. 2007; Arning et al. 2008; Zhang et al. 2008a,b), and in illnesses such as HIV-encephalopathy (Miura et al. 2003) or in meningitis (Gianinazzi et al. 2003) for example.

On the other hand, apoptosis is an important aspect of normal brain development. Neural stem/progenitor cells (NPCs) are present in the developing and adult central nervous system, and are capable to undergo apoptosis during brain development (Peng et al. 2005). In addition, the olfactory bulb is one of the few structures in the adult mammalian CNS that contains a continuous supply of newly generated neurons in the subventricular zone. Therefore, the balance between the supply of new cells and apoptosis might determine olfactory function (Mori et al. 2005).

Further elucidation of the apoptotic pathways and their relationships to the situation commonly referred to as "stress", under which all living organisms are constantly subjected to a variety of physical, chemical or biological damaging factors, will have implications for a better knowledge of the physiology and pathophysiology of the CNS, and in the finding of neuroprotective treatments for neurodegenerative diseases.

5.2 Neuronal Apoptosis

5.2.1 Extrinsic Pathway of Apoptosis in Neurons

The extrinsic pathway of apoptosis is related to the activation of TNF superfamily of cell membrane receptors. Upon activation, TNF-alpha receptor (TNF-R) can engage apoptotic or survival pathways (Fig. 5.4).

The neuronal loss associated with inflammation in Alzheimer's disease could be linked to age-related changes in TNF-R expression (Patel and Brewer, 2008). In this disease, beta amyloid protein plays a pivotal role, and beta amyloid-induced activation of TNF-alpha cascades has been also proposed. These age-related changes underlying Alzheimer's disease could involve a change in susceptibility to TNF-alpha through TNF receptor 1 (TNF-R1) and receptor 2 (TNF-R2) expression, that could be the basis for the effects of beta-amyloid protein. In this study, aging affects TNF-R expression in cultured adult rat cortical neurons, being old neurons more susceptible to beta amyloid toxicity, and this was exacerbated in the presence of TNF-alpha. In addition to aging, there are other numerous conditions that can lead to apoptosis involving TNF-alpha signalling.

TNF-alpha elicits toxicity in retinal ganglion cells, and involvement of caspase 8 was demonstrated by the employment of the caspase 8 inhibitor Z-IETD-FMK (Fuchs et al. 2005). In PC12 cells and cortical neurons, treatment with actinomycin D increases the susceptibility to TNF-alpha-induced cell death (Gozzelino et al. 2008). In these two cell types, TNF-alpha induced cell death through the activation of caspase 8, generation of tBid and activation of caspases 9 and 3. In addition, survival Bcl-x(L) protein levels, but not those proapoptotic members of

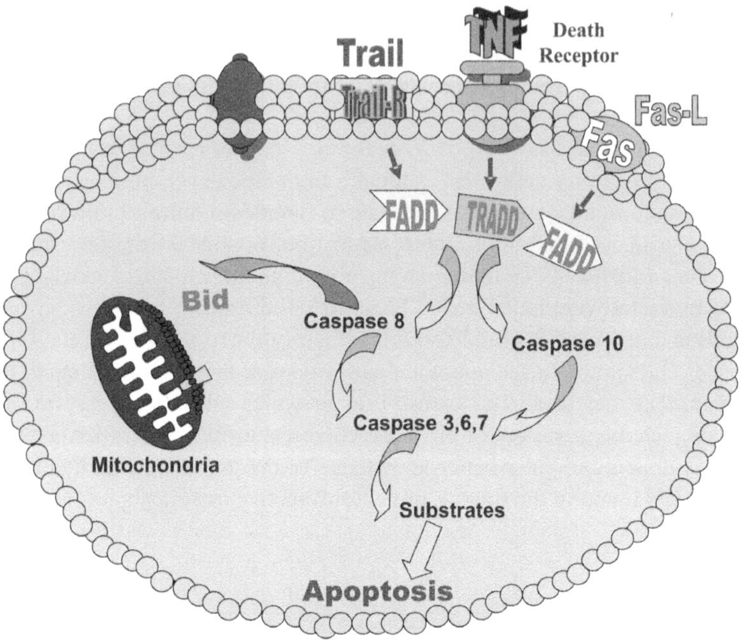

Fig. 5.4 Extrinsic pathways of apoptosis. Brief description of the chain of events activated by death ligands action on their membrane-bound death-receptors and consequent activation of apoptosis (See also Plate 6 in the Color Plate Section on page 227)

Bcl-2, Bax and Bak, are reduced in this study. Finally, an overexpression of Bcl-x(L) fully protects cells against TNF-alpha-induced cell death.

An increase in the expression of TNF-alpha in spinal cord, with increased levels of plasma homocysteine, was found in copper/zinc superoxide dismutase SOD1(G93A) mice model of amyotrophic lateral sclerosis. These mice depict mutations in SOD1 (Zhang et al. 2008c). The increase in TNF-alpha expression was associated to cleavage of caspase 3, increased level of poly(ADP-ribose)polymerase (PARP) and down-regulation of anti-apoptotic protein Bcl-2. These effects were counteracted by treatment with folic acid and vitamin B12. In relation to these studies, caspase 3 activation mediated via TNF-R1 has been observed in cerebellar granule neurons too (Taylor et al. 2005).

Further mediation of apoptosis of neurons through TNF-alpha and involvement of activation of both caspase 8 and caspase 3 has been shown in ischemic stroke, a situation where neurons in the core are rapidly committed to die, whereas neuron death in the slowly developing penumbra is more amenable to therapeutic intervention (Kaushal and Schlichter 2007). The process was mediated by activated microglia, which became neurotoxic leading to neuronal death through apoptotic mechanisms. In this line, TNF-induced apoptosis in primary rat hippocampal neurons and SH-SY5Y human neuroblastoma cells has been shown to occur following focal cerebral ischemia; in this study, TNF effect involves the proteolytic cleavage of caspase 8 and 3 (Yu et al. 2006b). Here, the authors show that gene transfer of A20, a gene that was originally identified as a primary response gene induced by TNF-alpha, resulted in reduction of infarct volume and improvement of neurological deficit, therefore proposing gene transfer as a promising approach to gene therapy for cerebral ischemia. Intervention of caspase 8, with increased activities of caspases 9 and 3 as well as increased levels of Bid, was found following limbic seizures, and accompanied by neuronal death in amygdala (Henshall et al. 2001).

Another pro-apoptotic protein, Bax, is involved in cell death. Using SH-SY5Y and SK-N-SH human neuroblastoma cells, Lee et al. (2008) showed that cell death was preceded by increases in Bax expression and mitochondrial dysfunction, such as collapse of mitochondrial membrane potential, release of cytochrome *c* from mitochondria into the cytoplasm, and increases in caspase 9 and 3 activities.

In addition to "natural occurring accidents", apoptosis can be activated in response to accumulation of chemical compounds, either synthesized in the body or entered by its consumption. Excitotoxicity is a condition caused by accumulation of excitatory aminoacids such as glutamate, as is closely related to overstimulation of Ca^{2+} signalling. Excitotoxicity can be a cause leading to TNF-alpha-mediated apoptosis too (Chaparro-Huerta et al. 2005). In PC12h cells glutamate evoked cytotoxicity through TNF-R1 and involved downstream activation of caspase 8 (Kogo et al. 2006).

A selective loss of dopaminergic neurons, which was mediated by apoptosis, was achieved by intranigral injection of lipopolysaccharide; the treatment evoked an increase in the expression of Bax, Fas and (TNF)-alpha, while expression of the anti-apoptotic gene Bcl-2 was decreased (Arimoto et al. 2006).

The activation of pro-apoptotic proteins Bax, caspase 9 and caspase 3 are supposed to underlie an aberrant outgrowth and decreased survival in locus coeruleus following cocaine exposure (Dey and Snow 2007). And participation of TNF-alpha was demonstrated by an increase in its expression in the presence of cocaine.

TRAIL, the TNF-related apoptosis-inducing ligand, is a member of the TNF family of cytokines that has been suggested to participate in apoptotic cell death, *via* the extrinsic pathway. As a death ligand, TRAIL was originally thought to target only tumor cells. However, whereas TRAIL is not typically present in CNS, apoptosis-inducing TRAIL receptors are found differently distributed on neurons (Bäurle et al. 2004; Aktas et al. 2007). In addition, emerging data show that TRAIL can be induced by immune stimuli on macrophage and microglia in the CNS. Upregulated TRAIL may then cause neuronal apoptosis through direct interaction with TRAIL receptors on neurons (Dörr et al. 2005; Huang et al. 2005). The involvement of TRAIL in cell death has been shown to involve an activation of caspases 8 and 3 (Murata et al. 2006).

TRAIL receptor 2 is highly expressed on human NPCs derived from fetal cortex, and it mediates caspase 3 activation and apoptosis (Peng et al. 2005). In SH-SY5Y human neuroblastoma cells TRAIL is involved in amyloid beta dependent neurotoxicity through activation of caspase 8 (Cantarella et al. 2008). And in both neuronal cell line and primary cortical neurons blockade of the TRAIL death receptor DR5 with a specific antibody completely prevented amyloid beta peptide neurotoxicity (Uberti et al. 2007).

By other side, apoptosis signal-regulating kinase 1 (ASK1) is an important regulator of stress-induced cell death. ASK1 is activated by oxidative stress, TNF and ER stress and activates the JNK-dependent intracellular death pathways. ASK1 is an evolutionarily conserved mitogen-activated protein 3-kinase that activates both JNK and p38 mitogen-activated protein kinases and plays an essential role in TNF-alpha-induced mitogen-activated protein kinase signalling (Matsuzawa et al. 2005; Zhao et al. 2007). The increased expression of several members of the TNF pathway and JNK activation of c-Jun ultimately result in neuronal apoptosis.

In a study, cocaine increased the levels of phosphorylated cJNK vía TNF-alpha signalling. The induction of cJNK may lead to caspase 3 activation and apoptosis in fetal locus coeruleus neurons (Dey and Snow 2007). On the other hand, employing a model of progressive cervical cord compression, Takenouchi et al. (2008) found increased levels of phosphorylated-ASK1, phosphorylated-JNK, and activated-caspase 3, all cell-specific markers that confirmed the presence of apoptosis signals. These findings were accompanied by secondary degeneration around the site of injury and chronic demyelination, with a resulting irreversible neurologic deficit.

Involvement of ASK1 in neuronal cell death and activation of JNK-signalling pathway can be observed as well in ischemia. These events may play an important role in dopaminergic neuronal death and thus, the JNK signalling may eventually emerge as a prime target for novel therapeutic approaches to treatment in ischemia and neurological disorders as Parkinson's disease (Pan et al. 2007). Participation of ASK1 in motor neurons apoptosis has been also demonstrated. Gemin5, a 170-kDa WD-repeat-containing protein, facilitates the activation of ASK1 and downstream signalling for JNK in apoptosis induced by hydrogen peroxide and TNF-alpha (Kim et al. 2007). In an animal model employed to study the pathogenic mechanisms leading to neurodegeneration in Parkinson's disease, coadministration of alpha-lipoic acid, a thiol antioxidant, abolished the activation of ASK1 and phosphorylation of downstream kinase JNK in substantia nigra, supporting their role in the pathogenesis of this disease (Karunakaran et al. 2007).

Activation of key members of the ASK1-mediated signalling pathway in a model of rabbit spinal cord ischemia and reperfusion was studied by Wang et al. (2007a). Increased levels of phosphorylation of ASK1 and JNK were observed after reperfusion of ischemic spinal cords, indicating that activation of ASK1 may play an important role in the apoptotic signalling mechanisms under this pathological situation. Furthermore, apoptosis through sequential phosphorylation of the ASK-1, MAPK kinase 4 and JNK leading to expression of c-Jun was demonstrated following treatment of SH-SY5Y cells with cadmium. The inhibitory action by N-acetyl-L-cysteine, a free radical scavenger, suggests the involvement of ROS. In addition, zDEVD and zVAD reduced apoptosis, indicating participation of caspase 3 (Kim et al. 2005b).

A crucial signal adaptor that mediates all intracellular responses from TNFR-1 is TRADD. TRADD contains a nuclear export and import sequence that allows shuttling between the nucleus and the cytoplasm. In the absence of export, TRADD is found within nuclear structures. In these structures, the TRADD death domain can activate an apoptosis pathway that is mechanistically distinct from its action at the membrane-bound TNF-R1 complex and can involve caspase 9 activation (Morgan et al. 2002; Bender et al. 2005). Therefore, nucleocytoplasmic shuttling of TRADD leads to the activation of distinct apoptosis mechanisms that connect the death receptor apparatus to nuclear events of apoptosis. TRADD induced FADD-mediated activation of caspase 8, which was blocked by caspases inhibitors (Bender et al. 2005). And the expression of TNFR-1 and TRADD regulates the receptor-linked downstream apoptotic cascades (Swarup et al. 2008).

Expression of Fas and FADD components of death receptor signalling were increased in amygdala neurons following seizure (Henshall et al. 2001). This study shows induced activities of caspases 8, 9, and 3 as well as cytochrome *c* release, DNA fragmentation, and neuronal death. Furthermore, up-regulation of genes involved in the TNF-alpha death pathway, caspase 8 and FADD, including that for TNF-alpha, was detected in neurons in a model of Niemann-Pick disease type C, an autosomal recessive neurovisceral storage disease that causes neurodegeneration (Wu et al. 2005).

In Alzheimer's disease, neuronal viability can be long-term affected by TRADD mediation of TNF-R1 signalling in response to oxidative or cytokine-promoted stresses. Increased TRADD expression was detected immunohistochemically in the hippocampus, with increased protein and mRNA expression of TRADD, TNF-R1, and activated JNK (Del Villar and Miller 2004).

Finally, ethacrynic acid can induce programmed cell death, which is initiated by membrane death receptors associated with TRADD and caspase 8 (Ding et al. 2007). The study shows an increase in the expression of executioner caspase 3 and caspase 6 downstream of caspase 8. The expression of caspase 9, which is associated with mitochondrial damage, proposes a crosstalk with apoptotic events associated to this organelle.

5.2.2 Intrinsic Pathway of Apoptosis in Neurons: Role of Mitochondria

Mitochondria play a central role in cell death by controlling cellular energy metabolism, production of ROS, and release of apoptotic factors into the cytosol. Loss of mitochondrial membrane integrity and release of apoptogenic factors are a key step in the signalling cascade leading to neuronal cell death in various neurological disorders. The most prominent pro-apoptotic factor released from mitochondria is cytochrome *c*, but the recently described mitochondrial proteins AIF, endonuclease G, Smac/DIABLO (second mitochondria-derived activator of caspases/direct IAP-binding protein with low pI) and Omi/HtrA2, also play roles in the regulation of apoptosis (van Gurp et al. 2003). The release of pro-apoptotic factors may occur with or without opening of the MPT pore (Fig. 5.5).

The activation of caspase 1, caspase 8, caspase 9, caspase 6 and caspase 3 was increased significantly in Neuro2A cells expressing Hippi, an interactor of

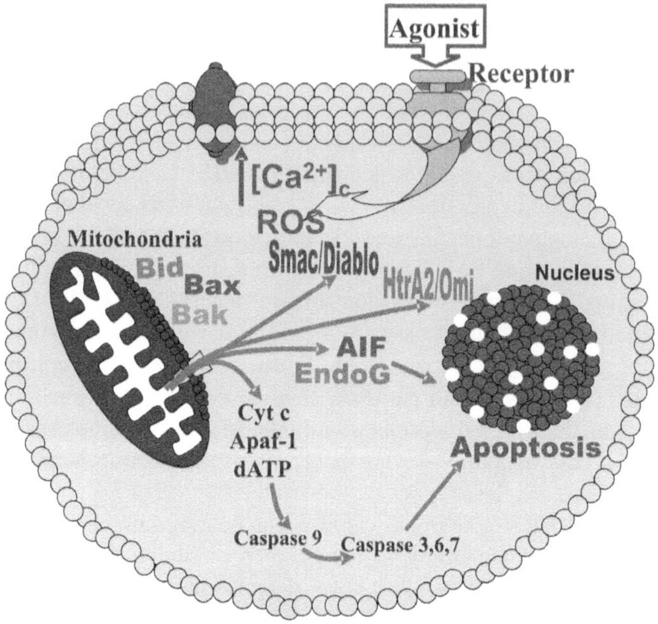

Fig. 5.5 Intrinsic pathways of apoptosis. Brief description of the mitochondrial-linked pathways of apoptosis (See also Plate 7 in the Color Plate Section on page 228)

huntingtin-interacting protein Hip1 (Majumder et al. 2006). In another study, it was observed a very early activation of cytosolic caspase 2 in primary cortical neurons subjected to acute serum-deprivation (Chauvier et al. 2005).

Mediation of apoptosis by mitochondria dysfunction pathway has been shown in human neuroblastoma SH-SY5Y cells too, where cytochrome *c* release and expression of Bcl-2 proteins were increased. In addition, nuclei condensation, DNA fragmentation, poly (ADP-ribose) polymerase cleavage and increased activity of caspase 3 were also observed (Liu et al. 2008). In relation to this study, collapse of mitochondrial membrane potential, with concomitant release of cytochrome *c* from mitochondria into the cytoplasm, and increases in caspase 9 and 3 activities accompanied with cell death were found in SH-SY5Y and SK-N-SH cells (Lee et al. 2008). And in hippocampal neurons, hyperhomocysteinemia induced cytochrome *c* release from mitochondria and caspase 3 and 9 activation. These effects were reduced by melatonin treatment, which then acted as an antipoptotic compound in this tissue (Baydas et al. 2005).

In mouse neuronal cells the proteasome inhibitor, N-benzyloxycarbonyl-Ile-Glu (O-t-butyl)-Ala-leucinal induced a decreased cell viability that involved a dose-dependent increase in caspase 3 and 7 activation and a mitochondrial dysfunction, manifested by the translocation of Bax from the cytoplasm to the mitochondria, membrane depolarization, and the release of cytochrome *c* and the AIF from mitochondria to the cytoplasm and nucleus, respectively (Papa et al. 2007).

Further investigations on cell death by mitochondria-mediated apoptotic pathway reported collapse of mitochondrial membrane potential, release of cytochrome *c* and activation of caspase 3, in response to neuromelanin in nigral dopamine neurons. The participation of neuromelanin in the pathogenesis of Parkinson's disease has been suggested (Nagatsu and Sawada, 2006). Similarly, reduction of mitochondrial membrane potential by cyanide induced apoptotic death in primary rat cortical cells, and was accompanied by the release of cytochrome *c* from mitochondria and elevated caspase 3 and 7 activity. Cyclosporin-A, an inhibitor of MPT, blocked the cyanide-induced apoptotic death (Prabhakaran et al. 2007). In a similar study, the herbicide paraquat decreased mitochondrial transmembrane potential and induced the release of cytochrome *c* from mitochondria in SY5Y cells. Increased activities of caspases 9 and 3 were also observed in this study, and the cells depicted nuclear condensation and DNA fragmentation (Yang and Tiffany-Castiglioni 2008). In this line, release of cytochrome *c*, activation of caspase 3 and upregulation of Bax leading to apoptotic death of striatal neurons was induced by kainic acid, suggesting a possible role in excitotoxicity (Wang et al. 2008). Moreover, cytochrome *c* release and caspase 3 and 9 activation was induced by hydrogen peroxide in brain-derived neural progenitor cells. Cell death process involved activation of JNK and was prevented by selenium, which shows antioxidant properties (Yeo and Kang 2007).

Interaction of pro-apoptotic member of the Bcl-2 family of proteins with mitochondria will also launch the mitochondrial apoptotic pathway. Phosphorilation of Bcl-2 at serine 24 was induced by hypoxia-ischemia, coincided with cytochrome *c* release from mitochondria and preceded caspase 3 activation (Hallin et al. 2006). A Bax-dependent loss of mitochondrial cytochrome *c* and caspase activation was observed in developing neurons deprived of trophic support, and underwent apoptosis (Barone et al. 2008). In another study, the release of cytochrome *c* and the activation of caspase 3 were prevented by curcuma oil, with antioxidant activity. The study supports the hypothesis of ROS inducing apoptosis via mitochondrial apoptosis pathway. The treatment suppressed the elevated protein level of Bax, and aided mitochondrial translocation and activation of Bcl-2, showing protection activity in cerebral stroke (Dohare et al. 2008). On the other hand, upregulation of Bax, the release of cytochrome *c* from mitochondria and an increase in caspase 9 and caspase 3 activities in retinal ganglion cells were related to the initiation apoptosis. The activation of caspase 8 and cleavage of Bid indicated a co-operation between the extrinsic and the intrinsic pathways of apoptosis (Das et al. 2006).

It has been also observed that Bax induced an increase of mitochondrial-derived ROS and the release of cytochrome *c* from mitochondria, with the subsequent activation of cytosolic caspases in neurons (Kirkland and Franklin 2007). An increase in Bax levels, reflected as a decrease in Bcl-2/Bax ratio, culminated in caspase 3 activation in neuronal-like PC12 cells treated with heroin hydrochloride (Cunha-Oliveira et al. 2007) and in human neuroblastoma SH-SY5Y cells treated with the neurotoxin 1-methyl-4-phenylpyridinium (Jung et al. 2007). Another study, reported increase in Bax/Bcl-2 ratio, mitochondrial release of cytochrome *c*, increase in cytosolic Smac/DIABLO, and activation of caspase9 and caspase 3 in SH-SY5Y cells treated with the garlic compounds diallyl sulfide and diallyl disulfide (Karmakar et al. 2007).

This pathway can be activated by stimuli related to the development of neurodegenerative diseases too. Cortical neurons exposed to amyloid protein developed an apototic process characterized by an increase in the amount of Bax protein, cytochrome *c* release from mitochondria and caspase 3 activation (Yu et al. 2006a). In the same study, the potassium channel blocker tetraethylammonium attenuated neuronal death and prevented the observed alterations, suggesting a role for potassium channel dysfunction in Alzheimer's disease.

Upregulation of another proapoptotic protein, Bim, accompanied by mitochondrial release of cytochrome *c* was detected in cerebrocortical neuron cultures treated with beta-amyloid peptide. Bim translocates to mitochondria and induces the release of cytochrome *c*. The apoptotic events were prevented by estrogens treatment, providing new understanding into the mechanisms contributing to estrogens neuroprotection (Yao et al. 2007).

Coupling of stress signals to the mitochondrial cell death pathways via the proapoptotic protein Bid was observed in hippocampal neurons in the presence of the glutamate receptor agonist N-methyl-D-aspartate (NMDA) (König et al. 2007).

Other pro-death molecules such as Omi/HtrA2 and AIF can be also released from the mitochondrion in addition to cytochrome *c* (Donovan et al. 2006). The mitochondrial flavoprotein AIF, a mitochondrial oxidoreductase, is the main mediator of caspase-independent apoptosis-like programmed cell death. Upon pathological permeabilization of the outer mitochondrial membrane, AIF is translocated to the nucleus, where it participates in chromatin condensation and is associated to large-scale DNA fragmentation (Krantic et al. 2007). Emerging evidence has suggested that the intramitochondrial protein AIF translocates to the nucleus and promotes caspase-independent cell death induced by glutamate toxicity, oxidative stress, hypoxia, or ischemia (Cao et al. 2007). However, the mechanism by which AIF is released from mitochondria after neuronal injury is not fully understood. It could be released into the cytoplasm to induce cell death in response to poly(ADP-ribose) (PAR) polymerase-1 (PARP-1) activation (Culmsee et al. 2005; Yu et al. 2006c). In addition, translocation of AIF could depend on changes in intracellular Ca^{2+} homeostasis and on calpain activity (Sanges et al. 2006).

Glutamate induces apoptotic cell death associated with activation of caspase 3 as well as upregulation and/or translocation of AIF from mitochondria to cytosol and nuclei in primary cortical cells (Zhang and Bhavnani 2006). A release of AIF from the mitochondria, associated to cleavage of Bid and the release of cytochrome *c*, was also increased in Neuro2A cells (Majumder et al. 2006). In addition, nuclear translocation of AIF and activation of executioner caspase 3 has been observed in neurons in the hippocampus, cerebellum, and cortex following infection of neonatal rats with Borna disease (Williams et al. 2008). On the other hand, hypoxia exhibited enhanced activation of the proapoptotic caspase 3 and increased mitochondrial release of AIF in cerebellar granule neurons (Russell et al. 2007). The pretreatment of cells with the caspase inhibitor ZVAD-FMK had little or no effect on AIF release and its subcellular translocation to the nucleus, suggesting caspase-independent AIF release. A similar result was observed in retinal neuronal cells subjected to diabetic retinopathy (Santiago et al. 2007).

Further studies show that thimerosal, an organic mercury compound that is commonly used as an antimicrobial preservative, induced death in neuroblastoma cells through apoptosis, and was associated with depolarization of mitochondrial membrane, generation of reactive oxygen species (ROS), and the release of cytochrome *c* and AIF from mitochondria to cytosol. In this study, caspase 9 and caspase 3 were also activated (Yel et al. 2005). Similarly, the fungal alkaloid militarinone A evoked apoptosis in murine neuroblastoma cell line N2a by nuclear translocation of AIF, activation of caspases and c-Jun/AP-1 transcription factor (Küenzi et al. 2008).

The release of second mitochondrial activator of caspases (Smac), Omi/HtrA2, and AIF which were delayed with respect to induction of Bax, loss of mitochondrial transmembrane potential and early release of cytochrome *c*, has been observed in primary neuronal cultures of murine cerebral cortex subjected to activation of AMPA subtype of receptor for l-glutamate, and could be a cause of excitotoxicity (Beart et al. 2007). Furthermore, excytotoxicity evoked by glutamate exposure of cultured hippocampal neurons led to swelling of mitochondria that was accompanied by the release of cytochrome *c*, Smac/DIABLO, Omi/HtrA2, and AIF (Shalbuyeva et al. 2006).

Smac/DIABLO was released from mitochondria in neural cells obtained from Niemann-Pick disease type C1 (NPC1) mouse brains. These cells sowed an elevated level of cholesterol in mitochondrial membranes, a situation that caused mitochondrial dysfunction and neuronal death (Huang et al. 2006). Furthermore, hypoxia-ischemia triggered the mitochondrial release of cytochrome *c* and Smac/DIABLO in neonatal rat brain (Russell et al. 2008) and in rat cerebellar granule neurons (Maycotte et al. 2008). And in neuronal cell death induced by seizures, cytosolic fraction of Smac/DIABLO was increased by 2 h (Li et al. 2006).

Further findings show an apoptosis associated mitochondrial release of cytochrome *c* and increased levels in cytosolic Smac/DIABLO in SH-SY5Y cells treated with the garlic compounds diallyl sulfide and diallyl disulfide (Karmakar et al. 2007). On the other hand, a release of Omi/HtrA2, accompanied by the release of cytochrome *c*, Smac/DIABLO and AIF, has been observed following exposure of cultured hippocampal neurons to excitotoxic glutamate (Shalbuyeva et al. 2006).

Other mitochondrial proteins and their relation to apoptosis have been also studied. Translocation of endonuclease G from mitochondria to the nucleus occurs in apoptotic processes. However, its implication has not been as widely studied as other mitochondrial proteins. Mitochondrion-specific nucleases AIF and endonuclease G translocated into the nuclei of apoptotic cells following noise exposure of cochlear hair cells (Han et al. 2006). And in Leber's hereditary optic neuropathy endonuclease G was released into the cytosol together with cytochrome *c* and AIF, and cells underwent caspase-independent apoptosis (Zanna et al. 2005). All these studies point out that changing the mitochondrial microenvironment can severely lead to the activation of apoptosis.

5.2.3 Endoplasmic Reticulum Stress and Apoptosis in Neurons

ER stress is caused by an accumulation of unfolded proteins in the ER lumen and alterations in Ca^{2+} homeostasis. The ER response is characterized by changes in specific proteins, causing translational attenuation, induction of ER chaperones and degradation of misfolded proteins. In case of prolonged or aggravated ER stress, cellular signals leading to cell death are activated (Lindholm et al. 2006). Apoptosis triggered by ER stress is associated with various pathophysiological conditions such as stroke and with neurodegenerative diseases such as Parkinson's disease and Alzheimer's disease. However, the mechanism by which ER stress induces neuronal apoptosis remains controversial (Fig. 5.6).

Sanges et al. (2006) reported the existence of an apoptotic pathway from the ER that is activated during the degenerative process in an animal model of retinitis pigmentosa, the rd1 mouse, and involved translocation of caspase 12. In a study performed on sympathetic neurons, the ER stress-inducing agent tunicamycin induced a neuronal apoptotic pathway that requires Bax activation and c-jun N-terminal kinase signalling (Smith and Deshmukh 2007). ER stress could be involved also in

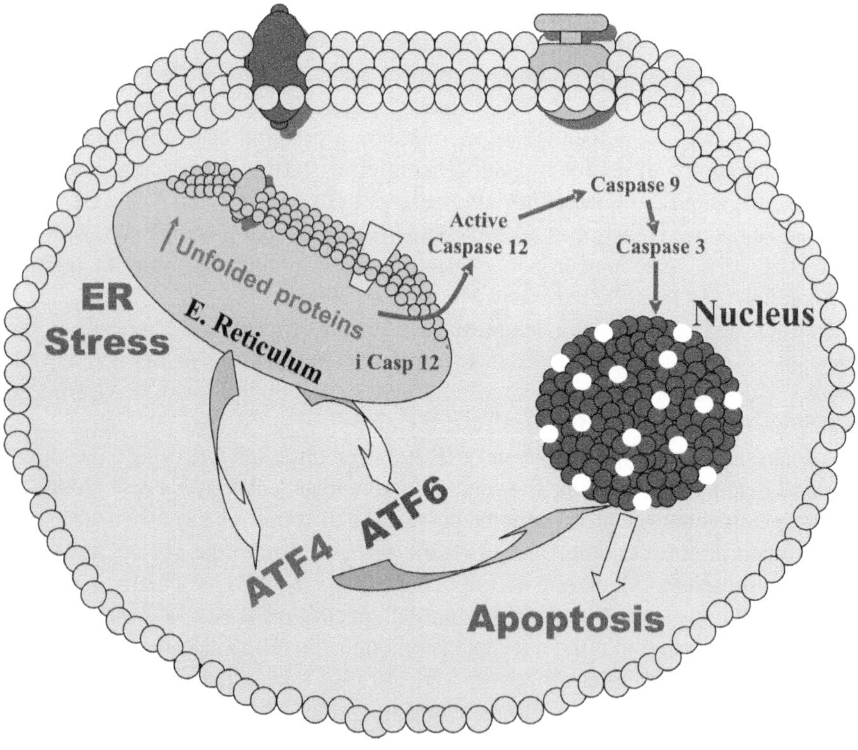

Fig. 5.6 Endoplasmic reticulum pathways of apoptosis. Brief description of apoptosis pathways coupled to endoplasmic reticulum stress (See also Plate 8 in the Color Plate Section on page 229)

apoptotic death induced by serum deprivation in neuronal cells. This study shows cleavage of procaspase 12 and apoptosis in the cell line HN9.10e, constituted by immortalized hippocampal neuroblasts (Voccoli et al. 2007). Caspase 12, which resides in the ER, is activated when the cells are exposed to excess levels of various stimuli that cause ER stress (Lee et al. 2005). An activation of caspase 12 via caspase 3, and regulation by protein kinase C, has been shown in the neuronal cell line Cas-12 (Lee et al. 2005). Cell death in neuronal PC6.3 cells is also related to ER stress, and involves cleavage of caspase 12 (Reijonen et al. 2008).

Caspase 12 activation was observed in the hippocampus in a time-dependent manner after recurrent febrile seizures (Chen et al. 2008), after cerebral ischemia-reperfusion (Zhang et al. 2008b) and in the cerebellum and the thalamus of thiamine (vitamin B1) deficient mice (Wang et al. 2007b). Furthermore, caspase 12 activation due to ER stress has been also shown in spinocerebellar ataxia type 14 (Lin and Takemoto 2007).

By another side, ER-mediated activation of apoptosis could involve the activation of certain transcription factors. An activation of pro-death pathways originating from the ER involving the induction of activating transcription factor (ATF4) was found in hypoxia-induced responses after stroke in cortical neurons (Halterman et al. 2008). Activation of ATF4 was also observed following transient forebrain ischemia, consistent with ER stress playing a pivotal role in post-ischemic neuronal death in the gerbil hippocampal CA1 subfield (Oida et al. 2008). And activation of ATF4, accompanied by cell death, was observed in cultured retinal ganglion cells after treatment with tunicamycin (an ER stress inducer) (Shimazawa et al. 2007). Furthermore, ATF4 translation and cleavage of the ATF6 protein was observed in borna disease virus-infected neonatal rats, implicating an imbalance between ER stress-mediated apoptosis and survival signalling as a critical determinant of neural cell fate in the disease (Williams and Lipkin 2006).

On the other hand, induction of activating transcription factor 6 (ATF6) has been observed in neuroblastoma cells under simulation of cerebral ischemia (Kudo et al. 2008), and in West Nile virus-mediated neuronal death, a hallmark of meningitis and encephalitis induced by the virus (Medigeshi et al. 2007). In the latter, premature cell death could represent a host defence mechanism to limit viral replication that might also be responsible for the widespread neuronal loss observed in the virus-infected neuronal tissue.

Finally, ER stress can engage the mitochondrial pathway of death, as it has been proposed that neurons release cytochrome *c* and that Bax plays a critical function in committing neurons to ER stress-induced apoptosis (Smith and Deshmukh 2007).

5.3 Apoptosis in Astrocytes

Astrocytes play many crucial roles in the CNS. In general, astrocytes support normal neuronal function by tightly regulating the extracellular environment with respect to ions and neurotransmitters and by providing energy substrates. Because astrocytes play a central role in maintaining neuronal viability both under normal conditions and during stress, dysfunction or loss of astrocytes can lead to neuronal death or dysfunction. Therefore, most studies have been focused onto the role that activated astrocytes and microglia can represent in neuronal apoptosis, due to their ability to release apoptosis activators towards surrounding neurons. Apoptosis-mediated neurotoxicity is mainly due to the release of neurotransmitters such as glutamate, ROS overproduction and the release of cytokines and other neuronal apoptosis mediators. However, like neurons do, astrocytes, and mainly those transformed into astrocyte malignant (glioma) cells, can undergo apoptosis too.

5.3.1 Extrinsic Pathway of Apoptosis in Astrocytes

There are very few studies focused onto the extrinsic pathway of apoptosis in normal astrocytes. Conversely, most of the studies have been directed towards the apoptotic events in malignant cells, because normal astrocytes do not show apoptosis through activation of TNF-alpha cascades. An elevated expression of TNF-alpha was detected in astrocytes distributed in the affected brain regions in the autosomal recessive neurovisceral storage disease Niemann-Pick disease type C, which involves neurodegeneration (Wu et al. 2005). In another study, trimethyltin induced a strong increase in TNF-R1 expression in astrocytes, suggesting that astrocytes may enhance trimethyltin -induced injury in the CNS via TNF-R1 (Figiel and Dzwonek 2007).

Rat astrocytes express all TRAIL receptor mRNAs and proteins. Thus, apoptosis-inducing TRAIL receptors are found differently distributed in oligodendrocytes and astrocytes, in addition to neurons. If, under pathologic circumstances, the CNS is inflamed, immune cells such as macrophages and T cells upregulate TRAIL upon activation and use this death ligand as a weapon, not only against tumor cells but also against neurons and oligodendrocytes within the inflamed CNS (Aktas et al. 2007). As for neurons, TRAIL aroused major interest due to its preferential toxic effect against malignant cells. However, TRAIL fails in inducing apoptosis of astrocytes (Cantarella et al. 2007). Nevertheless, there are few studies revealing that the TRAIL system can also induce death of non-transformed cells.

Programmed cell death, demonstrated by apoptosis-specific caspase 3 protease activity, was observed in murine astrocytes. The study implicated TRAIL and TNF-R in apoptosis signalling (Rubio et al. 2003). TRAIL induced apoptosis in glioma cells too, with the involvement of caspase 8 activation (Okhrimenko et al. 2005). On the other hand, modulation of TRAIL expression by oxidative stress was shown in LN215 cells, an astroglioma cell line (Kwon and Choi 2006). Furthermore, hydrogen peroxide increased caspase 8 expression and cell death in a time- and dose-dependent manner, and was blocked upon treatment with a TRAIL-specific antagonistic protein. These findings suggest that oxidative stress sensitizes human astroglial cells to TRAIL-induced cell death (Kwon et al. 2008).

The previous observations are in agreement with those obtained by Son et al. (2007), who observed a rapid apoptosis in TRAIL-resistant glioma cells, but not in human astrocytes, with an increase in the processing of procaspase 3 in response to silibinin, a flavonoid isolated from *Silybum marianum*. In relation to these observations, the proteolytic processing of procaspase 3 by TRAIL was enhanced by rottlerin and induced rapid apoptosis (Kim et al. 2005a). Another study shows that TRAIL induced cell death in human glioma cells, resulting in cleavage and activation of the apoptotic initiator caspase 8 and the effector caspase 3, as well as cleavage of Bid and the downstream initiator caspase 9 (Nagane et al. 2007). Crosstalk with mitochondrial pathway occurred, as TRAIL treatment of cells augmented the release of cytochrome *c* from the mitochondria into the cytosol, as well as AIF.

On the other hand, the selective increase of phosphorylated ASK1 and concomitant upregulation of the TNF-R that occurs in motor neurons of transgenic mice overexpressing amyotrophic lateral sclerosis-linked SOD1 mutants has not been observed in reactive astrocytes (Veglianese et al. 2006). Infection of adult hippocampus-derived rat progenitors with an adenovirus encoding the constitutively active form of ASK1 induced a marked depletion of glial cells, revealing that ASK1 acts as a potent inhibitor of glial-specific gene transcription (Faigle et al. 2004). Additionally, activation of ASK1 in cultured primary astrocytes by 15-deoxy-delta(12-14)-prostaglandin J(2) and its involvement in the activation of JNK was observed (Lennon et al. 2002). Despite these initial investigations, to date little attention has been paid to the activation of TNF-R and its interaction with JNK through ASK1 in astrocytes.

Finally, human astrocytes express Fas, which is known to recruit FADD and caspase 8 to induce apoptosis. However, astrocytes are resistant to Fas-induced apoptosis (Song et al. 2006).

5.3.2 Intrinsic Pathway of Apoptosis in Astrocytes. Role of Mitochondria

As in neurons, mitochondria are central to both apoptotic and necrotic cell death, as well as to normal physiological function in astrocytes. Caspases are essential components of the mammalian cell death machinery. Caspase 9 is positioned at the apex of the pro-apoptotic signalling cascade induced by cytochrome *c* release from mitochondria. Sequential release of cytochrome *c* from mitochondria, and activation of caspase 9 and 3, was observed in astrocytes from neonatal rat brain after unilateral focal ischemia. Apoptotic cell death was finally accomplished by PARP-1 cleavage, which functions as endogenous caspase 3 substrate (Benjelloun et al. 2003).

The mitochondria appear to be a main target in cobalt toxicity in CNS. Cobalt is suspected to cause memory deficit in humans and was reported to induce neurotoxicity in animal models. In this study, cobalt caused apoptosis related to the loss of mitochondrial membrane potential and release from the mitochondria of apoptogenic factors, e.g. AIF, in primary cultures of mouse astrocytes. Release of AIF involved opening of the MPT (Karovic et al. 2007). The release of AIF from mitochondria and its nuclear translocation was studied in hippocampi from neonatal rats infected with Borna disease virus. AIF expression was increased in astrocytes as did executioner caspase 3 activation (Williams et al. 2008). Similarly, expression of AIF and its translocation to the nucleus was detected in motor neurons and astrocytes during amyotrophic lateral sclerosis pathogenesis (Oh et al. 2006). In another study, mitochondrial membrane potential loss and subsequent release of pro-apoptotic factors cytochrome *c* and AIF were observed in MPP(+)-induced astrocytic apoptosis (Zhang et al. 2007). Similarly, death of mouse astrocytes resulted upon activation of PARP-1, and was preceded by NAD(+) depletion, mitochondrial membrane depolarization and MPT opening. This was accompanied by translocation of AIF from mitochondria to nucleus (Alano et al. 2004).

On other side, human cortical neurons from Down's syndrome patients showed increased Bax, cytoplasmic translocation of cytochrome *c* and AIF, and active caspases 3 and 7, consistent with activation of the apoptotic mitochondrial death pathway (Helguera et al. 2005). Furthermore, Bax expression and caspase 3 activation were triggered by hydrogen peroxide in rat astrocytes. The cells finally exhibited cell shrinkage, nuclear condensation and marked DNA fragmentation, and the process was inhibited by melatonin (Juknat et al. 2005). Similarly, increased endogenous Bad was translocated from the cytoplasm to mitochondria to induce apoptosis in astrocytes in primary cultures (Chen et al. 2005).

OSU-03012, a phosphoinositide-dependent kinase-1 inhibitor, caused a dose-dependent induction of death to a greater extent in primary human glioma cells than in non transformed astrocytes. This inhibitor promoted the release of cathepsin B from the lysosomal compartment and a Bid-dependent release of AIF from mitochondria (Yacoub et al. 2006). In this line, a specific overexpression of the proapoptotic protein Bax led to apoptosis and cell death in glioma cell lines through activation of both caspase 8 and 9, and activated downstream caspase 3 (Kaliberov et al. 2004).

In opposition to neuronal studies, investigations regarding the involvement of endonuclease G, Smac/DIABLO and Omi/HtrA2 in the regulation of apoptosis in astrocytes are currently lacking. Conversely, there are numerous findings about the other mediators of the intrinsic pathway for apoptosis mentioned above.

Interesting results have arisen from studies focused onto the anti-cancer actions of certain compounds. Alkylphosphocholines are candidate anticancer agents. These substances induced the formation of large vacuoles and typical features of apoptosis in human glioma cell lines, and promote caspase 7 activation, PARP-1 processing and cytochrome *c* release from mitochondria (Naumann et al. 2004). Capsaicin (8-methyl-N-vanillyl-6-nonenamide), the major pungent ingredient of red pepper, has been reported to possess anti-carcinogenic and anti-mutagenic activities. Treatment of human glioblastoma A172 cells with capsaicin inhibited cell growth and induced apoptosis through down-regulation of Bcl-2 and activation of caspase-3 (Gil and Kang 2008). On the other hand, saponins have exhibited broad anti-cancer and pro-apoptotic activity.

Ardipusilloside III induces apoptosis in U251MG cells, a human glioblastoma cell line. Exposure of cells to this compound resulted in time-dependent Bad dephosphorylation and cleavage as well as activation of caspase 8 and caspase 3 (Lin et al. 2008). Therefore, both the intrinsic pathway of apoptosis, mediated by Bad dephosphorylation and cleavage, and the extrinsic pathway of apoptosis, mediated by caspase 8 and caspase 3 activation, were involved in ardipusilloside III-induced apoptosis. The microtubule-binding drug taxol induced apoptosis in human glioblastoma T98G and U87MG cells through mitochondrial release of cytochrome *c* and Smac into the cytosol and activation of calpain, caspase 9, and caspase 3 (Das et al. 2008).

By another side, triptolide, a derivate from the traditional Chinese herb, *Tripterygium wilfordii*, sensitizes cancer cells to apoptosis. Triptolide showed dose-dependent inhibition of cell proliferation and induction of apoptosis in the glioma cell lines, U251MG and U87MG. It also increased the Bax/Bcl-2 proteins ratio (Lin et al. 2007). An up-regulation of Bax and down-regulation of Bcl-2 proteins lead to apoptosis in astrocytes, and is prevented by p73, a protein that is implicated in apoptosis and cell cycle control (Saunders et al. 2005).

In another study addition of bacterial lipopolysaccharides to rat astrocytes in primary culture promoted loss of ATP and mitochondrial membrane potential followed by the mitochondrial release of cytochrome *c*, activation of caspase 3 and morphological evidence of apoptosis (Palomba et al. 2007). Similarly, release of mitochondrial cytochrome *c* and fragmentation of caspase 3 was observed in astrocytes incubated in Kreb's buffer deprived of oxygen and glucose (resembling ischemia). Supply of exogenous pyruvate restored the morphological integrity of post-ischemic astrocytes and prevented gliosis (Sharma et al. 2003).

Cytochrome *c* release or apoptosome-driven caspase 9 activation has been reported in glioblastoma, an astrocytic brain tumour. And involvement of caspase 7 was also shown, which was inhibited by Bcl-2 expression (Stegh et al. 2007).

Finally, Bax translocation to the mitochondria and cytochrome *c* release was induced by nitric oxide in mouse astrocytes. Several features of apoptosis were additionally observed, including chromatin condensation and phosphatidylserine exposure on the outer leaflet of the plasma membrane (Yung et al. 2004).

5.3.3 Endoplasmic Reticulum Stress and Apoptosis in Astrocytes

The clearest results showing involvement of ER stress in astrocyte apoptosis come from the studies performed by Liu et al. (2004), employing Moloney murine leukemia virus that induce progressive paralysis and immune deficiency. The virus triggers apoptosis in immortalized astrocytes, C1 cells, and primary cultured astrocytes. This apoptosis is caused by ER stress resulting from accumulation of the viral envelope preprotein gPr80(env). Cleavage of procaspase 12 was also detected in primary cultured astrocytes infected with the virus. ER stress was followed by mitochondrial stress, detected as mitochondrial transmembrane potential dissipation, cleavage of procaspase 9, and induction of activated caspase 3. In another study, infection with the murine retrovirus causes apoptosis in astrocytes via ER and caspase 8 activation. ER stress was reduced by treatment with Z-IETD-FMK, a specific inhibitor of caspase 8 enzymatic activity (Liu et al. 2006). Caspase-12 induction in neurons in both the cortex and hippocampus as well as in astrocytes in response to traumatic brain injury has been proposed (Larner et al. 2004).

On the other hand ER stress transducers IRE1, PERK and ATF6 are well known to transduce signals from the ER to the cytoplasm and nucleus when unfolded proteins are accumulated in the ER. An up-regulation of ATF-6beta was observed in HIV-infected cortex, suggesting that ER stress response is activated in this process (Lindl et al. 2007).

5.4 Astrocytes and Neuroprotection

Astrocytes face the synapses, send end-foot processes that enwrap the brain capillaries, and form an extensive network interconnected by gap junctions. Astrocytes express several membrane proteins and enzymes that are critical for uptake of glutamate at the synapses, ammonia detoxification, buffering of extracellular K^+, and volume regulation. They also participate in detection, propagation, and modulation of excitatory synaptic signals, provide metabolic support to the active neurons, and contribute to functional hyperemia in the active brain tissue. Due to their close apposition to neurons, coupling in glial networks may be functionally important in determining neuronal vulnerability (Benarroch 2005).

Cell survival is a critical issue in the onset and progression of neurodegenerative diseases and following pathological events including ischemia and traumatic brain injury. Oxidative stress is the main cause of cell damage in such pathological conditions. Astrocytes play an important role in the homeostasis of the CNS both in normal conditions and after ischemic injury, as they are crucial for neuronal metabolic, antioxidant, and trophic support, as well as normal synaptic function (Ouyang and Giffard 2004). In the setting of stress, such as during cerebral ischemia, altered glial function can further lead to oxidative stress and excitotoxicity, and may contribute to the initiation or progression of neuronal death and may therefore compromise the ability of neurons to survive in neurodegenerative diseases.

When the nervous system is subjected to stressful stimuli, reactive gliosis often occurs. Associated with neuronal injuries caused by many CNS insults is an activation of glial cells (particularly astrocytes and microglia), termed gliosis, at the sites of injury. The activation of mechanisms that result in activated glia causes secondary neuronal damage. For example, the swelling of astrocytes is observed during and several seconds after brain ischemia. Then ischemia stimulates sequential morphological and biochemical changes in glia and induces its proliferation.

Reactive astrocytes demonstrate hypertrophy of astrocyte processes than then present a stellate morphology, increased glial fibrillary acidic protein (GFAP) immunoreactivity, increased number of mitochondria as well as elevated enzymatic and non-enzymatic antioxidant activities.

On the other hand, the ability of astrocytes to protect neurons against the toxic action of free radicals depends on their specific energy metabolism, high glutathione level, increased antioxidant enzyme activity (catalase, superoxide dismutase, glutathione peroxidase) and overexpression of antiapoptotic Bcl-2 gene. Astrocytes produce cytokines (TNF-alpha, IL-1, IL-6) involved in the initiation and maintaining of immunological response in the CNS. An in addition, in astrocytes, like in neurons, ischemia induces the expression of immediate early genes: *c-fos, c-jun, fos B, jun B, jun D, Krox-24, NGFI-B* and others. The protein products of these genes modulate the expression of different proteins, both destructive ones and those involved in the neuroprotective processes (Gabryel and Trzeciak 2001).

Opposing functions of activated glia, namely neuroprotection or neurotrophy versus neurodestruction or neurotoxicity, have been observed in a number of experimental models of neurotrauma and neurodegenerative diseases. Therefore, the regulation of astrocyte apoptosis is essential to physiological and pathological processes in the CNS. Nevertheless, the mechanism(s) involved in the determination of which function is executed by activated glia under a given set of conditions still remains to be elucidated.

Estrogen involvement in neuroprotection is widely accepted, and astrocytes act as contributing cells that result in estrogen-mediated neuroprotection (Bains et al. 2007). In this estudy it was shown that glial cells express significant levels of estrogen receptors, and it is postulated an indirect mechanism of estrogen-mediated neuroprotection through glial cell interaction, because removal of glial cells from the cultures significantly reduced the neuroprotective effects of estrogens. It was also shown that estradiol stimulates neurite growth and required the release of apolipoprotein E by glial cells in the cultures (Struble et al. 2007).

When high levels of glutamate accumulate, astrocytes participate in effectively clearance of the neurotransmitter from the extracellular space. An interesting association between astrocytes and T cells has been reported in which T cell-derived glutamate elicits the release of neuroprotective thiols (cysteine, glutathione, and cysteinyl-glycine) and lactate from astrocytes, and reduces neuronal apoptosis induced by oxidative stress in primary neuronal cultures (Garg et al. 2008). The mechanism could rely on T cells secreted cytokines that restore glutamate clearance capacity of astrocytes under oxidative conditions.

Another study shows that astrocytes protect neurons from glutamate toxicity via EAAT-1 and EAAT-2 glutamate transporters (Du et al. 2007). The importance of glutamate uptake by astrocytes in neuroprotection is further supported by the recent observations of Weller et al. (2008) and Wu et al. (2008), in which overexpression of excitatory amino acid transporters in astrocytes enhances neuroprotection.

Activation of a purinergic receptor signalling pathway in astrocytes significantly increases the resistance of astrocytes and neurons to oxidative stress. In a study, neuroprotection is mediated by a Ca^{2+}-dependent increase in mitochondrial metabolism (Wu et al. 2007a). Another study proposes an implication of astrocytes as the primary mediators of interleukin-1beta induced neuronal death, which occurs via generation of ROS by astrocytes (Sharma et al. 2007). Here, flavonoids significantly decreased the release of ROS from astrocytes. This decrease was accompanied by an increase in expression of superoxide dismutase and thioredoxin mediators associated with protection against oxidative stress.

Japanese encephalitis virus (JEV) infection induces increased astrogliosis and confers neuroprotection (Mishra et al. 2007); this elevated physiological state of astrocyte was characterized by the expression of GFAP, the glutamate aspartate transporter, glutamate transporter-1 and ceruloplasmin.

Interesting is the role of astrocytes in neuroprotection by the release of neurotrophins and their participation in the inhibition of apoptosis in neurons. Their expression in the CNS after various insults is thus of major therapeutic consequence (Aharoni et al. 2005; Sharma 2005). An increase in transcript levels of growth factors, such as nerve growth factor and ciliary neurotrophin factor, and of brain-derived neurotrophic factor was observed in JEV- infected astrocytes (Mishra et al. 2007). Released brain-derived neurotrophic factor by astrocytes was also neuroprotective against excitotoxic insult evoked by NMDA on transcortical neurons (Bemelmans et al. 2006). The protective function of neurotrophic factors against death signalling has been further studied by Murata et al. (2006). The glial cell line-derived neurotrophic factor inhibits TRAIL that, as previously described, mediates activation of caspase 8 and apoptosis. Swarup et al. (2008) suggest that the engagement of TNFR-1 and TRADD following JEV infection plays a crucial role in glial activation also and influences the outcome of viral pathogenesis.

Moreover, it has been suggested that CNS glial cells could interfere with Schwann cells (SC) migration, survival, maturation, and remyelination in the CNS. However, in one study, rat mixed CNS glial induced proliferation of SC (Lisak et al. 2006). The authors probe how soluble products of CNS glial cells have the potential to increase proliferation of SC and their resistance to cytokine-mediated death because of the inhibition TNF-alpha-induced SC death.

Interesting is as well the ability of astrocytes to release biochemical compounds that activate the survival biochemical routes in neurons. Hypothalamic astrocytes secrete TGF-beta, which attenuates neuronal death by a mechanism of protection that appears to involve phosphorylation of MKK4, JNK, c-Jun(Ser63), and enhancement of AP-1 binding (Mahesh et al. 2006).

On the other hand, astrocytes can release an array of substances, other than neurotrophic factors, with implications in neuroprotection. Glutathione (GSH) is one of the major antioxidants in the brain. GSH is secreted by astrocytes and this extracellular GSH is used by neurons to maintain and increase their intracellular GSH levels (Pope et al. 2008). Neuroprotection by enhancing glutathione export from as astrocytes is supported in the study by Shih et al. (2006), where cysteine-dependent glutathione production modulates both neuroprotection from oxidative stress and cell proliferation. Furthermore, increased expression and release of glutathione peroxidase by astrocytes in response to thrombin shows a protective effect towards surrounding neurons (Ishida et al. 2006).

Guanine derivates have been implicated in many relevant extracellular roles, such as modulation of glutamate transmission, thus protecting neurons against excitotoxic damage. Guanine derivatives are spontaneously released to the extracellular space from cultured astrocytes and may act as trophic factors, glutamate receptors blockers or glutamate transport modulators, therefore promoting neuroprotection. The positive effects of astrocyte-released guanosine were mediated via K^+ channels activation, and depended on extracellular Ca^{2+} levels and the modulation of the PKA, PKC, MEK and/or PI-3K pathways (Oleskovicz et al. 2008). Furthermore, activation of PI-3K/Akt signal pathways by astrocyte-conditioned medium that was added to neurons in culture promoted neuronal survival subjected to prolonged or excessive exposure to corticosterone, which leads to neuronal damage in the brain (Zhu et al. 2006).

S100B is a Ca^{2+}-binding protein expressed and secreted by astrocytes, which has been implicated in glial-neuronal communication, and appears to protect hippocampal neurons against toxic concentrations of glutamate (Tramontina et al. 2006). In addition, it has been shown that astrocytic S100B can reduce apoptosis in ethanol-treated rat fetal rhombencephalic neurons. It appears that these neuroprotective effects are linked to activation of the PI-3K pathways, and increases in the formation of pAkt and the up-regulation of two downstream NF-kappaB-dependent pro-survival genes: XIAP and Bcl-2 (Druse et al. 2007).

On the other hand, l-serine plays an essential role in neuronal survival. Glutamate, hydrogen peroxide, IL-1-beta, and sodium nitroprusside all induce the secretion of l-serine in astrocytes, which may be needed for neuronal survival during brain insults such as ischemic stroke (Yamagata et al. 2006).

Finally, recent studies indicate that transplanted neural stem/progenitor cells (NSPs) can interact with the environment of the CNS and stimulate protection and regeneration of host cells exposed to oxidative stress. The presence of NSPs significantly improved cell viability by interfering with production of free radicals and increasing the expression of neuroprotective factors. This process was accompanied by elevated expression of ciliary neurotrophic factor and vascular endothelial growth factor in a network of NSPs and local astrocytes. In addition, enhanced growth factor secretion stimulated a robust upregulation of the antioxidant enzyme superoxide dismutase 2 in neurons and resulted in their improved survival (Madhavan et al. 2008). On the other hand, human amniotic cells can express the marker of neural progenitor cells and differentiate into neuron, dopaminergic neuron, astrocyte and oligodendrocyte upon transplantation into the brain, and participated in neuroprotection. This action maybe related to the increased level of the neurotrophins brain-derived neurotrophic factor and glial cell line-derived neurotrophic factor (Kong et al. 2008).

5.5 Conclusion

Apoptosis is an integral part of the life cycle. It is a cell suicide program leading mainly to selective elimination of an organism's useless cells. In this process, the dying cell is an active participant. Under physiological conditions, apoptosis is most often found during normal cell turnover and tissue homeostasis, embryogenesis, induction and maintenance of immune tolerance, development of the nervous system, and endocrine-dependent tissue atrophy. Additionally, this process is essential in the homeostasis of cell number in organs in order to determine the cell numbers of the whole organism.

Apoptotic cells are characterized by loss of cell membrane phospholipids asymmetry, condensation of chromatin, reduction of nuclear size, intranucleosomal DNA cleavage, shrinkage of the cell, membrane blebbing and breakdown of the cell into apoptotic bodies, which are subsequently phagocytosed by other cells. There are myriads of reasons and ways for a cell to die; among them apoptosis is a specific form. Programmed cell death or apoptosis was initially considered a physiological process; however, it has also been suggested to contribute to pathology. Although the etiology of neurodegenerative diseases may be distinct, different diseases display a similar pathogenesis. Recent studies show that cells of the CNS in a degenerative state undergo apoptosis, being a hallmark of all forms of neurodegenerative diseases an impairment of neuronal functions, and in many cases of neuronal cell death.

Acknowledgments This work was supported by grant PRI08A018 funded by Junta de Extremadura and FEDER.

References

Aharoni R, Eilam R, Domev H et al. (2005) The immunomodulator glatiramer acetate augments the expression of neurotrophic factors in brains of experimental autoimmune encephalomyelitis mice. Proc Natl Acad Sci USA 102:19045–19050

Aktas O, Schulze-Topphoff U, Zipp F (2007) The role of TRAIL/TRAIL receptors in central nervous system pathology. Front Biosci 12:2912–2921

Alano CC, Ying W, Swanson RA (2004) Poly(ADP-ribose) polymerase-1-mediated cell death in astrocytes requires NAD+ depletion and mitochondrial permeability transition. J Biol Chem 279:18895–18902

Alexandrov PN, Zhao Y, Pogue AI et al. (2005) Synergistic effects of iron and aluminum on stress-related gene expression in primary human neural cells. J Alzheimers Dis 8:117–127

Araque A, Carmignoto G, Haydon PG (2001) Dynamic signaling between astrocytes and neurons. Annu Rev Physiol 63:795–813

Arimoto T, Choi DY, Lu X et al. (2006) Interleukin-10 protects against inflammation-mediated degeneration of dopaminergic neurons in substantia nigra. Neurobiol Aging 28:894–906

Arning L, Monté D, Hansen W et al. (2008) ASK1 and MAP2K6 as modifiers of age at onset in Huntington's disease. J Mol Med 86:485–490

Bains M, Cousins JC, Roberts JL (2007) Neuroprotection by estrogen against MPP+-induced dopamine neuron death is mediated by ERalpha in primary cultures of mouse mesencephalon. Exp Neurol 204:767–776

Barone MC, Desouza LA, Freeman RS (2008) Pin1 promotes cell death in NGF-dependent neurons through a mechanism requiring c-Jun activity. J Neurochem 106:734–745

Basu A, Castle VP, Bouziane M et al. (2006) Crosstalk between extrinsic and intrinsic cell death pathways in pancreatic cancer: synergistic action of estrogen metabolite and ligands of death receptor family. Cancer Res 66:4309–4318

Bäurle J, Frischmuth S, Kranda K (2004) TRAIL-related death receptors in normal Lurcher and weaver mutant mouse brain. Neurosci Lett 372:46–51

Baydas G, Reiter RJ, Akbulut M et al. (2005) Melatonin inhibits neural apoptosis induced by homocysteine in hippocampus of rats via inhibition of cytochrome *c* translocation and

caspase-3 activation and by regulating pro- and anti-apoptotic protein levels. Neuroscience 135:879–886
Beart PM, Lim ML, Chen B et al. (2007) Hierarchical recruitment by AMPA but not staurosporine of pro-apoptotic mitochondrial signaling in cultured cortical neurons: evidence for caspase-dependent/independent cross-talk. J Neurochem 103:2408–2427
Bemelmans AP, Husson I, Jaquet M et al. (2006) Lentiviral-mediated gene transfer of brain-derived neurotrophic factor is neuroprotective in a mouse model of neonatal excitotoxic challenge. J Neurosci Res 83:50–60
Benarroch EE (2005) Neuron-astrocyte interactions: partnership for normal function and disease in the central nervous system. Mayo Clin Proc 80:1326–1338
Bender LM, Morgan MJ, Thomas LR et al. (2005) The adaptor protein TRADD activates distinct mechanisms of apoptosis from the nucleus and the cytoplasm. Cell Death Differ 12:473–481
Benjelloun N, Joly LM, Palmier B et al. (2003) Apoptotic mitochondrial pathway in neurones and astrocytes after neonatal hypoxia-ischaemia in the rat brain. Neuropathol Appl Neurobiol 29:350–360
Bezprozvanny I, Mattson MP (2008) Neuronal calcium mishandling and the pathogenesis of Alzheimer's disease. Trends Neurosci 31:454–463
Brustovetsky T, Antonsson B, Jemmerson R et al. (2005) Activation of calcium-independent phospholipase A (iPLA) in brain mitochondria and release of apoptogenic factors by BAX and truncated BID. J Neurochem 94:980–994
Burke RE (2008) Programmed cell death and new discoveries in the genetics of parkinsonism. J Neurochem 104:875–890
Cantarella G, Di Benedetto G, Pezzino S et al. (2008) TRAIL-related neurotoxicity implies interaction with the Wnt pathway in human neuronal cells in vitro. J Neurochem 105:1915–1923
Cantarella G, Lempereur L, D'Alcamo MA et al. (2007) Trail interacts redundantly with nitric oxide in rat astrocytes: potential contribution to neurodegenerative processes. J Neuroimmunol 182:41–47
Cao G, Xing J, Xiao X et al. (2007) Critical role of calpain I in mitochondrial release of apoptosis-inducing factor in ischemic neuronal injury. J Neurosci 27:9278–9293
Chao PL, Fan SF, Chou YH et al. (2007) N-acetylcysteine attenuates arsenite-induced oxidative injury in dorsal root ganglion explants. Ann NY Acad Sci 1122:276–288
Chaparro-Huerta V, Rivera-Cervantes MC, Flores-Soto ME et al. (2005) Proinflammatory cytokines and apoptosis following glutamate-induced excitotoxicity mediated by p38 MAPK in the hippocampus of neonatal rats. J Neuroimmunol 165:53–62
Chauvier D, Lecoeur H, Langonné A et al. (2005) Upstream control of apoptosis by caspase-2 in serum-deprived primary neurons. Apoptosis 10:1243–1259
Chen J, Qin J, Liu X et al. (2008) Nitric oxide-mediated neuronal apoptosis in rats with recurrent febrile seizures through endoplasmic reticulum stress pathway. Neurosci Lett 443:134–139
Chen XQ, Lau LT, Fung YW (2005) Inactivation of bad by site-specific phosphorylation: the checkpoint for ischemic astrocytes to initiate or resist apoptosis. J Neurosci Res 79:798–808
Cheng HY, Hsieh MT, Wu CR et al. (2008) Schizandrin protects primary cultures of rat cortical cells from glutamate-induced excitotoxicity. J Pharmacol Sci 107:21–31
Cregan SP, Fortin A, MacLaurin JG et al. (2002) Apoptosis-inducing factor is involved in the regulation of caspase-independent neuronal cell death. J Cell Biol 158:507–517
Culmsee C, Landshamer S (2006) Molecular insights into mechanisms of the cell death program: role in the progression of neurodegenerative disorders. Curr Alzheimer Res 3:269–283
Culmsee C, Zhu C, Landshamer S et al. (2005) Apoptosis-inducing factor triggered by poly(ADP-ribose) polymerase and Bid mediates neuronal cell death after oxygen-glucose deprivation and focal cerebral ischemia. J Neurosci 25:10262–10272
Cunha-Oliveira T, Rego AC, Garrido J et al. (2007) Street heroin induces mitochondrial dysfunction and apoptosis in rat cortical neurons. J Neurochem 101:543–554
Das A, Banik NL, Ray SK (2008) Retinoids induced astrocytic differentiation with down regulation of telomerase activity and enhanced sensitivity to taxol for apoptosis in human glioblastoma T98G and U87MG cells. J Neurooncol 87:9–22

Das A, Garner DP, Del Re AM et al. (2006) Calpeptin provides functional neuroprotection to rat retinal ganglion cells following Ca^{2+} influx. Brain Res 1084:146–157

Del Río P, Montiel T, Massieu L (2008) Contribution of NMDA and non-NMDA receptors to in vivo glutamate-induced calpain activation in the rat striatum Relation to neuronal damage. Neurochem Res 33:1475–1483

Del Villar K, Miller CA (2004) Down-regulation of DENN/MADD a TNF receptor binding protein correlates with neuronal cell death in Alzheimer's disease brain and hippocampal neurons. Proc Natl Acad Sci USA 101:4210–4215

Dempsey PW, Doyle SE, He JQ et al. (2003) The signaling adaptors and pathways activated by TNF superfamily. Cytokine Growth Factor Rev 14:193–209

Dey S, Snow DM (2007) Cocaine exposure in vitro induces apoptosis in fetal locus coeruleus neurons through TNF-alpha-mediated induction of Bax and phosphorylated c-Jun NH(2)-terminal kinase. J Neurochem 103:542–556

Ding D, Jiang H, Wang P et al. (2007) Cell death after co-administration of cisplatin and ethacrynic acid. Hear Res 226:129–139

Dohare P, Varma S, Ray M (2008) Curcuma oil modulates the nitric oxide system response to cerebral ischemia/reperfusion injury. Nitric Oxide 19:1–11

Donovan M, Doonan F, Cotter TG (2006) Decreased expression of pro-apoptotic Bcl-2 family members during retinal development and differential sensitivity to cell death. Dev Biol 291:154–169

Dörr J, Roth K, Zurbuchen U et al. (2005) Tumor-necrosis-factor-related apoptosis-inducing-ligand (TRAIL)-mediated death of neurons in living human brain tissue is inhibited by flupirtine-maleate. J Neuroimmunol 167:204–209

Druse MJ, Gillespie RA, Tajuddin NF et al. (2007) S100B-mediated protection against the pro-apoptotic effects of ethanol on fetal rhombencephalic neurons. Brain Res 1150: 46–54

Du Y, Chen CP, Tseng CY et al. (2007) Astroglia-mediated effects of uric acid to protect spinal cord neurons from glutamate toxicity. Glia 55:463–472

Faigle R, Brederlau A, Elmi M et al. (2004) ASK1 inhibits astroglial development via p38 mitogen-activated protein kinase and promotes neuronal differentiation in adult hippocampus-derived progenitor cells. Mol Cell Biol 24:280–293

Figiel I, Dzwonek K (2007) TNFalpha and TNF receptor 1 expression in the mixed neuronal-glial cultures of hippocampal dentate gyrus exposed to glutamate or trimethyltin. Brain Res 1131:17–28

Fuchs C, Forster V, Balse E et al. (2005) Retinal-cell-conditioned medium prevents TNF-alpha-induced apoptosis of purified ganglion cells. Invest Ophthalmol Vis Sci 46:2983–2991

Gabryel B, Trzeciak HI (2001) Role of astrocytes in pathogenesis of ischemic brain injury. Neurotox Res 3:205–221

Garg SK, Banerjee R, Kipnis J (2008) Neuroprotective immunity: T cell-derived glutamate endows astrocytes with a neuroprotective phenotype. J Immunol 180:3866–3873

Gianinazzi C, Grandgirard D, Imboden H et al. (2003) Caspase-3 mediates hippocampal apoptosis in pneumococcal meningitis. Acta Neuropathol 105:499–507

Gil YG, Kang MK (2008) Capsaicin induces apoptosis and terminal differentiation in human glioma A172 cells. Life Sci 82:997–1003

Goodall EF, Morrison KE (2006) Amyotrophic lateral sclerosis (motor neuron disease): proposed mechanisms and pathways to treatment. Expert Rev Mol Med 8:1–22

Gozzelino R, Sole C, Llecha N et al. (2008) BCL-X(L) regulates TNF-alpha-mediated cell death independently of NF-kappaB FLIP and IAPs. Cell Res 10:1020–1036

Guthmann F, Wissel H, Schachtrup C et al. (2005) Inhibition of TNFalpha in vivo prevents hyperoxia-mediated activation of caspase 3 in type II cells. Respir Res 21:6–10

Hallin U, Kondo E, Ozaki Y et al. (2006) Bcl-2 phosphorylation in the BH4 domain precedes caspase-3 activation and cell death after neonatal cerebral hypoxic-ischemic injury. Neurobiol Dis 21:478–486

Halterman MW, De Jesus C, Rempe DA et al. (2008) Loss of c/EBP-beta activity promotes the adaptive to apoptotic switch in hypoxic cortical neurons. Mol Cell Neurosci 38:125–137
Han JM, Chang BJ, Li TZ et al. (2007) Protective effects of ascorbic acid against lead-induced apoptotic neurodegeneration in the developing rat hippocampus in vivo. Brain Res 1185:68–74
Han W, Shi X, Nuttall AL (2006) AIF and endoG translocation in noise exposure induced hair cell death. Hear Res 211:85–95
Hans G, Malgrange B, Lallemend F et al. (2005) Beta-carbolines induce apoptosis in cultured cerebellar granule neurons via the mitochondrial pathway. Neuropharmacology 48:105–117
Hara MR, Snyder SH (2007) Cell signaling and neuronal death. Annu Rev Pharmacol Toxicol 47:117–141
Harukuni I, Bhardwaj A (2006) Mechanisms of brain injury after global cerebral ischemia. Neurol Clin 24:1–21
Helguera P, Pelsman A, Pigino G (2005) ets-2 promotes the activation of a mitochondrial death pathway in Down's syndrome neurons. J Neurosci 25:2295–2303
Hellmich HL, Eidson K, Cowart J et al. (2008) Chelation of neurotoxic zinc levels does not improve neurobehavioral outcome after traumatic brain injury. Neurosci Lett 440:155–159
Henshall DC, Bonislawski DP, Skradski SL et al. (2001) Cleavage of bid may amplify caspase-8-induced neuronal death following focally evoked limbic seizures. Neurobiol Dis 8:568–580
Hu H, Jiang C, Schuster T et al. (2006) Inorganic selenium sensitizes prostate cancer cells to TRAIL-induced apoptosis through superoxide/p53/Bax-mediated activation of mitochondrial pathway. Mol Cancer Ther 5:1873–1882
Huang Y, Erdmann N, Zhao J et al. (2005) The signaling and apoptotic effects of TNF-related apoptosis-inducing ligand in HIV-1 associated dementia. Neurotox Res 8:135–148
Huang Z, Hou Q, Cheung NS et al. (2006) Neuronal cell death caused by inhibition of intracellular cholesterol trafficking is caspase dependent and associated with activation of the mitochondrial apoptosis pathway. J Neurochem 97:280–291
Hsu MJ, Hsu CY, Chen BC et al. (2007) Apoptosis signal-regulating kinase 1 in amyloid beta peptide-induced cerebral endothelial cell apoptosis. J Neurosci 27:5719–5729
Ishida Y, Nagai A, Kobayashi S et al. (2006) Upregulation of protease-activated receptor-1 in astrocytes in Parkinson disease: astrocyte-mediated neuroprotection through increased levels of glutathione peroxidise. J Neuropathol Exp Neurol 65:66–77
Jin Z, El-Deiry WS (2006) Distinct signaling pathways in TRAIL- versus tumor necrosis factor-induced apoptosis. Mol Cell Biol 26:8136–8148
Juknat AA, Méndez Mdel V, Quaglino A et al. (2005) Melatonin prevents hydrogen peroxide-induced Bax expression in cultured rat astrocytes. J Pineal Res 38:84–92
Jung TW, Lee JY, Shim WS et al. (2007) Rosiglitazone protects human neuroblastoma SH-SY5Y cells against MPP+ induced cytotoxicity via inhibition of mitochondrial dysfunction and ROS production. J Neurol Sci 253:53–60
Kaliberov S, Stackhouse MA, Kaliberova L et al. (2004) Enhanced apoptosis following treatment with TRA-8 anti-human DR5 monoclonal antibody and overexpression of exogenous Bax in human glioma cells. Gene Ther 11:658–667
Karmakar S, Banik NL, Patel SJ et al. (2007) Garlic compounds induced calpain and intrinsic caspase cascade for apoptosis in human malignant neuroblastoma SH-SY5Y cells. Apoptosis 12:671–684
Karovic O, Tonazzini I, Rebola et al. (2007) Toxic effects of cobalt in primary cultures of mouse astrocytes Similarities with hypoxia and role of HIF-1alpha. Biochem Pharmacol 73:694–708
Karunakaran S, Diwakar L, Saeed U et al. (2007) Activation of apoptosis signal regulating kinase 1 (ASK1) and translocation of death-associated protein Daxx in substantia nigra pars compacta in a mouse model of Parkinson's disease: protection by alpha-lipoic acid. FASEB J 21:2226–2236
Kaushal V, Schlichter LC (2007) Mechanisms of microglia-mediated neurotoxicity in a new model of the stroke penumbra. J Neurosci 28:2221–2230
Keane RW, Davis AR, Dietrich WD (2006) Inflammatory and apoptotic signaling after spinal cord injury. J Neurotrauma 3:335–344

Keeble JA, Gilmore AP (2007) Apoptosis commitment – translating survival signals into decisions on mitochondria. Cell Res 17:976–984

Kim EK, Noh KT, Yoon JH et al. (2007) Positive regulation of ASK1-mediated c-Jun NH(2)-terminal kinase signaling pathway by the WD-repeat protein Gemin5. Cell Death Differ 14:1518–1528

Kim EH, Kim SU, Choi KS (2005a) Rottlerin sensitizes glioma cells to TRAIL-induced apoptosis by inhibition of Cdc2 and the subsequent downregulation of survivin and XIAP. Oncogene 24:838–849

Kim SD, Moon CK, Eun SY et al. (2005b) Identification of ASK1 MKK4 JNK c-Jun and caspase-3 as a signaling cascade involved in cadmium-induced neuronal cell apoptosis. Biochem Biophys Res Commun 328:326–334

Kirkland RA, Franklin JL (2007) Bax affects production of reactive oxygen by the mitochondria of non-apoptotic neurons. Exp Neurol 204:458–461

Kogo J, Takeba Y, Kumai T et al. (2006) Involvement of TNF-alpha in glutamate-induced apoptosis in a differentiated neuronal cell line. Brain Res 1122:201–208

Kong XY, Cai Z, Pan L et al. (2008) Transplantation of human amniotic cells exerts neuroprotection in MPTP-induced Parkinson disease mice. Brain Res 1205:108–115

König HG, Rehm M, Gudorf D et al. (2007) Full length Bid is sufficient to induce apoptosis of cultured rat hippocampal neurons. BMC Cell Biol 8:7

Krantic S, Mechawar N, Reix S et al. (2007) Apoptosis-inducing factor: a matter of neuron life and death. Prog Neurobiol 81:179–196

Krieger C, Duchen MR (2002) Mitochondria Ca^{2+} and neurodegenerative disease. Eur J Pharmacol 447:177–188

Kudo T, Kanemoto S, Hara H et al. (2008) A molecular chaperone inducer protects neurons from ER stress. Cell Death Differ 15:364–375

Küenzi P, Kiefer S, Koryakina A et al. (2008) Promotion of cell death or neurite outgrowth in PC-12 and N2a cells by the fungal alkaloid militarinone A depends on basal expression of p53. Apoptosis 13:364–376

Kuffler SW (1984) Physiology of neuroglial cells. In: Kuffler SW, Nicholls JG, Martin AR (eds) From neuron to brain, 2nd ed. Sinauer Associates Inc. Publishers, Sunderland, Massachusetts, USA, p 324

Kwon D, Choi IH (2006) Hydrogen peroxide upregulates TNF-related apoptosis-inducing ligand (TRAIL) expression in human astroglial cells and augments apoptosis of T cells. Yonsei Med J 47:551–557

Kwon D, Choi K, Choi C et al. (2008) Hydrogen peroxide enhances TRAIL-induced cell death through up-regulation of DR5 in human astrocytic cells. Biochem Biophys Res Commun 372:870–874

Lacour P, Heimrich B, Pröls F (2007) Induction of cellular stress and chaperone activation in organotypic slice cultures of hippocampus. J Neurosci Methods 166:24–31

Larner SF, Hayes RL, McKinsey DM (2004) Increased expression and processing of caspase-12 after traumatic brain injury in rats. J Neurochem 88:78–90

Lee DY, Lee KS, Lee HJ et al. (2008) Kynurenic acid attenuates MPP(+)-induced dopaminergic neuronal cell death via a Bax-mediated mitochondrial pathway. Eur J Cell Biol 87:389–397

Lee W, Kim DH, Boo JH et al. (2005) ER stress-induced caspase-12 activation is inhibited by PKC in neuronal cells. Apoptosis 10:407–415

Lennon AM, Ramaugé M, Dessouroux A et al. (2002) MAP kinase cascades are activated in astrocytes and preadipocytes by 15-deoxy-Delta(12–14)-prostaglandin J(2) and the thiazolidinedione ciglitazone through peroxisome proliferator activator receptor gamma-independent mechanisms involving reactive oxygenated species. J Biol Chem 277:29681–29685

Li T, Lu C, Xia Z et al. (2006) Inhibition of caspase-8 attenuates neuronal death induced by limbic seizures in a cytochrome c-dependent and Smac/DIABLO-independent way. Brain Res 1098:204–211

Lin D, Takemoto DJ (2007) Protection from ataxia-linked apoptosis by gap junction inhibitors. Biochem Biophys Res Commun 362:982–987
Lin H, Zhang X, Cheng G et al. (2008) Apoptosis induced by ardipusilloside III through BAD dephosphorylation and cleavage in human glioblastoma U251MG cells. Apoptosis 13:247–257
Lin J, Chen LY, Lin ZX et al. (2007) The effect of triptolide on apoptosis of glioblastoma multiforme (GBM) cells. J Int Med Res 35:637–643
Lindholm D, Wootz H, Korhonen L (2006) ER stress and neurodegenerative diseases. Cell Death Differ 13:385–392
Lindl KA, Akay C, Wang Y et al. (2007) Expression of the endoplasmic reticulum stress response marker BiP in the central nervous system of HIV-positive individuals. Neuropathol Appl Neurobiol 33:658–669
Lisak RP, Bealmear B, Nedelkoska L et al. (2006) Secretory products of central nervous system glial cells induce Schwann cell proliferation and protect from cytokine-mediated death. J Neurosci Res 83:1425–1431
Liu N, Kuang X, Kim HT et al. (2004) Possible involvement of both endoplasmic reticulum- and mitochondria-dependent pathways in MoMuLV-ts1-induced apoptosis in astrocytes. J Neurovirol 10:189–198
Liu N, Scofield VL, Qiang W et al. (2006) Interaction between endoplasmic reticulum stress and caspase 8 activation in retrovirus MoMuLV-ts1-infected astrocytes. Virology 348:398–405
Liu X, Shibata T, Hisaka S et al. (2008) DHA Hydroperoxides as a potential inducer of neuronal cell death: a mitochondrial dysfunction-mediated pathway. J Clin Biochem Nutr 43:26–33
Madhavan L, Ourednik V, Ourednik J (2008) Neural stem/progenitor cells initiate the formation of cellular networks that provide neuroprotection by growth factor-modulated antioxidant expression. Stem Cells 26:254–265
Mahesh VB, Dhandapani KM, Brann DW (2006) Role of astrocytes in reproduction and neuroprotection. Mol Cell Endocrinol 246:1–9
Majumder P, Chattopadhyay B, Mazumder A et al. (2006) Induction of apoptosis in cells expressing exogenous Hippi a molecular partner of huntingtin-interacting protein Hip1. Neurobiol Dis 22:242–256
Manfredi G, Xu Z (2005) Mitochondrial dysfunction and its role in motor neuron degeneration in ALS. Mitochondrion 5:77–87
Matsuzawa A, Ichijo H (2008) Redox control of cell fate by MAP kinase: physiological roles of ASK1-MAP kinase pathway in stress signaling. Biochim Biophys Acta 1780:1325–1336
Matsuzawa A, Saegusa K, Noguchi T et al. (2005) ROS-dependent activation of the TRAF6-ASK1-p38 pathway is selectively required for TLR4-mediated innate immunity. Nat Immunol 6:587–592
Mattson MP (2007) Calcium and neurodegeneration. Aging Cell 6:337–350
Maycotte P, Blancas S, Morán J (2008) Role of inhibitor of apoptosis proteins and Smac/DIABLO in staurosporine-induced cerebellar granule neurons death. Neurochem Res 33:1534–1540
Medigeshi GR, Lancaster AM, Hirsch AJ et al. (2007) West Nile virus infection activates the unfolded protein response leading to CHOP induction and apoptosis. J Virol 81:10849–10860
Micheau O, Tschopp J (2003) Induction of TNF receptor I-mediated apoptosis via two sequential signaling complexes. Cell 114:181–190
Mishra MK, Koli P, Bhowmick S et al. (2007) Neuroprotection conferred by astrocytes is insufficient to protect animals from succumbing to Japanese encephalitis. Neurochem Int 50:764–773
Miura Y, Koyanagi Y, Mizusawa H (2003) TNF-related apoptosis-inducing ligand (TRAIL) induces neuronal apoptosis in HIV-encephalopathy. J Med Dent Sci 50:17–25
Morgan M, Thorburn J, Pandolfi PP et al. (2002) Nuclear and cytoplasmic shuttling of TRADD induces apoptosis via different mechanisms. J Cell Biol 157:975–984
Mori K, Kaneko YS, Nakashima A et al. (2005) Peripheral lipopolysaccharide induces apoptosis in the murine olfactory bulb. Brain Res 1039:116–129
Muranyi M, Li PA (2005) Bongkrekic acid ameliorates ischemic neuronal death in the cortex by preventing cytochrome c release and inhibiting astrocyte activation. Neurosci Lett 384:277–281

Murata T, Tsuboi M, Hikita K et al. (2006) Protective effects of neurotrophic factors on tumor necrosis factor-related apoptosis-inducing ligand (TRAIL)-mediated apoptosis of murine adrenal chromaffin cell line tsAM5D. J Biol Chem 281:22503–22516

Nadeau PJ, Charette SJ, Toledano MB et al. (2007) Disulfide bond-mediated multimerization of Ask1 and its reduction by thioredoxin-1 regulate H_2O_2-induced c-Jun NH_2-terminal kinase activation and apoptosis. Mol Biol Cell 18:3903–3913

Nagai H, Noguchi T, Takeda K et al. (2007) Pathophysiological roles of ASK1-MAP kinase signaling pathways. J Biochem Mol Biol 40:1–6

Nagane M, Cavenee WK, Shiokawa Y (2007) Synergistic cytotoxicity through the activation of multiple apoptosis pathways in human glioma cells induced by combined treatment with ionizing radiation and tumor necrosis factor-related apoptosis-inducing ligand. J Neurosurg 106:407–416

Nagatsu T, Sawada M (2006) Molecular mechanism of the relation of monoamine oxidase B and its inhibitors to Parkinson's disease: possible implications of glial cells. J Neural Transm Suppl 71:53–65.

Nagy K, Székely-Szüts K, Izeradjene K et al. (2006) Proteasome inhibitors sensitize colon carcinoma cells to TRAIL-induced apoptosis via enhanced release of Smac/DIABLO from the mitochondria. Pathol Oncol Res 12:133–142

Nakka VP, Gusain A, Mehta SL et al. (2008) Molecular mechanisms of apoptosis in cerebral ischemia: multiple neuroprotective opportunities. Mol Neurobiol 37:7–38

Naumann U, Wischhusen J, Weit S, Rieger J, Wolburg H, Massing U, Weller M (2004) Alkylphosphocholine-induced glioma cell death is BCL-X(L)-sensitive caspase-independent and characterized by massive cytoplasmic vacuole formation. Cell Death Differ 11:1326–1341

Nicholls DG, Budd SL (2000) Mitochondria and neuronal survival. Physiol Rev 80:315–360

Oh YK, Shin KS, Kang SJ (2006) AIF translocates to the nucleus in the spinal motor neurons in a mouse model of ALS. Neurosci Lett 406:205–210

Oida Y, Shimazawa M, Imaizumi K et al. (2008) Involvement of endoplasmic reticulum stress in the neuronal death induced by transient forebrain ischemia in gerbil. Neuroscience 151:111–119

Okhrimenko H, Lu W, Xiang C et al. (2005) Protein kinase C-epsilon regulates the apoptosis and survival of glioma cells. Cancer Res 65:7301–7309

Oleskovicz SP, Martins WC, Leal RB et al. (2008) Mechanism of guanosine-induced neuroprotection in rat hippocampal slices submitted to oxygen-glucose deprivation. Neurochem Int 52:411–418

Ouyang YB, Giffard RG (2004) Changes in astrocyte mitochondrial function with stress: effects of Bcl-2 family proteins. Neurochem Int 45:371–379

Palomba L, Amadori A, Cantoni O (2007) Early release of arachidonic acid prevents an otherwise immediate formation of toxic levels of peroxynitrite in astrocytes stimulated with lipopolysaccharide/interferon-gamma. J Neurochem 103:904–913

Pan J, Wang G, Yang HQ et al. (2007) K252a prevents nigral dopaminergic cell death induced by 6-hydroxydopamine through inhibition of both mixed-lineage kinase 3/c-Jun NH2-terminal kinase 3 (JNK3) and apoptosis-inducing kinase 1/JNK3 signaling pathways. Mol Pharmacol 72:1607–1618

Papa L, Gomes E, Rockwell P (2007) Reactive oxygen species induced by proteasome inhibition in neuronal cells mediate mitochondrial dysfunction and a caspase independent cell death. Apoptosis 12:1389–1405

Patel JR, Brewer GJ (2008) Age-related changes to tumor necrosis factor receptors affect neuron survival in the presence of beta-amyloid. J Neurosci Res 86:2303–2313

Penas C, Guzmán MS, Verdú E (2007) Spinal cord injury induces endoplasmic reticulum stress with different cell-type dependent response. J Neurochem 102:1242–1255

Peng H, Huang Y, Duan Z et al. (2005) Cellular IAP1 regulates TRAIL-induced apoptosis in human fetal cortical neural progenitor cells. J Neurosci Res 82:295–305

Pope SA, Milton R, Heales SJ (2008) Astrocytes protect against copper-catalysed loss of extracellular glutathione. Neurochem Res 33:1410–1418

Prabhakaran K, Ghosh D, Chapman GD et al. (2008) Molecular mechanism of manganese exposure-induced dopaminergic toxicity. Brain Res Bull 76:361–367
Prabhakaran K, Li L, Zhang L et al. (2007) Upregulation of BNIP3 and translocation to mitochondria mediates cyanide-induced apoptosis in cortical cells. Neuroscience 150:159–167
Reijonen S, Putkonen N, Nørremølle A et al. (2008) Inhibition of endoplasmic reticulum stress counteracts neuronal cell death and protein aggregation caused by N-terminal mutant huntingtin proteins. Exp Cell Res 314:950–960
Resende R, Ferreiro E, Pereira C et al. (2008) Neurotoxic effect of oligomeric and fibrillar species of amyloid-beta peptide 1-42: involvement of endoplasmic reticulum calcium release in oligomer-induced cell death. Neuroscience 155:725–737
Rubio N, Martin-Clemente B, Lipton HL (2003) High-neurovirulence GDVII virus induces apoptosis in murine astrocytes through tumor necrosis factor (TNF)-receptor and TNF-related apoptosis-inducing ligand. Virology 311:366–375
Russell JC, Blue ME, Johnston MV et al. (2007) Enhanced cell death in MeCP2 null cerebellar granule neurons exposed to excitotoxicity and hypoxia. Neuroscience 150:563–574
Russell JC, Whiting H, Szuflita N et al. (2008) Nuclear translocation of X-linked inhibitor of apoptosis (XIAP) determines cell fate after hypoxia ischemia in neonatal brain. J Neurochem 106:1357–1370
Sanges D, Comitato A, Tammaro R et al. (2006) Apoptosis in retinal degeneration involves cross-talk between apoptosis-inducing factor (AIF) and caspase-12 and is blocked by calpain inhibitors. Proc Natl Acad Sci USA 103:17366–17371
Santiago AR, Cristóvão AJ, Santos PF et al. (2007) High glucose induces caspase-independent cell death in retinal neural cells. Neurobiol Dis 25:464–472
Saunders M, Eldeen MB, Del Valle L et al. (2005) p73 modulates HIV-1 Tat transcriptional and apoptotic activities in human astrocytes. Apoptosis 10:1419–1431
Scorrano L, Stanley JK (2003) Mechanisms of cytochrome c release by proapoptotic BCL-2 family members. Biochem Biophys Res Commun 304:437–444
Shalbuyeva N, Brustovetsky T, Bolshakov A et al. (2006) Calcium-dependent spontaneously reversible remodeling of brain mitochondria. J Biol Chem 281:37547–37558
Shang L, Liu J, Zhu Q et al. (2006) Gypenosides protect primary cultures of rat cortical cells against oxidative neurotoxicity. Brain Res 1102:163–174
Sharma HS (2005) Neuroprotective effects of neurotrophins and melanocortins in spinal cord injury: an experimental study in the rat using pharmacological and morphological approaches. Ann NY Acad Sci 1053:407–421
Sharma P, Karian J, Sharma S et al. (2003) Pyruvate ameliorates post ischemic injury of rat astrocytes and protects them against PARP mediated cell death. Brain Res 992:104–113
Sharma V, Mishra M, Ghosh S et al. (2007) Modulation of interleukin-1beta mediated inflammatory response in human astrocytes by flavonoids: implications in neuroprotection. Brain Res Bull 73:55–63
Sharma V, Tewari R, Sk UH et al. (2008) Ebselen sensitizes glioblastoma cells to Tumor Necrosis Factor (TNFalpha)-induced apoptosis through two distinct pathways involving NF-kappaB downregulation and Fas-mediated formation of death inducing signaling complex. Int J Cancer 123:2204–2212
Schneider-Brachert W, Tchikov V, Neumeyer J et al. (2004) Compartmentalization of TNF receptor 1 signaling: internalized TNF receptosomes as death signaling vesicles. Immunity 21:415–448
Schuster N, Bender H, Rössler OG et al. (2003) Transforming growth factor-beta and tumor necrosis factor-alpha cooperate to induce apoptosis in the oligodendroglial cell line OLI-neu. J Neurosci Res 73:324–333
Shih AY, Erb H, Sun X et al. (2006) Cystine/glutamate exchange modulates glutathione supply for neuroprotection from oxidative stress and cell proliferation. J Neurosci 26:10514–10523
Shimazawa M, Ito Y, Inokuchi Y et al. (2007) Involvement of double-stranded RNA-dependent protein kinase in ER stress-induced retinal neuron damage. Invest Ophthalmol Vis Sci 48:3729–3736

Smith MI, Deshmukh M (2007) Endoplasmic reticulum stress-induced apoptosis requires bax for commitment and Apaf-1 for execution in primary neurons. Cell Death Differ 14:1011–1019

Sofo V, Salmeri FM, Di Bella P et al. (2005) Short communication: impairment of membrane markers on peripheral blood mononuclear cells and imbalance of cytokine secretion in the pathogenesis of multiple sclerosis active phases. J Interferon Cytokine Res 25:661–665

Son YG, Kim EH, Kim JY et al. (2007) Silibinin sensitizes human glioma cells to TRAIL-mediated apoptosis via DR5 up-regulation and down-regulation of c-FLIP and surviving. Cancer Res 67:8274–8284

Song JH, Bellail A, Tse MC et al. (2006) Human astrocytes are resistant to Fas ligand and tumor necrosis factor-related apoptosis-inducing ligand-induced apoptosis. J Neurosci 26:3299–3308

Stegh AH, Kim H, Bachoo RM et al. (2007) Bcl2L12 inhibits post-mitochondrial apoptosis signaling in glioblastoma. Genes Dev 21:98–111

Struble RG, Nathan BP, Cady C et al. (2007) Estradiol regulation of astroglia and apolipoprotein E: an important role in neuronal regeneration. Exp Gerontol 42:54–63

Sui Y, Stehno-Bittel L, Li S et al. (2006) CXCL10-induced cell death in neurons: role of calcium dysregulation. Eur J Neurosci 23:957–964

Swarup V, Ghosh J, Das S et al. (2008) Tumor necrosis factor receptor-associated death domain mediated neuronal death contributes to the glial activation and subsequent neuroinflammation in Japanese encephalitis. Neurochem Int 52:1310–1321

Takenouchi T, Setoguchi T, Yone K et al. (2008) Expression of apoptosis signal-regulating kinase 1 in mouse spinal cord under chronic mechanical compression: possible involvement of the stress-activated mitogen-activated protein kinase pathways in spinal cord cell apoptosis. Spine 33:1943–1950

Taylor DL, Jones F, Kubota ES et al. (2005) Stimulation of microglial metabotropic glutamate receptor mGlu2 triggers tumor necrosis factor alpha-induced neurotoxicity in concert with microglial-derived Fas ligand. J Neurosci 25:2952–2964

Tramontina F, Tramontina AC, Souza DF et al. (2006) Glutamate uptake is stimulated by extracellular S100B in hippocampal astrocytes. Cell Mol Neurobiol 26:81–86

Uberti D, Ferrari-Toninelli G, Bonini SA et al. (2007) Blockade of the tumor necrosis factor-related apoptosis inducing ligand death receptor DR5 prevents beta-amyloid neurotoxicity. Neuropsychopharmacology 32:872–880

Van Gurp M, Festjens N, van Loo G et al. (2003) Mitochondrial intermembrane proteins in cell death. Biochem Biophys Res Commun 304:487–497

Veglianese P, Lo Coco D, Bao Cutrona M et al. (2006) Activation of the p38MAPK cascade is associated with upregulation of TNF alpha receptors in the spinal motor neurons of mouse models of familial ALS. Mol Cell Neurosci 31:218–231

Verkhratsky A, Orkand RK, Kettenmann H (1998) Glial calcium: homeostasis and signalling function. Physiol Rev 78:99–141

Voccoli V, Mazzoni F, Garcia-Gil M et al. (2007) Serum-withdrawal-dependent apoptosis of hippocampal neuroblasts involves Ca^{2+} release by endoplasmic reticulum and caspase-12 activation. Brain Res 1147:1–11

Volterra A, Meldolesi J (2005) Astrocytes from brain glue to communication elements: the revolution continues. Nat Rev Neurosci 6:626–640

Wang P, Cao X, Nagel DJ et al. (2007a) Activation of ASK1 during reperfusion of ischemic spinal cord. Neurosci Lett 415:248–252

Wang X, Wang B, Fan Z et al. (2007b) Thiamine deficiency induces endoplasmic reticulum stress in neurons. Neuroscience 144:1045–1056

Wang Y, Han R, Liang ZQ et al. (2008) An autophagic mechanism is involved in apoptotic death of rat striatal neurons induced by the non-N-methyl-D-aspartate receptor agonist kainic acid. Autophagy 4:214–226

Wang Y, Singh R, Lefkowitch JH et al. (2006) Tumor necrosis factor-induced toxic liver injury results from JNK2-dependent activation of caspase-8 and the mitochondrial death pathway. J Biol Chem 281:15258–15267

Weller ML, Stone IM, Goss A et al. (2008) Selective overexpression of excitatory amino acid transporter 2 (EAAT2) in astrocytes enhances neuroprotection from moderate but not severe hypoxia-ischemia. Neuroscience 155:1204–1211

Williams BL, Hornig M, Yaddanapudi K et al. (2008) Hippocampal poly(ADP-Ribose) polymerase 1 and caspase 3 activation in neonatal bornavirus infection. J Virol 82:1748–1758

Williams BL, Lipkin WI (2006) Endoplasmic reticulum stress and neurodegeneration in rats neonatally infected with borna disease virus. J Virol 80:8613–8626

Wu J, Holstein JD, Upadhyay G et al. (2007a) Purinergic receptor-stimulated IP3-mediated Ca^{2+} release enhances neuroprotection by increasing astrocyte mitochondrial metabolism during aging. J Neurosci 27:6510–6520

Wu JY, Niu FN, Huang R et al. (2008) Enhancement of glutamate uptake in 1-methyl-4-phenylpyridinium-treated astrocytes by trichostatin A. Neuroreport 19:1209–1212

Wu KL, Hsu C, Chan JY (2007b) Impairment of the mitochondrial respiratory enzyme activity triggers sequential activation of apoptosis-inducing factor-dependent and caspase-dependent signaling pathways to induce apoptosis after spinal cord injury. J Neurochem 101:1552–1566

Wu YP, Mizukami H, Matsuda J et al. (2005) Apoptosis accompanied by up-regulation of TNF-alpha death pathway genes in the brain of Niemann-Pick type C disease. Mol Genet Metab 84:9–17

Yacoub A, Park MA, Hanna D et al. (2006) OSU-03012 promotes caspase-independent but PERK-cathepsin B- BID- and AIF-dependent killing of transformed cells. Mol Pharmacol 70:589–603

Yamagata K, Shoji Y, Terashima T et al. (2006) Glutamate reduces secretion of l-serine in astrocytes isolated from stroke-prone spontaneously hypertensive rats. Neuroscience 143:729–737

Yang S, Thor AD, Edgerton S et al. (2006) Caspase-3 mediated feedback activation of apical caspases in doxorubicin and TNF-alpha induced apoptosis. Apoptosis 11:1987–1997

Yang W, Tiffany-Castiglioni E (2008) Paraquat-induced apoptosis in human neuroblastoma SH-SY5Y cells: involvement of p53 and mitochondria. J Toxicol Environ Health A 71:289–299

Yao M, Nguyen TV, Pike CJ (2007) Estrogen regulates Bcl-w and Bim expression: role in protection against beta-amyloid peptide-induced neuronal death. J Neurosci 27:422–1433

Yel L, Brown LE, Su K et al. (2005) Thimerosal induces neuronal cell apoptosis by causing cytochrome c and apoptosis-inducing factor release from mitochondria. Int J Mol Med 16:971–977

Yeo JE, Kang SK (2007) Selenium effectively inhibits ROS-mediated apoptotic neural precursor cell death in vitro and in vivo in traumatic brain injury. Biochim Biophys Acta 1772:1199–1210

Yoshida I, Monji A, Tashiro K (2006) Depletion of intracellular Ca^{2+} store itself may be a major factor in thapsigargin-induced ER stress and apoptosis in PC12 cells. Neurochem Int 48:696–702

Yu HB, Li ZB, Zhang HX (2006a) Role of potassium channels in Abeta(1–40)-activated apoptotic pathway in cultured cortical neurons. J Neurosci Res 84:1475–1484

Yu L, Miao H, Hou Y et al. (2006b) Neuroprotective effect of A20 on TNF-induced postischemic apoptosis. Neurochem Res 31:21–32

Yu SW, Andrabi SA, Wang H et al. (2006c) Apoptosis-inducing factor mediates poly(ADP-ribose) (PAR) polymer-induced cell death. Proc Natl Acad Sci USA 103:18314–18319

Yung HW, Bal-Price AK, Brown GC et al. (2004) Nitric oxide-induced cell death of cerebrocortical murine astrocytes is mediated through p53- and Bax-dependent pathways. J Neurochem 89:812–821

Zanna C, Ghelli A, Porcelli AM et al. (2005) Caspase-independent death of Leber's hereditary optic neuropathy cybrids is driven by energetic failure and mediated by AIF and Endonuclease G. Apoptosis 10:1997–1007

Zha J, Shu H (2002) Molecular mechanism of signaling by tumor necrosis factor. Sci China C Life Sci 45:113–119

Zhang H, Li Q, Graham RK et al. (2008a) Full length mutant huntingtin is required for altered Ca^{2+} signaling and apoptosis of striatal neurons in the YAC mouse model of Huntington's disease. Neurobiol Dis 31:80–88

Zhang H, Song LC, Jia CH (2008b) Effects of ATP sensitive potassium channel opener on the mRNA and protein expressions of caspase-12 after cerebral ischemia-reperfusion in rats. Neurosci Bull 24:7–12

Zhang S, Zhou F, Ding JH et al. (2007) ATP-sensitive potassium channel opener iptakalim protects against MPP-induced astrocytic apoptosis via mitochondria and mitogen-activated protein kinase signal pathways. J Neurochem 103:569–579

Zhang X, Chen S, Li L et al. (2008c) Folic acid protects motor neurons against the increased homocysteine inflammation and apoptosis in SOD1 G93A transgenic mice. Neuropharmacology 54:1112–1119

Zhang Y, Bhavnani BR (2006) Glutamate-induced apoptosis in neuronal cells is mediated via caspase-dependent and independent mechanisms involving calpain and caspase-3 proteases as well as apoptosis inducing factor (AIF) and this process is inhibited by equine estrogens. BMC Neurosci 15:7–49

Zhao Y, Conze DB, Hanover JA et al. (2007) Tumor necrosis factor receptor 2 signaling induces selective c-IAP1-dependent ASK1 ubiquitination and terminates mitogen-activated protein kinase signaling. J Biol Chem 282:7777–7782

Zheng L, Bidere N, Staudt D et al. (2006) Competitive control of independent programs of tumor necrosis factor receptor-induced cell death by TRADD and RIP1. Mol Cell Biol 26:3505–3513

Zhu ZH, Yang R, Fu X et al. (2006) Astrocyte-conditioned medium protecting hippocampal neurons in primary cultures against corticosterone-induced damages via PI3-K/Akt signal pathway. Brain Res 1114:1–10

Chapter 6
Apoptotic Events in Blood Cells

J.A. Rosado

Abstract Cells undergo apoptotic events during development, tissue homeostasis or disease and are subsequently cleared by phagocytes, inducing changes in the immune response. Lymphocyte apoptosis is responsible for the homeostasis of immune cells and plays an essential role in the elimination of autoreactive lymphocytes. Apoptosis also modulates neutrophil life span, regulating the balance between their function as effectors of the immune system and the clearance of potentially harmful cells. Programmed mammalian red blood cells death, or eryptosis, is a special form of apoptosis that shows all features of apoptosis, except nuclear condensation. Similarly, platelets, small, anuclear cytoplasmic fragments, develop apoptotic events that regulates platelet life span. Apoptotic events, which are enhanced in mature megakaryocytes, the platelet precursors, have been proposed as a major force driving proplatelet formation and platelet release. In addition to apoptosis, platelets undergo apoptotic-like events, including the rapid and reversible activation of caspase 3, that are essential for platelet Ca^{2+} signalling and aggregation independently of programmed cell death. Finally elevated apoptosis and enhanced oxidative stress in peripheral blood cells, such as platelets and lymphocytes, have been proposed as a biomarker for Alzheimer disease. This chapter describes the physiological and pathological implications of apoptosis in blood cells.

Keywords Erythrocytes · Platelets · Neutrophils · Eryptosis · Microparticles

6.1 Introduction

Mamalian blood consists of a suspension of cells, representing 40–45% of total volume, in liquid medium called plasma. In an adult man, blood is about 1/12th of the body weight and represents about 5–6 l. Blood performs a number of important functions, some of them associated to blood cells and others to plasma components.

J.A. Rosado (✉)
Department of Physiology, University of Extremadura, Avda. Universidad s/n, 10071 Cáceres, Spain
e-mail: jarosado@unex.es

G.M. Salido, J.A. Rosado (eds.), *Apoptosis: Involvement of Oxidative Stress and Intracellular Ca^{2+} Homeostasis*, DOI 10.1007/978-1-4020-9873-4_6,

For instance, oxygen is carried by means of the haemoglobin, a protein contained in the erythrocytes that is also involved in the transport of carbon dioxide (CO_2) from the tissues to the lungs. The defense of the organism is mediated by the immune system, where the leucocytes are the main components. Plasma also transports nutritive substances, including glucose, amino acids, lipoproteins, hormones and ions, as well as excreted material derived mostly from the metabolic activity.

There are different types of cells in blood, which can be grouped in erythrocytes and leucocytes. In addition, blood contains platelets, cell fragments derived from megakaryocytes. Red blood cells or erythrocytes are biconcave cells with a diameter between 6 and 8 μm and 1.5–1.8 μm thick. Erythrocytes are the most numerous blood cells with a concentration in blood ranging 4–6 millions/mm^3. In mammals, including human, erythrocytes are devoid of a nucleus; however, in other vertebrates, including birds, reptilians, amphibians and fishes, erythrocytes have a nucleus. Erythrocytes are involved in the transport of oxygen and, at least partially, CO_2, since is mostly carried by plasma, by means of the protein haemoglobin. Haemoglobin is an iron-containing protein that transports oxygen from the lungs to the body tissues. Haemoglobin is composed of four subunits, each containing a heme (iron-containing) group, which is responsible for oxygen transport. The heme group of haemoglobin binds with oxygen in the lungs releasing it in the body capillaries, where the oxygen pressure is low, and takes CO_2, which is transported back to the lungs to be exhaled. Haemoglobin binds or release oxygen depending on the relative concentration of each gas in its environment, i.e. around the red blood cell (Schroeder 1963).

In addition to lacking nucleus, erythrocytes do not have mitochondria and produce energy through glycolysis, an anaerobic pathway that converts glucose into pyruvate with the production of a relatively small amount of adenosine triphosphate (ATP). It should be noted that glucose uptake by erythrocytes is not regulated by the pancreatic hormone insulin.

Erythrocytes are continuously generated in the red bone marrow or large bones, at a rate of 2 million/s, through a process named erythropoiesis. During embryo and fetal stages erythrocytes are mostly produced in the liver. Erythropoiesis can be stimulated by the renal hormone erythropoietin, the major regulator of red blood cell production that promotes the differentiation and development of erythrocytes and initiates the production of hemoglobin (Moritz et al. 1997). Erythrocytes develop from stem cells to mature erythrocytes in about 7 days and live a total of about 120 days. Immature erythrocytes released by the bone marrow are known as reticulocytes. Reticulocytes differ from mature erythrocytes in that they have a more convoluted shape, and, despite they are devoid of nucleus, they present residual RNA (Koepke and Koepke 1986).

Aged erythrocytes are cleared by phagocytosis in the spleen, liver and bone marrow (Kelton 1987). The aging erythrocyte shows changes in its plasma membrane, making it susceptible to recognition by phagocytes. Several of the cellular components are recirculated after phagocytosis of red cells. This is the case of the heme group of hemoglobin, which is broken down into Fe^{3+} and biliverdin. Iron is recycled associated to transferrin and used in the synthesis of new haemoglobin

(Harding and Stahl 1983; Callus et al. 1996) and biliverdin is further metabolized by reduction into bilirubin, which is released into the plasma and recirculated to the liver. When erythrocytes are hemolyzed hemoglobin is bound to a plasma protein named haptoglobin which is not excreted by the kidney, then saving its components. Cellular heme levels are finely controlled by the balance between heme biosynthesis and catabolism by the enzyme heme oxygenase (Kelton 1987).

Leucocytes or white cells are a different type of blood cells. Leukocytes are a family of cells that belongs the immune system involved in the defense the body against pathogens, tumoral cells and foreign materials. The number of leucocytes in blood is normally between 4×10^3 and 11×10^3 cells per mm^3, making up approximately 1% of blood in a healthy adult (Klenerman et al. 2002). The number of leukocytes in the blood is often used as an indicator of acute or chronic disease. An increase in the leukocyte count in blood is known as leukocytosis. It might occur in response to a wide variety of situations, including viral, bacterial, fungal, or parasitic infection, neoplasia and hemorrhage. Leukocytosis can also be an indication of neoplastic processes concerning leukocytes, named as leukemias, where the number of leukocytes is higher than normal. On the other hand, a reduction in the number of circulating leukocytes in blood is known as leukopenia, which might affect to all types of leukocytes or just to one type.

Several different types of leukocytes can be found in circulation, which are all derived from a multipotent hematopoietic stem cell in the bone marrow. In addition to the bone marrow, lymphocytes can also been produced in other lymphatic tissues. There are two types of leucocytes according to the presence of different staining granules in the cytoplasm. The granulocytes, also known as polymorphonuclear leucocytes, are leucocytes with different granules stained by different colorants, including neutrophils, basophils and eosinophils. Their names arrive from the staining characteristics of their cytosolic granules on hematoxylin and eosin cytological preparations. Neutrophil granules stain a neutral pink whereas basophilic granules stain dark blue and eosinophilic granules stain bright red. On the other hand, agranulocytes, also named mononuclear leucocytes, are characterized by the lack of apparent granules. This class includes lymphocytes and monocytes or macrophages.

Neutrophils are the most common type of leukocytes, comprising about 33–70% of all white blood cells, with mean concentrations of 4.4×10^3 cells/mm^3. Neutrophils are small cells, about 9–10 μm in diameter. Neutrophils are highly dynamic phagocytic cells that constitute the first line of defense of the innate immune system. These cells engulf and degrade bacteria through the discharge of the contents of their abundant cytoplasmic into the phagocytic vacuole containing the pathogens (Segal 2005). Following activation of neutrophils by bacterial by-products or other agents with immunological activity, including glycolipids, lipopolysaccharide and methylated DNA, neutrophils perform a number of specialized functions such as chemotaxis (a process by which neutrophils migrate to sites of bacterial ingress or tissue damage), phagocytosis, and the generation of reactive oxygen species, including superoxide ion (Cicchetti et al. 2002). Neutrophils migrate toward the source of chemoattractants or chemotatic agents, such as bacterial products, C5a, a product of the complement cascade, leukotriene B_4, a products of phospholipid metabolism, and chemokines such as interleukin-8 (Zachariae 1993), in order to eliminate invading pathogens or cellular debris. A single neutrophil can phagocyte between 3 and 20 bacteria before it gets inactivated and dies. Circulating basophils are similar to tissue mast cells and, together with mast cells, are the central players in the allergic response. Immunoglobulin (Ig) E a known factor in allergic inflammation, stimulates the basophils to release histamine, a major mediator in the allergic response associated with inflammation. Basophils are also specialized in the secretion of the anticoagulant agent heparin, bradikinin, a potent but short-lived agent that induces arteriolar dilation and increased capillary permeability, and serotonin (5-hydroxytryptamine), a neurotransmitter and a peripheral signal mediator involved in mediating post-injury vasoconstriction (Bochner and Schleimer 2001).

Basophils are the less abundant leucocytes in blood, representing fewer than 0.5% of blood leukocytes. These cells are associated to inflammatory reactions associated to allergy, Thus upon activation by IgE basophils, as well as mast cells in the tissues, release a number of inflammatory factors, including histamine, bradikinin, serotonin and heparin, mentioned above, that initiate a cascade of vascular and tissue local reactions. Recent studies have also proposed the hypothesis that basophils could contribute to the development of type 2 immunity by releasing interleukin-4, an agent that is important in T cell polarization and the recruitment of other effector cells such as eosinophils or neutrophils (Min and Paul 2008).

Eosinophils are granulocytes produced in the bone marrow before they are release into circulation that represents about 1–6% of white blood cells. Eosinophils are recognized as a proinflammatory leukocytes involved in the defense against parasitic infection and is also implicated in allergic diseases, including bronchial asthma, allergic rhinitis, and atopic dermatitis (Kita and Gleich 1996). Eosinophils act against parasitic infections through the release of a number of agents, including hydrolytic enzymes, reactive oxygen species such as superoxide, and a polypeptide named major basic protein. The major basic protein is a 117-amino acid enzyme with potent activity against helminths, bacteria and mammalian cells, in fact this enzyme is associated to immune hypersensitivity reactions. In addition, eosinophil major basic protein also stimulates the release of histamine from mast cells and basophils, activates neutrophils and alveolar macrophages, and is involved in epithelial cell damage and bronchospasm in asthma (Wasmoen et al. 1988; Vanhaesebroeck and Alessi 2000). In fact, eosinophils have been shown to accumulate in tissues affected by allergic reactions, such as peribronchial tissue in patients suffering from asthma. Eosinophils produce large amounts of the major basic protein, the sulphidopeptide leukotrienes (LTC4/D4 and E4) and platelet activating factor, which are believed to cause bronchospasm during asthma. Furthermore, in support of the role of eosinophils in asthma, it has long been known that one of the most relevant effects of glucocorticoids in asthma is their ability to reduce the blood and airway eosinophilia (Corrigan and Kay 1992; Wardlaw et al. 2000).

Lymphocytes are small cells involved in the immune defense of the body. There are two types of lymphocytes named B lymphocytes (or simply B cells) and T lymphocytes (T cells). In mammals, B cells are produced and matured in the bone marrow, but in birds maturation takes place in an organ called the bursa of Fabricus.

B cells are responsible for the humoral immunity and make antibodies that attack bacteria and toxins. Most B cells differentiate resulting in the so called "plasma cells" that are specialized to produce high amounts of antibodies to fight against infections while a minority of B cells mature into memory B cells. Plasma cells derived from a single B cell produce the same antibody which is directed against the antigen that stimulated the B cell to mature. On the other hand, T cells are responsible of the cellular immunity and attack body cells themselves when they have been infected by viruses or have become tumoral. T cells are produced in the bone marrow and immature T cells migrate to the thymus gland, where they differentiate into various types of mature T cells and become active in response to several factors, including a hormone called thymosin. Lymphocytes secrete a potent set of small proteins, known as lymphokines, such as interleukins or the granulocyte-macrophage colony-stimulating factor, that modulate the functional activities of many cell types, including all cells of the hematopoietic system to regulate their growth and differentiation, and other different cells (Miyajima et al. 1988; Paul 1989).

Finally, circulating monocytes are produced in the bone marrow from hematopoietic stem cells and represents about 5–10% of circulating white blood cells. They are the largest leukocytes (10–20 μm diameter) with phagocytic capacity and differentiation potentials, so that they are thought to be committed precursors in transit from the bone marrow to ultimate sites of activity (Godon 1995). Circulating monocytes differentiate into a variety of tissue phagocytes, such as macrophages, dendritic cells, osteoclasts, cells of microglia (located in the central nervous), and hepatic Kupffer cells (Miyamoto et al. 2001). Peripheral blood monocytes have been demonstrated to participate in the normal tissue renewal and a potential approach for tissue regeneration through cell transplantation therapies using monocyte-derived multipotential cells has been proposed (Kuwana et al. 2003; Seta and Kuwana 2007).

In addition to red and white cells, blood contains cells, or strictly speaking cellular fragments, named platelets, that play an essential role in hemostasia. Platelets are irregularly-shaped, colorless cells that, along with other substances, form clots to stop bleeding. The platelet size is about 20% of the diameter of red blood cells. Platelets were the last blood cell to be identified. They were first detected by Donné (1842), although the existence of platelet plugs was recognized some years latter by Bizzonero (1882), and the same year Hayem (1882) recognized that thrombocytopenia would lead to impaired hemostasis (Gresele et al. 2003).

Platelets are anucleated cells that remain as quite as possible in blood, as they circulate along the intact endothelium, during their 10–12 day lifespan. Ultrastructurally, platelets posses an internal membrane system, named the open canalicular system, that is continuous with, and part of, the plasma membrane and a highly dynamic cytoskeleton, consisting of tubulin microtubules and actin filaments, that support the discoid shape of resting platelets and provide a contractile system involved in shape changes, pseudopod formation, internal contraction and secretion during platelet activation (Mezzano et al. 1984; Fox 2001; Rosado and Sage 2000a,b). Platelets store a number of activators of hemostasia in secretory granules that are classified into two groups on the base of their different ultrastructural properties, dense granules and alpha granules. Dense granules (also called dense bodies) show a highly condensed core that contains serotonin, Ca^{2+}, ATP, ADP, and pyrophosphate, which allows these granules to be readily detected by whole-mount electron microscopy (Dell'Angelica et al. 2000). Alpha granules contain platelet factor IV, von Willebrand factor, thrombospodin, fibrinogen and different immunoglobulins that might take by endocytosis (Harrison and Cramer 1993). Secretion of dense and alpha granules is a critical event in the formation of the hemostatic plug.

For a long time since the discovery of platelets there was considerable disagreement about their origin. Wright's work in the early 1900s using stained sections of tissue and supravital preparations, reported the first clear evidence that platelets derived from megakaryocytes (Wright 1906; Mustard et al. 2001) but it were Humphrey (1955), on the basis of studies with fluorescent antibodies, and Kinoshita and Ohno (1961), developing a technique for placing windows in bone that allowed the direct observation of platelet production from megakaryocytes, who provided compelling pieces of evidence that platelets were derived from a precursor cell known as megakaryocyte. The process of megakaryocyte generation in the bone marrow and subsequent maturation and differentiation leading to platelet production is still little understood. The process is known as megakaryopoiesis and involves a number of unique biological features, including an increase in the nuclear DNA content (endoreplication) and fractionation of cytoplasm and membranes into platelets that will be released into circulation. The primary regulator of platelet generation is thrombopoietin, an acidic glycoprotein produced primarily in the liver, bone marrow and kidney (Kaushansky 2005).

Accumulating evidence indicate that circulating platelets exert a number of vital functions not only associated to wound repair and hemostasia but also in the innate immune response, and metastatic tumor cell biology. In fact, it has been reported, both in experimental animal models of metastatic cancer and patients with tumors, that higher platelet counts in circulation are associated to an unfavorable prognosis (Gupta and Massague 2004) probably associated to adverse cardiovascular events (Thaulow et al. 1991).

6.2 Apoptosis and Leucocytes

Leucocytes are associated to apoptotic events in two possible manners. First of all, these, as well as other, cells can undergo apoptotic processes that determine their life span, and, furthermore, leucocytes, due to their function in the immune system, continuously monitor tissue cells and induce apoptosis in those that undergo significant antigenic changes.

A role for lymphocytes in the development of apoptotic events in cell tissues was first demonstrated in 1977 by Don et al. (1977). In an study performed in a DBA/2 mouse mastocytoma cell line they found that addition of splenic lymphocytes from mice previously immunized against the tumour enhanced the development of apoptosis compared with the addition of lymphocytes from normal (not immunized) mice. Immunized lymphocytes were found to be firmly attached to the surface of the mastocytoma cells, suggesting that cellular immune attack might be the responsible for the induction of apoptosis.

Lymphocytes, including cytotoxic T lymphocytes and natural killer cells, are able to recognize cells infected by viruses as well as transformed cells and induce the development of apoptotic events in them through the release of death receptor–ligands or perforin, a pore-forming 67-kDa protein, expressed and secreted during lymphocyte differentiation (Bolitho et al. 2007). The secretory granules of cytotoxic lymphocytes contain perforin and a number of pro-apoptotic factors that are released to destroy target cells. Perforin has been reported to be important for the release of granzymes, a family of serine proteases that are released by cytotoxic T lymphocytes and natural killer cells to induce apoptosis in target cells. Granzymes have been shown to induce cleaving of caspases, including caspase 3 (Metkar et al. 2003), an effector caspase that activates caspase-activated DNase. In addition, granzymes cleave the pro-apoptotic protein Bid, which recruits other proapoptotic proteins, named Bax and Bak, to the mitochondria, to change the mitochondrial membrane permeability and induce membrane depolarisation, thus inducing the release of cytochrome *c*, one of initial steps in the activation of the caspase cascade. In addition, granzymes have been reported to show some cleaving activity and promote the development of apoptosis in a manner independent of the activation of caspases (Sebbagh et al. 2005).

The importance of perforin and granzymes secretion is revealed by a number of manifestations and diseases. For instance, in patients with Chediak-Higashi, Grisceli type 2 or X-linked lymphoproliferative syndromes, which are associated to defective granule secretion, lymphocytes show impaired cytotoxicity, which results in a reduced immunological capability (Menasche et al. 2005). Similar findings have been reported in perforin- or granzymes-deficient animal models (Bots and Medema 2006).

Lymphocytes have been demonstrated to induce apoptosis in macrophages a process that might be involved in a mechanism by which immunologically stimulated cells are removed to reduce immune-mediated injury of normal tissues. Lymphocyte-induced apoptosis in macrophages has been reported to be mediated by the release of cytokines that might directly induce endogenous nitric oxide synthase in macrophages, resulting in excessive production of nitric oxide and the development of apoptosis pathways in macrophages (Choi et al. 2008).

A number of studies have provided evidence that leucocytes are involved in the inflammatory process associated with alcohol-induced liver injury. Alcohol-associated liver damage is a consequence of the immunologic response of the liver to cellular injury, which plays a key role in the pathogenesis of chronic liver disease in alcoholics. This effect is mostly mediated through the secretion of cytokines, especially TNF alpha, by macrophages/Kupffer cells, which is able to cause liver cell apoptosis through the TNF alpha receptor or Fas/CD95 which is expressed by liver cells (Batey and Wang 2002). In addition, interleukin 1, interleukin 6 and a number of chemokines, facilitate the development of inflammatory processes of the liver during alcohol-derived cell damage (Laso et al. 2005).

In addition, it has long been demonstrated that lymphocytes can recognize cells undergoing apoptosis. An early study in thymocytes reported that macrophages bind to apoptotic thymocytes, a phenomenon that involves an interaction between a thymocyte carbohydrate moiety and a lectin-like macrophage surface receptor (Duvall et al. 1985). Cells undergoing apoptosis have been reported to be recognized and phagocytosed by macrophages before their degradation, thus reducing the inflammatory reaction and protecting tissues from the damaging effects that the secretion of inflammatory factors might induce (Falasca et al. 1996).

6.2.1 Apoptotic Agents in Leucocytes

Among the earliest cell death events resembling apoptosis was described in human glucocorticoid-sensitive and -insensitive lymphoid cell lines. Exposure to different concentrations of glucocorticoids induces a number of morphological changes in dying cells, including rounding up of cells, condensation of nuclear chromatin and detachment of chromatin from the nuclear matrix, although the intracellular events leading to cell death differ from those reported for apoptotic phenomena (Blewitt et al. 1983). In mature human T lymphocytes apoptosis induced by dexamethasone, a synthetic glucocorticosteroid, was reversed either by progression in the cell cycle or by occupancy of the T cell receptor, thus suggesting that glucocorticoids are involved in the regulation of the proliferative or apoptotic response of antigen-activated human T lymphocytes (Tuosto et al. 1994). By the same time Migliorati et al. (1994) reported that dexamethasone-induced apoptosis was prevented by treatment with interleukin-2 and interluekin-4, a process that involves, at least for interleukin-4, induction of c-jun. Interestingly, dexamethasone (methyl in position 16 alpha) has been shown to be a more efficient apoptotic inducer than betamethasone (methyl in position 16 beta) or triamcinolone (hydroxyl in position 16), thus suggesting that structural differences at position 16 of the steroid nucleus is associated with a different pro-apoptotic activity by glucocorticoids (Perrin-Wolff et al. 1995).

Lymphocytes can also undergo apoptosis when exposed to ultraviolet radiation. In 1968, a study by Binet et al. (1968) presented a therapeutic strategy to treat patients with chronic lymphatic leukaemia by exposing their blood to short-wave ultraviolet radiation in an extracorporeal circulation. Since a similar treatment was found to induce lymphopenia in previously healthy goats, further studies were performed to investigate this approach, in particular the effects on lymphocyte functions. In 1983, Gunn et al. (1983) reported that exposure to short-wave (254 nm) ultraviolet radiation decreased the response of mononuclear cells to mitogen stimulation. The irradiated lymphocytes underwent apoptosis and necrosis during the first day of tissue culture both in absence or presence of mitogen stimulation. A more recent study has revealed that the oxygen radical scavenger trolox protects cells from irradiation as demonstrated by the increased viability and thymidine incorporation observed by a postirradiation incubation with trolox. These findings indicate oxidation events triggered by radiation are responsible for apoptosis induced in the human lymphocytic leukemia MOLT-4 cell line (McClain et al. 1995).

The fungal toxin and immunomodulating agent gliotoxin, which is able to attenuate proliferation of T and B cells, has been reported to induce apoptotic events, including DNA fragmentation, in macrophages, T and B cells. This phenomenon has been attributed to the synthesis of IP_3 (Waring 1990), which is likely associated to the generation of Ca^{2+} signals. Different chemical agents used in medicine have been shown to induce apoptotic effects in leucocytes resulting in suppression of the immune response. This is the case of certain volatile anesthetics can modulate lymphocyte function during surgery, which might have negative effects in the postoperative immune response. The anesthetics sevoflurane and isoflurane have been demonstrated to induce apoptosis in T lymphocytes through an increased mitochondrial membrane permeability and caspase 3 activation, but independently of death receptor signaling (Loop et al. 2005). Antitumoral drugs, such as hydroxyurea or adriamycin, have also been shown to induce apoptosis in lymphocytes obtained from chronic myelogenous leukemia patients. These agents stimulate a number of morphological changes, including membrane blebbing, cell shrinkage, chromatin condensation, and DNA fragmentation, which are all characteristics of the apoptotic phenomenon (Anand et al. 1995). Long-term administration of macrolides, including clarithromycin, azithromycin or josamycin, leads to a reduction in the number of lymphocytes by inducing apoptosis through Fas-Fas ligand pathway (Ishimatsu et al. 2004). In addition, clarithromycin and azithromycin have been demonstrated to enhance apoptosis of activated lymphocytes by down-regulation of Bcl-xL (Mizunoe et al. 2004).

In addition to radiation or chemical compounds, biological agents have also been reported to induce apoptosis in lymphocytes. The Newcastle disease virus has been found to induce both apoptotic and necrotic in chicken peripheral blood lymphocytes as compared with controls. Infected cells showed a number of morphological and biochemical changes characteristics of apoptosis, including apoptotic bodies, margination of chromatin and extensive DNA fragmentation (Lam and Vasconcelos 1994). The human immunodeficiency virus type 1 (HIV-1) infection in humans has been associated with rapid turnover of $CD4^+$ T cells due to a number of factors, including the induction of lymphocyte apoptosis (Ameisen and Capron 1991). The apoptotic actions of HIV-1 have been demonstrated to be enhanced by coinfection with the hepatitis C virus (Núñez et al. 2006). Macrophages have been reported as a reservoir for HIV-1, which induces apoptosis and cell death in these cells through the activation of the phosphatidylinositol 3-kinase/Akt/FOXO3 pathway (Cui et al. 2008). The influenza A virus has also been shown to induce apoptosis in peripheral blood monocytes and lymphocytes, a phenomenon that might underlie the transient but severe leucopenia detected after virus infection. Leucopenia due to apoptosis after influenza A virus infection might be mediated by induction of cytokine stimulation, viral induction of the Fas-FasL signalling (Nichols et al. 2001). A similar apoptotic action has been reported the rabies virus. Annexin V-fluorescein isothiocyanate staining of mice splenocytes and thymocytes revealed apoptotic events in these cells after day 6 of infection, which occurred by the same time as rabies virus antigen become detectable in the brain. Thus it has been suggested that the infection of neurons might trigger apoptosis of lymphoid cells leading to suppression of the immune system (Kasempimolporn et al. 2001).

By contrast, many viruses interfere with programmed cell death, preventing apoptosis of infected cells as a direct effect of viral infection. This is the case of the Epstein-Barr virus infects human B lymphocytes maintaining cell division. As a result, the lymphoblastoid cell lines derived from Epstein-Barr virus infection are relatively resistant to apoptosis (Spender et al. 1999). In addition, the herpes simplex viruses-1 inhibits the oligonucleosomal DNA fragmentation characteristic of apoptosis induced by $CD8^+$ cytotoxic T lymphocytes, without having any significant effect on the membrane manifestations of apoptosis, such as phosphatidylserine exposure. This effect of herpes simplex viruses-1 infection might be involved in the mechanism of immune suppression induced by the virus (Jerome et al. 1998).

In addition, a number of pathological cardiovascular situations have a direct relationship with apoptotic processes. This is the case of advanced atherosclerosis, where apoptosis of macrophages, due to stressed endoplasmic reticulum, involving the proteins *S*ignal *T*ransducer and *A*ctivator of *T*ranscription-1 (STAT1) and calmodulin kinase II and changes in cytosolic Ca^{2+} concentration ($[Ca^{2+}]_c$; Lim et al. 2008), contributes to the development of plaque necrosis (Tabas 2005). Plaque necrosis, in turn, induces plaque disruption and arterial thrombosis, which are the proximate causes of acute cardiovascular events (Tabas 2005; Schrijvers et al. 2007). Macrophage apoptosis occurs at all the stages of atherosclerosis. Its consequences on lesion progression mostly depend on how efficiently apoptotic cells are cleared by phagocytes, mainly macrophages. In the early stages apoptotic cells are efficiently cleared but clearance of apoptotic macrophages in advanced lesions is reduced, leading to necrosis, inflammation, and thrombosis, which are thought to play a key role in the clinical consequences (Tabas 2005).

6.3 Eryptosis

Eryptosis or apoptosis-like death of erythrocytes is a phenomenon characterized by phosphatidylserine exposure and erythrocyte shrinkage, which are both typical features of apoptosis in nucleated cells. Erythrocyte injury such as osmotic shock (Lang et al. 2006), oxidative stress or energy depletion induces the activation of cyclooxygenase, which, in turn, leads to the synthesis of prostaglandin E_2 (Lang et al. 2005). Prostaglandin E_2 activates a Ca^{2+} permeable cation channel thus increasing $[Ca^{2+}]_c$ and activating a number of Ca^{2+}-dependent processes, including Ca^{2+} sensitive K^+ channels, thus leading to hyperpolarization and cell shrinkage. The rise in $[Ca^{2+}]_c$ activates the enzyme scramblase, a protein responsible for the transport of phospholipids between the two monolayers of the cell membrane (Zwaal et al. 2005), shifting phosphatidylserine from the inner to the outer layer of the plasma membrane. Scramblase has been shown to be sensitized by ceramide, which is produced by sphingomyelinase following osmotic stress. Phosphatidylserine externalization is a common feature of apoptosis that induce macrophages to engulf and degrade the affected red blood cells. Furthermore, erythrocytes exposing phosphatidylserine may adhere to the vascular wall and thus interfere with microcirculation (Lang et al. 2006). Among the channels involved in ion leakage, resulting in Ca^{2+} and N^{+} entry, during eryptosis, TRPC6 has been show to play a relevant role as revealed in studies performed in $TRPC6^{-/-}$ erythrocytes, where $[Ca^{2+}]_c$, cell-shrinkage, and phosphatidylserine externalization were significantly smaller than in wildtype cells (Föller et al. 2008). TRPC6 proteins have been reported to form both store-operated and diacylglycerol-activated non-capacitative Ca^{2+} permeable channels (Jardin et al. 2008).

The hyperosmotic conditions that exist in the renal inner medulla might be expected to induce eryptosis due to osmotic stress; however, the time erythrocytes are exposed to the environment of renal medulla is usually too short to induce eryptosis (Lang et al. 2006). Eryptosis is inhibited by erythropoietin, which protect red blood cells from apoptosis, thus extending the life span of circulating erythrocytes (Polenakovic and Sikole 1996).The relevance of the anti-apoptotic actions of erythropoietin is revealed by the observation that the number of erythrocytes exposing phosphatidylserine in circulation in patients with end stage renal disease has been shown to be significantly decreased after 3 h of treatment with erythropoietin (Myssina et al. 2003).

Eryptosis might be involved in physiological processes, such as the limitation of erythrocyte survival. Erythrocyte ageing, as well as oxidative stress, is associated to an increase of $[Ca^{2+}]_c$ (Romero and Romero 1999; Damonte et al. 1992), which accelerate erythrocyte death by eryptosis and clearance from circulation.

In addition to osmotic or oxidative stress, other agents and conditions trigger premature eryptosis thus favouring the development of anaemia. This is the case of the cytotoxic drug cisplatin, a platinum-based chemotherapy compound used for the treatment of several types of cancers, including sarcomas, some carcinomas, lymphomas and germ cell tumours, induces suicidal death or eryptosis of red blood cells, leading to the development of anaemia. Phosphatidilserine exposure after cisplatin administration has been shown to be reduced in the absence of extracellular Ca^{2+}. Cisplatin moderately decreased the cellular concentration of ATP, which, as mentioned above, is well known to induce eryptosis (Mahmud et al. 2008). A similar effect has been reported for azathioprine, a widely used immunosuppressive drug that has been shown to induce anaemia. Azathioprine-induced anaemia has long been attributed to bone marrow suppression; however, recent studies have provided evidence supporting that azathioprine administration might result in anaemia by accelerating eryptosis (Geiger et al. 2008). Zn^{2+} has also been reported to induce eryptosis through the stimulation of ceramide formation and an increase of $[Ca^{2+}]_c$ (Kiedaisch et al. 2008), as well as gold, which has been shown to induce eryptosis through the elevation in $[Ca^{2+}]_c$ (Sopjani et al. 2008). Vanadate, which has been shown to interfere with a wide variety of enzymes including Ca^{2+} ATPases (Rosado et al. 2001), induces eryptosis probably mediated through increase of $[Ca^{2+}]_c$. The effect of vanadate might contribute to the development of anaemia in chronic renal failure, where vanadate has been shown to be accumulated in blood due to reduced excretion (Föller et al. 2008). Retinoid acid, have previously been shown to confer protection against malaria by fostering the phagocytosis of parasitized erythrocytes, trigger phosphatididylserine exposure and cell shrinkage of erythrocytes, typical characteristics of eryptosis. This effect is mimicked by the specific retinoic acid receptor agonist 4-(E-2-[5,6,7,8-tetrahydro-5,5,8,8-tetramethyl-2-naphthalenyl]-1-propenyl) benzoic acid. Eryptosis induced by retinoic acid might explain the accelerated clearance of parasitized erythrocytes from circulating blood following treatment with retinoids (Niemoeller et al. 2008).

6.4 Apoptosis in Platelets

Platelets are small cellular fragments of approximately $0.5 \times 3.0\,\mu m$ that respond to a number of stimuli with changes in the shape from a smooth disc to an irregular form with multiple projections. Shape change, as well as other intracellular pathways, is mediated by reorganisation of the actin cytoskeleton, which is a highly dynamic cellular component that consists of two major structures: a cytoplasmic actin network and a membrane-associated cytoskeleton (Ma and Abrams 1999).

Despite they are cellular fragments, platelets express a number of components of the apoptotic machinery, including caspases-3, and -9 (Shcherbina and Remold-O'Donnel 1999; Wolf et al. 1999; Leytin et al. 2006; Ben Amor et al. 2006a,b). Platelets also express caspases-8 and 10; however, caspase 12 has not been found in platelets, despite it is highly expressed in mature megakaryocytes (Kerrigan et al. 2004). Several proapoptotic proteins, including Bid, a member of the "BH3 domain only" subclass of the Bcl-2 family proposed to connect surface death receptors with Bcl-2 or Bax (Wang et al. 1996), the protein Bax (Dale and Friese 2006; Leytin et al. 2006) and Bak (Mason et al. 2007); however only a weak expression of Bcl-2 has been reported in human platelets (Leytin et al. 2006).

Apoptotic events in platelets share several pathways with those reported in nucleated cells. As reported in previous chapters, pro-apoptotic proteins have been reported to translocate to mitochondria where they insert as apparent integral membrane proteins (Gross et al. 1998) inducing permeabilisation of the outer mitochondrial membrane and releasing multiple intermembrane space proteins (Kluck et al. 1999), including cytochrome *c* (Crompton 1999). Cytochrome *c* release results in activation of initiator caspase 9 and subsequent activation of effector caspases, leading to phosphatidylserine exposure.

In human platelets the physiological agonist thrombin has been shown to activate several apoptotic events, including enhanced expression of pro-apoptotic Bax and Bak proteins associated to translocation of Bid, Bax and Bak to the mitochondria, reduction in the expression of the anti-apoptotic Bcl-2 protein, mitochondrial inner transmembrane potential depolarization, cytochrome *c* release, caspase 9 activation and subsequent caspase 3 activation and phosphatidylserine externalization (Leytin et al. 2006; Lopez et al. 2007, 2008). Activation of caspases-3 and -9 by thrombin requires their translocation to the newly polymerized actin filament network, a process that depends on functional protein kinase C but is independent on rises in

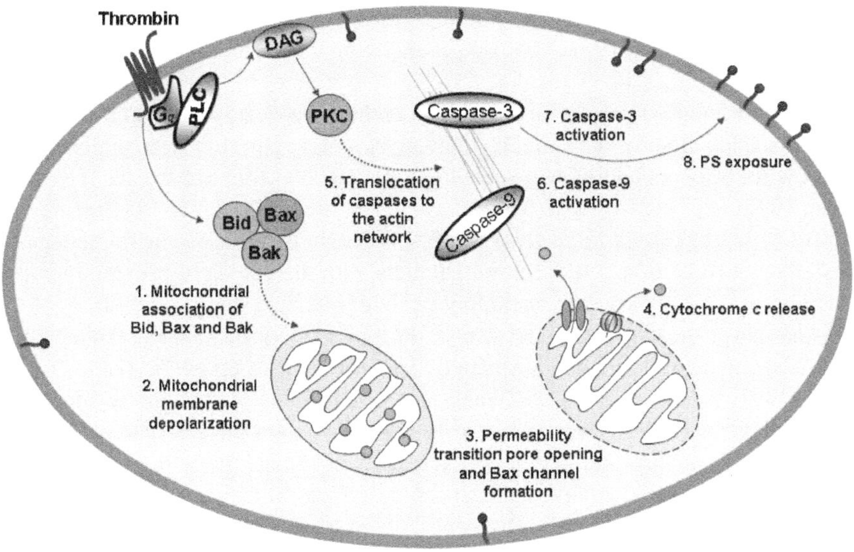

Fig. 6.1 Schematic representation of the apoptotic events activated by thrombin in human platelets. Platelet stimulation with the physiological agonist thrombin results in mitochondrial association of the pro-apoptotic proteins Bid, Bax and Bak, which might lead to the formation and opening of the Bax channel and the permeability transition pore, resulting in mitochondrial membrane depolarization and cytochrome *c* release. Released cytochrome *c* induces the activation of caspases-9, which, in turn, activates the executioner caspase 3, a process that requires association of caspases to the newly polymerized actin cytoskeleton. Caspase activation results in phosphatidylserine externalization and a number of morphological changes characteristics of apoptosis (See also Plate 9 in the Color Plate Section on page 230)

$[Ca^{2+}]_i$ (Fig. 6.1; Ben Amor et al. 2006a,b). A number of apoptotic events have also been observed after stimulation with ADP but not after treatment with the thromboxane A2 analogue, U46619, which has been reported to induce "apoptotic-like events" in platelets in the absence of phosphatidylserine externalization (Tonon et al. 2002). Apoptosis in platelets induces the release of microparticles, platelet membrane fractions, which play a role in the processes of inflammation, coagulation and vascular function, involved in the development of cardiovascular diseases (VanWijk et al. 2003).

Microparticles were first reported by Wolf (1967) as "platelet dust". Microparticles are membrane fractions derived from platelets and other cell types, including endothelial cells, vascular smooth muscle cells, leukocytes and erythrocytes, smaller than 0.1 μm in diameter, which suspended in plasma are able to promote coagulation (VanWijk et al. 2003). Microparticles have been reported to be involved both in physiological and pathological processes. For instance, they play an important role in inflammation, coagulation and vascular function. Two major processes have been involved in the formation of microparticles, cell activation and apoptosis, with clear differences in the processes thought to be involved in microparticle formation depending on the origin. Platelets activated by thrombin, the Ca^{2+} ionophore A23187,

ADP plus collagen or shear stress can generate and release microparticles (Sims et al. 1989; Tans et al. 1991; Miyazaki et al. 1996; Taube et al. 1999). In these cells, microparticle formation is associated to the activation of the glycoprotein IIb–IIIa, the main fibrinogen receptor on the platelet surface. Especially relevant is the amino acid sequence arg–gly–asp (RGD) of the glycoprotein IIb–IIIa, thus, platelet treatment with RGD-containing peptides blocks the release of microparticles (Gemmell et al. 1993).

Apoptosis has an especial interest in the process of megakaryocyte fragmentation and platelet formation, which shows parallels with the course of programmed cell death in a number of events, including ruffling, blebbing and condensation of the plasma and nuclear membranes, remodelling of the cytoskeletal network, typical DNA fragmentation and cell shrinkage and formation of apoptotic bodies (Zauli et al. 1997; Ogilvy et al. 1999; Wang and Wang 1999). Although the molecular and cellular mechanisms through which megakaryocytes differentiate and mature remain poorly understood apoptotic-like events have been presented as necessary for platelet formation. Apoptosis has been described in mature megakaryocytes in bone marrow, although most cells appear to round up rather than undergo fragmentation (Radley and Haller 1983). One of the major regulators of cytoskeletal organization and apoptosis is the protein Bcl-xL (Boise et al. 1993). Bcl-xl is the dominant regulator of apoptosis, with cell death repressor activity, whose level has been reported to be reduced in mature cultured megakaryocytes (Sanz et al. 2001). Bcl-xL levels has been shown to be finely regulated in mature megakaryocytes in vivo, thus, deregulated, high level expression of BclxL impairs the ability of the cells to fragment into platelets (Kaluzhny et al. 2002). In addition, expression of Bcl-2, another anti-apoptotic protein, has been shown to be decreased in a megakaryoblastic cell line (Sanz et al. 2001).

Some authors have suggested that megakaryocyte apoptosis does not occur until full megakaryocyte maturation occurs (Falcieri et al. 2000; Sanz et al. 2001), while others have reported that megakaryocyte apoptosis and maturation are intimately associated (Zauli et al. 1997; De Botton et al. 2002; Yang et al. 2002). In fact, it has been demonstrated that addition of caspase inhibitors delays platelet formation and postpones megakaryocyte polyploidization (De Botton et al. 2002; Yang et al. 2002). Furthermore, irregular patterns of megakaryocyte apoptosis have been shown to be linked with megakaryocyte-associated diseases including immune thrombocytopenic purpura (Houwerzijl et al. 2004).

Recent studies have suggested that stimulation of platelets by agonists results in the activation of a number of proteins of the apoptotic machinery, leading to physiological functions rather than to the development of apoptotic events, that are required for full platelet activation. Studies in our laboratory have reported that the physiological agonist thrombin induces a bimodal activation of the pro-apoptotic caspase 3 at a wide range of agonist concentrations from 0.01 to 1 U/ml. The early caspase 3 activation was found within 1 min of stimulation with thrombin at low concentrations (0.01 U/ml) or 1 U/ml. The early increase in caspase 3 activity was found to be independently of the activation of the upstream caspases-8, -9, and -10 and mitochondrial-dependent apoptosis. In contrast, the delayed increase in caspase

3 activity, which reached a maximum after 1 h of stimulation with 1 U/mL thrombin is more likely an apoptotic event that occurs downstream of caspase 9 activation and mitochondrial cytochrome *c* release. Apoptotic (delayed) caspase 3 activation was not observed when platelets were stimulated with low concentrations of thrombin. In addition, phosphatidylserine externalization after treatment with thrombin was associated to the delayed activation of caspase 3, thus paralleling the activation of caspase 9, and was not detected when platelets were stimulated with low concentrations of thrombin (Rosado et al. 2006). In agreement, studies by other groups have reported that thrombin induces platelet activation and apoptosis at different concentrations, thus at high thrombin concentrations generated during blood coagulation a number of platelets became apoptotic, which indicates that hypercoagulable states may be associated with increased numbers of apoptotic platelets (Leytin et al. 2007). Early caspase 3 activation in human platelets was found to require protein kinase C activity, a serine/threonine kinase that has been shown to amplify apoptotic signalling via activation of the caspase cascade (Kanthasamy et al. 2003). In addition, inhibition of PKC has also been reported to attenuate apoptosis induced by different factors (Imamdi et al. 2004; Zhou et al. 2005). We further observed that thrombin-induced early caspase 3 activation is required for platelet aggregation. The role of early caspase 3 activation by agonists in cellular functions was further demonstrated in the non-related pancreatic acinar cells, where impairment of the activation of caspase 3 attenuated amylase secretion stimulated by the physiological agonist cholecystokinin octapeptide (Rosado et al. 2006). In addition, a role for caspases in non-apoptotic processes has been demonstrated in different cell types, including neurones, where activation of caspase 3 can be measured in normally functioning cells. There is a body of evidence supporting the possibility that caspases, including caspase 3, may be involved in synaptic plasticity modulation in normal brain functioning. For instance, caspase 3 is likely to be involved in the long-term potentiation phenomenon (Gulyaeva 2003). Furthermore, caspase 3 enzymatic activity has been shown to be detected in astrocytes at 10 h and 1 day excitotoxic damage in postnatal rats. Caspase 3 is co-localized with caspase-cleaved fragments of glial fibrillary acidic protein (CCP-GFAP). At longer survival times, when astroglial hypertrophy was observed, caspase 3 activation has not been found to be correlated with GFAP cleavage, but instead was associated with de novo expression of the cytoskeletal protein vimentin (Acarin et al. 2007). In ovarian cancer cells the extracellular matrix protein laminin-10/11, through the interaction with the integrin β_1, triggers the cleavage of $iPLA_2$ via a caspase 3 dependent pathway, resulting in the formation of truncated active $iPLA_2$. The activation of $iPLA_2$ increases the production of lysophosphatidic acid, which plays a an anti-apoptotic role and allows cell survival, and arachidonic acid, which is involved in cell migration (Zhao et al. 2006).

Therefore, apoptosis plays an important role in platelet biology from platelet generation to death, both mediating platelet formation and limiting platelet life span. Especially relevant is the antagonistic balance between Bcl-xL and Bak in the delimitation of platelet life span as demonstrated by genetic ablation or pharmacological inactivation of Bcl-xL, which has been shown to reduce platelet half-life and induce thrombocytopenia in mice. This effect is overcome by deletion of Bak. In support of the role of Bak in limiting life span, platelets from Bak-deficient mice live longer than normal. Thus, there is a body of evidence supporting that platelets, if not involved in haemostatic processes, are genetically programmed to die by apoptosis (Mason et al. 2007). In addition, as demonstrated in other cell types, a number of proteins of the apoptotic machinery appear to be involved in physiological platelet processes.

Acknowledgments This work was supported by grant BFU2007-60104 funded by the Spanish Ministry of Science and Innovation.

References

Acarin L, Villapol S, Faiz M et al. (2007) Caspase-3 activation in astrocytes following postnatal excitotoxic damage correlates with cytoskeletal remodeling but not with cell death or proliferation. Glia 55:954–965

Ameisen JC, Capron A (1991) Cell dysfunction and depletion in AIDS: the programmed cell death hypothesis. Immunol Today 12:102–105

Anand S, Verma H, Kumar L et al. (1995) Induction of apoptosis in chronic myelogenous leukemia lymphocytes by hydroxyurea and adriamycin. Cancer Lett 88:101–105

Batey RG, Wang J (2002) Molecular pathogenesis of T lymphocyte-induced liver injury in alcoholic hepatitis. Front Biosci 7:1662–1675

Ben Amor N, Pariente JA, Salido GM et al. (2006a) Caspases 3 and 9 are translocated to the cytoskeleton and activated by thrombin in human platelets. Evidence for the involvement of PKC and the actin filament polymerization. Cell Signal 18:1252–1261

Ben Amor N, Pariente JA, Salido GM et al. (2006b) Thrombin-induced caspases 3 and 9 translocation to the cytoskeleton is independent of changes in cytosolic calcium in human platelets. Blood Cells Mol Dis 36:392–401

Binet JL, Villeneuve B, van Rapenbusch R et al. (1968) Extracorporeal radiation of the blood by ultraviolet rays. Nouv Rev Fr Hematol 8:733–744

Bizzonero J (1882) Über einen formbestandtheil des blutes und dessen rolle bei der thrombose und der blutgerinnung. Virchow's Arch Path Anat Physiol Klin Med 90:261–332

Blewitt RW, Abbott AC, Bird CC (1983) Mode of cell death induced in human lymphoid cells by high and low doses of glucocorticoid. Br J Cancer 47:477–486

Bochner BS, Schleimer RP (2001) Mast cells, basophils, and eosinophils: distinct but overlapping pathways for recruitment. Immunol Rev 179:5–15

Boise LH, Gonzalez-Garcia M, Postema CE et al. (1993) Bcl-x, a bcl-2-related gene that functions as a dominant regulator of apoptotic cell death. Cell 74:597–608

Bolitho P, Voskoboinik I, Trapani JA et al. (2007) Apoptosis induced by the lymphocyte effector molecule perforin. Curr Opin Immunol 19:339–347

Bots M, Medema JP (2006) Granzymes at a glance. J Cell Sci 119:5011–5014

Callus BA, Iacopetta BJ, Kühn LC et al. (1996) Effects of overexpression of the transferrin receptor on the rates of transferring recycling and uptake of non-transferrin-bound iron. Eur J Biochem 238:463–469

Cicchetti G, Allen PG, Glogauer M (2002) Chemotactic signaling pathways in neutrophils: from receptor to actin assembly. Crit Rev Oral Biol Med 13:220–228

Choi KS, Song EK, Yim CY (2008) Cytokines secreted by IL-2-activated lymphocytes induce endogenous nitric oxide synthesis and apoptosis in macrophages. J Leukoc Biol doi: 10.1189/jlb.1007701

Corrigan CJ, Kay AB (1992) T cells and eosinophils in the pathogenesis of asthma. Immunol Today 13:501–507
Crompton M (1999) The mitochondrial permeability transition pore and its role in cell death. Biochem J 341:233–249
Cui M, Huang Y, Zhao Y et al. (2008) Transcription factor FOXO3a mediates apoptosis in HIV-1-infected macrophages. J Immunol 180:898–906
Dale GL, Friese P (2006) Bax activators potentiate coated-platelet formation. J Thromb Haemost 4:2664–2669
Damonte G, Guida L, Sdraffa A (1992) Mechanisms of perturbation of erythrocyte calcium homeostasis in favism. Cell Calcium 13:649–658
De Botton S, Sabri S, Daugas E et al. (2002) Platelet formation is the consequence of caspase activation within megakaryocytes. Blood 100:1310–1317
Dell'Angelica EC, Mullins C, Caplan S et al. (2000) Lysosome-related organelles. FASEB J 14:1265–1278
Don MM, Ablett G, Bishop CJ et al. (1977) Death of cells by apoptosis following attachment of specifically allergized lymphocytes in vitro. Aust J Exp Biol Med Sci 55:407–417
Donné A (1842) De l'origine des globules du sang, de leur mode de formation et luer fin. CR Acad Sci 14:366
Duvall E, Wyllie AH, Morris RG (1985) Macrophage recognition of cells undergoing programmed cell death (apoptosis). Immunology 56:351–358
Falasca L, Bergamini A, Serafino A et al. (1996) Human Kupffer cell recognition and phagocytosis of apoptotic peripheral blood lymphocytes. Exp Cell Res 224:152–162
Falcieri E, Bassini A, Pierpaoli S et al. (2000) Ultrastructural characterization of maturation, platelet release, and senescence of human cultured megakaryocytes. Anat Rec 258:90–99
Föller M, Kasinathan RS, Koka S et al. (2008) TRPC6 contributes to the Ca^{2+} leak of human erythrocytes. Cell Physiol Biochem 21:183–192
Föller M, Sopjani M, Mahmud H et al. (2008) Vanadate-induced suicidal erythrocyte death. Kidney Blood Press Res 31:87–93
Fox JE (2001) Cytoskeletal proteins and platelet signaling. Thromb Haemost 86:198–213
Geiger C, Föller M, Herrlinger KR et al. (2008) Azathioprine-induced suicidal erythrocyte death. Inflamm Bowel Dis doi:10.1002/ibd.20433
Gemmell CH, Sefton MV, Yeo EL (1993) Platelet-derived microparticle formation involves glycoprotein IIb–IIIa. Inhibition by RGDS and a Glanzmann's thrombasthenia defect. J Biol Chem 268:14586–14589
Godon S (1995) The macrophage. Bioessays 17:977–986
Gresele P, Page C, Fuster V et al. (2003) Platelets in thrombotic and nonthrombotic disorders – pathophysiology, pharmacology and therapeutics. Leukemia 17:478–479
Gross A, Jockel J, Wei MC et al. (1998) Enforced dimerization of BAX results in its translocation, mitochondrial dysfunction and apoptosis. EMBO J 17:3878–3885
Gulyaeva NV (2003) Non-apoptotic functions of caspase-3 in nervous tissue. Biochemistry (Mosc) 68:1171–1180
Gunn A, Scrimgeour D, Potts RC et al. (1983) The destruction of peripheral-blood lymphocytes by extracorporeal exposure to ultraviolet radiation. Immunology 50:477–485
Gupta GP, Massague J (2004) Platelets and metastasis revisited: a novel fatty link. J Clin Invest 114:1691–1693
Harding C, Stahl P (1983) Transferrin recycling in reticulocytes: pH and iron are important determinants of ligand binding and processing. Biochem Biophys Res Commun 113:650–658
Harrison P, Cramer EM (1993) Platelet alpha-granules. Blood Rev 7:52–62
Hayem G (1882) Sur le méchanisme de l'arrêt des hémorrhagies. CR Acad Sci 95:18
Houwerzijl EJ, Blom NR, van der Want JJL et al. (2004) Ultrastructural study shows morphologic features of apoptosis and para-apoptosis in megakaryocytes from patients with idiopathic thrombocytopenic purpura. Blood 103:500–506

Humphrey JH (1955) Origin of blood platelets. Nature 176:38
Imamdi R, de Graauw M, van de Water B (2004) Protein kinase C mediates cisplatin-induced loss of adherens junctions followed by apoptosis of renal proximal tubular epithelial cells. J Pharmacol Exp Ther 311:892–903
Ishimatsu Y, Kadota J, Iwashita T et al. (2004) Macrolide antibiotics induce apoptosis of human peripheral lymphocytes in vitro. Int J Antimicrob Agents 24:247–253
Jardin I, Redondo PC, Salido GM et al. (2008) Phosphatidylinositol 4,5-bisphosphate enhances store-operated calcium entry through hTRPC6 channel in human platelets. Biochim Biophys Acta 1783:84–97
Jerome KR, Tait JF, Koelle DM et al. (1998) Herpes simplex virus type 1 renders infected cells resistant to cytotoxic T-lymphocyte-induced apoptosis. J Virol 72:436–441
Kaluzhny Y, Yu G, Sun S et al. (2002) BclxL overexpression in megakaryocytes leads to impaired platelet fragmentation. Blood 100:1670–1678
Kanthasamy AG, Kitazawa M, Kanthasamy A et al. (2003) Role of proteolytic activation of protein kinase C delta in oxidative stress-induced apoptosis. Antioxid Redox Signal 5:609–620
Kasempimolporn S, Tirawatnapong T, Saengseesom W et al. (2001) Immunosuppression in rabies virus infection mediated by lymphocyte apoptosis. Jpn J Infect Dis 54:144–147
Kaushansky K (2005) The molecular mechanisms that control thrombopoiesis. J Clin Invest 115:3339–3347
Kelton JG (1987) Platelet and red cell clearance is determined by the interaction of the IgG and complement on the cells and the activity of the reticuloendothelial system. Transfus Med Rev 1:75–84
Kerrigan SW, Gaur M, Murphy RP et al. (2004) Caspase-12: a developmental link between G-protein-coupled receptors and integrin alphaIIbbeta3 activation. Blood 104:1327–1334
Kiedaisch V, Akel A, Niemoeller OM et al. (2008) Zinc-induced suicidal erythrocyte death. Am J Clin Nutr 87:1530–1534
Kinoshita R, Ohno S (1961) Biodynamics of thrombopoiesis. In: Blood platelets. Henry Ford Hospital International Symposium 1960, Little, Brown & Co, Boston
Kita H, Gleich GJ (1996) Chemokines active on eosinophils: potential roles in allergic inflammation. J Exp Med 183:2421–2426
Klenerman P, Cerundolo V, Dunbar PR (2002) Tracking T cells with tetramers: new tales from new tools. Nat Rev Immunol 2:263–272
Kluck RM, Esposti MD, Perkins G et al. (1999) The pro-apoptotic proteins, Bid and Bax, cause a limited permeabilization of the mitochondrial outer membrane that is enhanced by cytosol. J Cell Biol 147:809–822
Koepke JF, Koepke JA (1986) Reticulocytes. Clin Lab Haematol 8:169–179
Kuwana M, Okazaki Y, Kodama H et al. (2003) Human circulating $CD14^+$ monocytes as a source of progenitors that exhibit mesenchymal cell differentiation. J Leukoc Biol 74:833–845
Lam KM, Vasconcelos AC (1994) Newcastle disease virus-induced apoptosis in chicken peripheral blood lymphocytes. Vet Immunol Immunopathol 44:45–56
Lang F, Lang KS, Lang PA et al. (2006) Osmotic shock-induced suicidal death of erythrocytes. Acta Physiol (Oxf) 187:191–198
Lang KS, Lang PA, Bauer C et al. (2005) Mechanisms of suicidal erythrocyte death. Cell Physiol Biochem 15:195–202
Laso FJ, Pastor I, Orfao A (2005) Immune system and alcoholic liver disease. Med Clin (Barc) 125:263–269
Leytin V, Allen DJ, Lyubimov E et al. (2007) Higher thrombin concentrations are required to induce platelet apoptosis than to induce platelet activation. Br J Haematol 136:762–764
Leytin V, Allen DJ, Mykhaylov S et al. (2006) Thrombin-triggered platelet apoptosis. J Thromb Haemost 4:2656–2663
Leytin V, Mykhaylov S, Starkey AF et al. (2006) Intravenous immunoglobulin inhibits anti-glycoprotein IIb-induced platelet apoptosis in a murine model of immune thrombocytopenia. Br J Haematol 133:78–82

Lim WS, Timmins JM, Seimon TA et al. (2008) Signal transducer and activator of transcription-1 is critical for apoptosis in macrophages subjected to endoplasmic reticulum stress in vitro and in advanced atherosclerotic lesions in vivo. Circulation 117:940–951

Loop T, Dovi-Akue D, Frick M et al. (2005) Volatile anesthetics induce caspase-dependent, mitochondria-mediated apoptosis in human T lymphocytes in vitro. Anesthesiology 102:1147–1157

Lopez JJ, Salido GM, Gómez-Arteta E et al. (2007) Thrombin induces apoptotic events through the generation of reactive oxygen species in human platelets. J Thromb Haemost 5:1283–1291

Lopez JJ, Salido GM, Pariente JA et al. (2008) Thrombin induces activation and translocation of Bid, Bax and Bak to the mitochondria in human platelets. J Thromb Haemost 6:1780–1788

Ma AD, Abrams CS (1999) Pleckstrin homology domains and phospholipid-induced cytoskeletal reorganisation. Thromb Haemost 82:399–406

Mahmud H, Föller M, Lang F (2008) Suicidal erythrocyte death triggered by cisplatin. Toxicology 249:40–44

Mason KD, Carpinelli MR, Fletcher JI et al. (2007) Programmed anuclear cell death delimits platelet life span. Cell 128:1173–1186

McClain DE, Kalinich JF, Ramakrishnan N (1995) Trolox inhibits apoptosis in irradiated MOLT-4 lymphocytes. FASEB J 9:1345–1354

Menasche G, Feldmann J, Fischer A et al. (2005) Primary hemophagocytic syndromes point to a direct link between lymphocyte cytotoxicity and homeostasis. Immunol Rev 203:165–179

Metkar SS, Wang B, Ebbs ML et al. (2003) Granzyme B activates procaspase-3 which signals a mitochondrial amplification loop for maximal apoptosis. J Cell Biol 160:875–885

Mezzano D, Aranda E, Foradori A et al. (1984) Kinetics of platelet density subpopulations in splenectomized mongrel dogs. Am J Hematol 17:373–382

Migliorati G, Nicoletti I, Nocentini G et al. (1994) Dexamethasone and interleukins modulate apoptosis of murine thymocytes and peripheral T-lymphocytes. Pharmacol Res 30:43–52

Min B, Paul WE (2008) Basophils and type 2 immunity. Curr Opin Hematol 15:59–63

Miyajima A, Miyatake S, Schreurs J et al. (1988) Coordinate regulation of immune and inflammatory responses by T cell-derived lymphokines. FASEB J 2:2462–2473

Miyamoto T, Ohneda O, Arai F et al. (2001) Bifurcation of osteoclasts and dendritic cells from common progenitors. Blood 98:2544–2554

Miyazaki Y, Nomura S, Miyake T et al. (1996) High shear stress can initiate both platelet aggregation and shedding of procoagulant containing microparticles. Blood 88:3456–3464

Mizunoe S, Kadota J, Tokimatsu I et al. (2004) Clarithromycin and azithromycin induce apoptosis of activated lymphocytes via down-regulation of Bcl-xL. Int Immunopharmacol 4:1201–1207

Moritz KM, Lim GB, Wintour EM (1997) Developmental regulation of erythropoietin and erythropoiesis. Am J Physiol 273:R1829–R1844

Mustard JF, Kinlough-Rathbone RL, Packham MA (2001) History of platelets. In: Gresele P, Page C, Fuster V, Vermylen J (eds) Platelets in thrombotic and non-thrombotic disorders. Pathophysiology, pharmacology and therapeutics. Cambridge University Press, Cambridge

Myssina S, Huber SM, Birka C, Lang PA, Lang KS, Friedrich B, Risler T, Wieder T, Lang F (2003) Inhibition of erythrocyte cation channels by erythropoietin. J Am Soc Nephrol 14:2750–2757

Nichols JE, Niles JA, Roberts NJ Jr (2001) Human lymphocyte apoptosis after exposure to influenza A virus. J Virol 75:5921–5929

Niemoeller OM, Foller M, Lang C et al. (2008) Retinoic acid induced suicidal erythrocyte death. Cell Physiol Biochem 21:193–202

Núñez M, Soriano V, López M et al. (2006) Coinfection with hepatitis C virus increases lymphocyte apoptosis in HIV-infected patients. Clin Infect Dis 43:1209–1212

Ogilvy S, Metcalf D, Print CG et al. (1999) Constitutive Bcl-2 expression throughout the hematopoietic compartment affects multiple lineages and enhances progenitor cell survival. Proc Natl Acad Sci USA 96:14943–14948

Paul WE (1989) Pleiotropy and redundancy: T cell-derived lymphokines in the immune response. Cell 57:521–524

Perrin-Wolff M, Bertoglio J, Bressac B et al. (1995) Structure-activity relationships in glucocorticoid-induced apoptosis in T lymphocytes. Biochem Pharmacol 50:103–110

Polenakovic M, Sikole A (1996) Is erythropoietin a survival factor for red blood cells? J Am Soc Nephrol 7:1178–1182.

Radley JM, Haller CJ (1983) Fate of senescent megakaryocytes in the bone marrow. Br J Haematol 53:277–287

Romero PJ, Romero EA (1999) Effect of cell ageing on Ca^{2+} influx into human red cells. Cell Calcium 26:131–137

Rosado JA, Lopez JJ, Gomez-Arteta E et al. (2006) Early caspase-3 activation independent of apoptosis is required for cellular function. J Cell Physiol 209:142–152

Rosado JA, Porras T, Conde M et al. (2001) Cyclic nucleotides modulate store-mediated calcium entry through the activation of protein-tyrosine phosphatases and altered actin polymerization in human platelets. J Biol Chem 276:15666–15675

Rosado JA, Sage SO (2000a) The actin cytoskeleton in store-mediated calcium entry. J Physiol 526:221–229

Rosado JA, Sage SO (2000b) A role for the actin cytoskeleton in the initiation and maintenance of store-mediated calcium entry in human platelets. Trends Cardiovasc Med 10:327–332

Sanz C, Benet I, Richard C et al. (2001) Antiapoptotic protein Bcl-x(L) is up-regulated during megakaryocytic differentiation of $CD34^+$ progenitors but is absent from senescent megakaryocytes. Exp Hematol 29:728–735

Schroeder WA (1963) The hemoglobins. Annu Rev Biochem 32:301–320

Sebbagh M, Hamelin J, Bertoglio J et al. (2005) Direct cleavage of ROCK II by granzyme B induces target cell membrane blebbing in a caspase-independent manner. J Exp Med 201:465–471

Schrijvers DM, De Meyer GR, Herman AG et al. (2007) Phagocytosis in atherosclerosis: molecular mechanisms and implications for plaque progression and stability. Cardiovasc Res 73:470–480

Segal AW (2005) How neutrophils kill microbes. Annu Rev Immunol 23:197–223

Seta N, Kuwana M (2007) Human circulating monocytes as multipotential progenitors. Keio J Med 56:41–47

Shcherbina A, Remold-O'Donnel E (1999) Role of caspase in a subset of human platelet activation responses. Blood 93:4222–4231

Sims PJ, Wiedmer T, Esmon CT et al. (1989) Assembly of the platelet prothrombinase complex is linked to vesiculation of the platelet plasma membrane. Studies in Scott syndrome: an isolated defect in platelet procoagulant activity. J Biol Chem 264:17049–17057

Sopjani M, Föller M, Lang F (2008) Gold stimulates Ca^{2+} entry into and subsequent suicidal death of erythrocytes. Toxicology 244:271–279

Spender LC, Cannell EJ, Hollyoake M et al. (1999) Control of cell cycle entry and apoptosis in B lymphocytes infected by Epstein-Barr virus. J Virol 73:4678–4688

Tabas I (2005) Consequences and therapeutic implications of macrophage apoptosis in atherosclerosis: the importance of lesion stage and phagocytic efficiency. Arterioscler Thromb Vasc Biol 25:2255–2264

Tans G, Rosing J, Thomassen MC et al. (1991) Comparison of anticoagulant and procoagulant activities of stimulated platelets and platelet-derived microparticles. Blood 77:2641–2648

Taube J, McWilliam N, Luddington R et al. (1999) Activated protein C resistance: effect of platelet activation, platelet-derived microparticles, and atherogenic lipoproteins. Blood 93:3792–3797

Thaulow E, Erikssen J, Sandvik L et al. (1991) Blood platelet count and function are related to total and cardiovascular death in apparently healthy men. Circulation 84:613–617

Tonon G, Luo X, Greco NJ et al. (2002) Weak platelet agonists and U46619 induce apoptosis-like events in platelets, in the absence of phosphatidylserine exposure. Thromb Res 107:345–350

Tuosto L, Cundari E, Gilardini Montani MS et al. (1994) Analysis of susceptibility of mature human T lymphocytes to dexamethasone-induced apoptosis. Eur J Immunol 24:1061–1065

Vanhaesebroeck B, Alessi DR (2000) The PI3K-PDK1 connection: more than just a road to PKB. Biochem J 346:561–576
VanWijk MJ, VanBavel E, Sturk A et al. (2003) Microparticles in cardiovascular diseases. Cardiovasc Res 59:277–287
Wang K, Yin XM, Chao DT et al. (1996) BID: a novel BH3 domain-only death agonist. Genes Dev 10:2859–2869
Wang TH, Wang HS (1999) Apoptosis: (2) characteristics of apoptosis. J Formos Med Assoc 98:531–542
Wardlaw AJ, Brightling C, Green R et al. (2000) Eosinophils in asthma and other allergic diseases. Br Med Bull 56:985–1003
Waring P (1990) DNA fragmentation induced in macrophages by gliotoxin does not require protein synthesis and is preceded by raised inositol triphosphate levels. J Biol Chem 265:14476–14480
Wasmoen TL, Bell MP, Loegering DA et al. (1988) Biochemical and amino acid sequence analysis of human eosinophil granule major basic protein. J Biol Chem 263:12559–12563
Wolf P (1967) The nature and significance of platelet products in human plasma. Br J Haematol 13:269–288
Wolf BB, Goldstein JC, Stennicke HR et al. (1999) Calpain functions in a caspase-independent manner to promote apoptosis-like events during platelet activation. Blood 94:1683–1692
Wright JH (1906) The origin and nature of the blood plates. Boston Med Surg J 4:178–195
Yang H, Miller WM, Papoutsakis ET (2002) Higher pH promotes megakaryocytic maturation and apoptosis. Stem Cells 20:320–328
Zachariae CO (1993) Chemotactic cytokines and inflammation. Biological properties of the lymphocyte and monocyte chemotactic factors ELCF, MCAF and IL-8. Acta Derm Venereol Suppl 181:1–37
Zauli G, Vitale M, Falcieri E et al. (1997) In vitro senescence and apoptotic cell death of human megakaryocytes. Blood 90:2234–2243
Zhao X, Wang D, Zhao Z et al. (2006) Caspase-3-dependent activation of calcium-independent phospholipase A2 enhances cell migration in non-apoptotic ovarian cancer cells. J Biol Chem 281:29357–29368
Zhou Y, Wang Q, Evers BM et al. (2005) Signal transduction pathways involved in oxidative stress-induced intestinal epithelial cell apoptosis. Pediatr Res 58:1192–1197
Zwaal RF, Comfurius P, Bevers EM (2005) Surface exposure of phosphatidylserine in pathological cells. Cell Mol Life Sci 62:971–988

Chapter 7
Apoptotic Events in Endothelial and Smooth Muscle Cells

J. Garcia-Estañ and N.M. Atucha

Abstract The normal function of the cardiovascular system depends on several interrelated mechanisms working both in the cells of the vascular vessels (endothelial and vascular smooth muscle cells) and in elements such as blood cells and humoral factors present in the blood. Many of the cardiovascular diseases are initiated by alterations in the cells of the vascular wall, which by different etiologic factors (endothelial dysfunction, atherosclerosis) induce proliferation and apoptosis of both endothelial cells and vascular smooth muscle cells. This apoptotic process is controlled by a diversity of cell signalling events, which originate either outside or inside the cells. Extracellular signals include cytokines, hormones, growth factors or nitric oxide that cross the plasma membrane or transduce to effect a response. The intracellular activation of the apoptotic machinery can be initiated by any stressful message to the cell with a resultant cell suicide. In the following lines, we will review the most important mechanisms that regulate apoptosis in endothelial and smooth muscle cells.

Keywords Endothelial cell · Smooth muscle cell · Cardiovascular disease · Cytokines · Angiotensin II · Atherosclerotic plaque

7.1 Introduction

Apoptosis is a word used in the Greek language apo for "apart" and ptosis for "fallen" to describe the dropping off or falling off of petals from flowers or shedding of leaves from trees. This name was first used by Kerr, Wyllie and Currie (1972), a group of researchers at the University of Edinburgh, when they described the distinct ultrastructural morphological changes that are characteristic of these dying cells. The name emphasizes that cell death (apoptosis) is the opposite of cell replication (mitosis), in which there is tight regulation of cell cycle and cell differentiation,

J. Garcia-Estañ (✉)
Department of Physiology, University of Murcia, Campus Espinardo, 30100 Murcia, Spain
e-mail: jgestan@um.es

G.M. Salido, J.A. Rosado (eds.), *Apoptosis: Involvement of Oxidative Stress and Intracellular Ca^{2+} Homeostasis*, DOI 10.1007/978-1-4020-9873-4_7,

survival of the fit test cell, and the optimum number of cells present for normal balance in the tissue environment. This term is not similar to necrosis (Reed 2000). Both necrosis and apoptosis are terms related to some morphological features of the dying cells. A necrotic cell swells, explodes, and release its cytosolic components which produces an inflammatory response to attract immune cells. Then, the necrotic cell is associated with an inflammatory response. In the apoptotic cell, however, there is shrinkage, loss of specialized surface features, ruffling, membrane blebbing, condensation and margination of nuclear chromatin and nuclear fragmentation. The cell fragmentation results in the formation of membrane-enclosed "apoptotic bodies" that are recognized by resident adjacent cells and rapidly eliminated. This process is typically completed very rapidly, usually in only 30–60 minutes. As no cytosolic contents are released during apoptosis, inflammation is not triggered. We now know that apoptosis is the main form of death cell, both in physiological and pathological processes (Raffetto et al. 2001).

7.2 Physiology of the Endothelium and Smooth Muscle

7.2.1 Endothelium

Only a few years ago, the endothelium was considered to be just a barrier to the movement of substances in the capillary walls. Now we know that apart from this function, the endothelial layer regulates blood coagulation and thrombosis, the tone of the smooth muscle layer, the proliferation and growth of the vascular wall cells, and leukocyte adhesion, among many others. Nowadays, the endothelium is considered to be a secretory organ, since several hemodynamic changes and hormonal signals induce it to release vasoactive and thromboregulatory substances, signalling and adhesion molecules and also growth factors (Cines et al. 1998). Therefore, in order to respond to all these stimuli, the surface of the endothelial cells is covered with a great variety of receptors, which makes the endothelium a very susceptible organ to disease and as such, endothelial malfunction or as it is widely known, endothelial dysfunction is involved in many diseases. Basically, the endothelium participates in three different functions:

1. Vascular tone: the endothelium synthesizes and releases vasodilatory and vasoconstrictor substances that affect the tone and structure of the blood vessels. The endothelium thus plays a pivotal role in arteriosclerosis, arterial hypertension, and in sepsis-induced hemodynamic shock.
2. Immunity and citotoxicity: the relations between endothelial cells and immunity cells, leucocytes and macrophages, are crucial to the understanding of pathologies of the connective system, vasculitis and also, sepsis.
3. Coagulation and fibrinolysis: these parts of the haemostatic process are of vital importance in order to maintain blood fluidity. The disequilibrium will produce either haemorrhage or thrombosis.

It was at the end of the 70s when Furchgott and Zawadzki (1980) discovered that acetylcholine and many other substances act on the endothelial cells to produce their vasodilatory or vasoconstrictor effects. The equilibrium between both of them affects the vascular tone and the development of important diseases such as arterial hypertension, cardiac insufficiency and artherioclerosis. The endothelial cells produce three main substances with vasodilatory action, nitric oxide (NO), endothelium-derived hyperpolarizing factor (EDHF), and prostacyclin (PGI2), which we will briefly review here.

The aminoacid L-arginine is converted into citrulline and NO by the action of a group of enzymes called nitric oxide synthases (NOS), an enzyme requiring several cofactors, tetrahidrobiopterin among others (Stamler et al. 1992). It is the first gas also described to be a neurotransmitter (Snyder 1992). Once released, its half-life is very short, some 3–5 seconds, then it is transformed into inactive forms, such as nitrite and nitrate, by the actions of oxyhemoglobin and oxygen free radicals. The synthase located in the endothelial cell is called endothelial NOS (eNOS) and it is known to be of a constitutive origin. Then, NO acts in a paracrine fashion in the cells in close proximity to its production and release. NO diffusses very easily to the subjacent layer, the smooth muscle cells, where acts on guanylate cyclase, its effector, to produce cGMP and then, relaxation (Loscalzo and Welch 1995; Buga et al. 1991).

The main actions of NO are cGMP-dependent: (a) smooth muscle relaxation, (b) platelet antiaggregant, (c) relaxation of the smooth muscle of the bronchi, uterus and intestine by way of the non adrenergic-non cholinergic neurotransmission (Sanders and Ward 1992), and (d) inhibition of the proliferation of the smooth muscle, which has an important impact on arteriosclerosis and other cardiovascular disease. There is an allosteric variant of the eNOS called inducible (iNOS) that it is activated by different infectious and inflammatory factors and produces much more NO, up to ten times more than the eNOS isoform. It is located in cells of the macrophage system, as well as in leucocytes and in many other cells. The big amount of NO produced in these circumstances plays an important role in the arterial vasodilation characteristic of sepsis (Parratt 1997).

The stimuli that elevate intracellular Ca^{2+} in the endothelial cells stimulate the production and release of NO. It is thought that this effect is mediated by a direct effect of Ca^{2+} releasing the eNOS enzyme from caveolin, a protein that anchors eNOS to the plasma membrane of the endothelial cell. Estrogens also stimulate NOS activity and this has been proposed as a mechanism that protects the cardiovascular system in the premenopausal woman. On the contrary, cholesterol upregulates caveolin, which decreases NO activity. Other important neuroendocrine stimuli of NO release and action are the vasoconstrictor catecholamines, vasopressin, histamine and endothelin. These substances act mainly by physical stimuli, which, in turn, induces NO release, since the shear stress of the blood stream is a very potent stimulus for the elevation of endothelial NO. Physical exercise is also a potent stimulus for the release of NO. Finally, the production of NO may be altered by deficiencies in some cofactors, such as tetrahydrobiopterin, or by alterations in L-arginine transport (Voetsch et al. 2004).

Apart of being a vasodilator, NO also reduces vascular permeability and monocyte- and lymphocyte-adhesion molecules synthesis. NO also reduces platelet aggregation, tissue oxidation, tissue inflammation, activation of thrombogenic factors, cell growth, proliferation and migration, inhibits proatherogenic and pro-inflammatory cytokines expression and it favours fibrinolysis. Nuclear factor kappa-B (NFkB) inhibitor is also expressed by NO. All these factors reduce atherogenesis and its complications. For this reason NO is considered the antiatherogenic molecule (Cooke and Tsao 1994; Baldwin 2001).

The EDHF seems to be important in small calibre vessels. Although the chemical identity of the factor(s) has not been determined, it is believed by many authors to be potassium ions; other researchers have shown evidence that an endocannabinoid (anandamide) might be involved by activating the CB1 receptor; however, it looks that the action of anandamide has now been shown to occur *via* a non-endothelium-dependent mechanism. In other cases, some arachidonic acid derivatives, the epoxyeicosatrienoic acids, have been found to mediate the vasodilation. These compounds are formed by epoxidation of any one of four double bonds of the arachidonic acid carbon backbone by cytochrome P450 epoxygenase enzymes. Finally, gap junctional coupling between the endothelium and smooth muscle has also been implicated in EDHF activity in many arteries, in such a way that the hyperpolarization would be transferred from the endothelium to the smooth muscle, but without the existence of a factor (Feletou and Vanhoutte 1988; Garland et al. 1995).

PGI2 is produced in endothelial cells from prostaglandin H2 by the action of the enzyme prostacyclin synthase, and it acts by way of the elevation in cAMP, sinergically with NO, at the neighbouring smooth muscle cells. It is synthesized in response to inflammatory mediators including interleukin 1 and platelet-derived and epidermal growth factors. It is also important in platelets, where PGI2 prevents the formation of the platelet plug involved in primary hemostasis.

In addition to the above mentioned substances, bradykinin is an endogenous vasodilator nonapeptide released from plasma globulins called kininogens. There are two kininogens, a high molecular weight form present in plasma and a low molecular weight form present in tissues. High molecular weight form kininogen is synthetized in the liver and released into plasma. Hydrolysis of plasma kininogen, catalyzed by plasma protease kallikrein gives bradykinin, whose half-life in plasma is less than one minute. Therefore, there are plasma and tissue kallikreins. The hydrolysis of low molecular weight kininogen by tissue kallikrein gives kallidin, a vasodilator decapeptide whose properties are quite similar. Bradykinin effects result from stimulation of B1 and B2 receptors and produces vasodilation and an increase in capillary permeability. Bradykinin is one of the most powerful known vasodilators. Its vasodilatator and hypotensive effects resulting from B2 receptor stimulation are, at least partially, the consequence of NO release. Vasodilation is particularly marked in capillaries, where reproduces symptoms of inflammation (Iñiguez et al. 2008).

The endothelial cell also participates in the formation of the potent vasoconstrictor angiotensin II (AII), the natural antagonist of NO, since it posseses angiotensin converting enzyme (ACE), which hydrolyzes angiotensin-I formed by the action of renin on liver-produced angiotensinogen. Through its binding at the AT1 receptor, AII causes vasoconstriction and induces prothrombogenic, oxidizing and antifybrinolitic effects. All these effects allow atherosclerosis to start, develop, progress and even, complicate. Also, AII stimulates endothelin converting enzyme that degrades big-endothelin releasing endothelin-I, the most powerful vasoconstrictor known on vessel walls. ACE not only produces AII but also participates in the degradation of bradykinin, which will reduce vasodilation by its own and by way of the lower bradykinin-stimulated NO production, so that the local increase of ACE induces vasoconstriction and vascular wall hypertrophy, vascular remodelling (Johnston 1992; Kawano et al. 2000).

Endothelins are also important vasoconstrictor hormones acting both as a paracrine and endocrine fashion (Levin 1996). They are vasoconstrictors (by way of receptor A) and also vasodilatory (through receptor B) and also act as growth factors in the cardiovascular system, with important roles related to ventricular hypertrophy and plaque formation. The main stimuli for the activation of the endothelin-1 gene are AII, catecholamines, growth factors, hypoxia, and also shear stress and thrombin. On the contrary, atrial natriuretic peptide and prostaglandins E2 and prostacyclin are inhibitory factors, but NO is its main natrural inhibitor.

As can be seen, depending on the balance of these two substances, NO and AII, a vasodilatation and antiatherosclerotic or vasoconstriction and atherogenic effect will prevail. It is not necessary the increased production of one or the other, the diminished synthesis of one will make normal amounts of the other prevail. The endothelium should maintain an adequate homeostasis so that the disease does not appear and this depends on the capacity it has of producing the protective molecules. All of the factors that augment vasoconstriction, such as oxygen free radicals, endothelin, AII, thromboxane A2, the plasminogen activator and some adhesion molecules and selectins induce an alteration in the normal endothelial homeostatic balance which initiates endothelial dysfunction, a term coined after many works demonstrated the importance of a normal endothelial function.

Today, when we talk about endothelial dysfunction, we are talking about a lower capacity of the endothelium to produce NO. But, it is not only the lower production of NO that damages the endothelial function, but also the conversion of an excess of oxygen free radicals, which takes place in several situations. In a situation of endothelial dysfunction, we can observe a greater platelet and leucocyte adhesion, vasoconstriction, aggregation and migration of smooth muscle cells, lipid deposition, altered anticoagulant and antiinflammatory function (Celermajer 1997).

Disturbed endothelial function may play a large role in cardiovascular disease, especially in atherosclerosis (Esper et al. 2006). Atherosclerosis results from excessive inflammatory and fibroproliferative responses to vascular insults and the earliest alterations in the vessel wall are formation of the fatty streak and monocyte adhesion. Many factors modulate endothelial cell function, such as cytokines, growth factors, lipids and enzymes, leading to lipid accumulation, vasoconstriction and promotion of thrombosis. Active endothelial participation is required for monocyte adhesion and migration to the sub-endothelium, a participation also stimulated by endothelin. Oxidised LDL by the endothelium and oxidant stress play a major role in impairing endothelial function, by reducing the bioavailability of NO and activating pro-inflammatory signalling pathways, such as NFkB. In fact, many studies have demonstrated that reduced NO activity is one of the earliest markers of endothelial dysfunction. Improved endothelial function is a clinical marker of atherogenic risk factor modification, as it has been shown with the improvement of endothelium and NO-dependent vasodilatory responses after therapies with cholesterol-lowering drugs. Also, increased shear stress from disturbed turbulent blood flow also activate the endothelium, increasing vasomotor dysfunction and promoting inflammation by up-regulating pro-atherogenic genes, just the contrary that occurs during normal laminar flow, where physiological shear stress promotes the expression of anti-atherosclerosis genes.

7.2.2 Vascular Smooth Muscle

It refers to the particular type of smooth muscle found within and composing the majority of the wall of blood vessels. Smooth muscle is composed of spindle-shaped fibers, with a diameter of 3–10 μm and lengths of several hundred μm, with a different structure to that of skeletal muscle, no visible striations, but with essentially the same contractile mechanisms as skeletal muscle. As in skeletal muscle, Ca^{2+} is very important to initiate and maintain force generation. Plasma membranes of the vascular smooth muscle cells (VSMCs) have pouchlike infoldings called *caveoli*, and Ca^{2+} is sequestered in the extracellular space near the *caveoli*, allowing rapid influx when channels are opened. Also different with the skeletal muscle, the VSMCs do not have neuromuscular junctions and the innervating nerves have bulbous swellings called varicosities which release neurotransmitters into the synaptic clefts. Also, vasoconstrictor and vasodilator substances act on cell receptors to stimulate contraction of VSMCs. The main difference with skeletal muscle is the presence of two enzymes that phosphorylate (kinase) or dephosphorylate (phosphatase) the myosin filament, exactly its light chain. Thus, the rise in intracellular Ca^{2+} produced by the release from the sarcoplasmic reticulum and from the extracellular space activates calmodulin which in turn activates myosin light chain kinase to activate myosin and interact with actin to produce shortening. Smooth muscle relaxes when intracellular Ca^{2+} levels drop, which causes kinase to inactivate or when the phosphatase enzyme is activated (Berridge 2008).

Vascular smooth muscle contracts or relaxes to both change the volume of blood vessels and the local blood pressure, a mechanism that is responsible to redistribution of the blood within the body to areas where it is needed (i.e. areas with temporarily enhanced oxygen consumption). Thus the main function of vascular smooth muscle tone is to regulate the caliber of the blood vessels in the body. Excessive vasoconstriction leads to hypertension, while excessive vasodilation as in shock, leads to hypotension. Arteries have much more smooth muscle within their walls than veins, thus their greater wall thickness. This is because they have to carry pumped blood away from the heart to all the organs and tissues that need the oxygenated blood. The endothelial lining of each is similar. Vascular smooth muscle is innervated primarily by the sympathetic nervous system through adrenergic receptors (adrenoceptors). Three types of adrenoceptors are present within vascular smooth muscle cells: α_1, α_2 and β_2. The main endogenous agonist of these cell receptors is norepinephrine (Webb 2003). Of course, receptors for all types of vasoconstrictor and vasodilators substances are present in their membranes (Overgaard and Dzavík 2008).

7.3 Cardiovascular Apoptosis

Apoptosis is a physiological event precisely regulated during the development of the cardiovascular system, contrary to what occurs during the adult life, where it is mainly a pathological event. During development, apoptosis plays an important role in the suppression of vestigial structures, the control of cell number and cardiovascular remodelling. Right ventricle suffers a physiological remodelling after birth, in which a number of cells die, in order to adapt to the new extrauterine pumping function (Colucci 1996). Apoptosis seems to be of importance also in the formation of atrial and atrioventricular nodes, by means of which a number of round or oval P cells present at birth die without signs of inflammation (James 1994). The autrioventricular connections are also eliminated in the postnatal stage and it has been suggested that the persistence of some of these connections may be related to Wolf Parkinson White syndrome and to auriculoventricular congenital blocks or the syndrome of long or prolonged QT (James et al. 1996).

In the adult life, apoptosis is now recognized to play an important role in ischemic and dilated cardiac diseases, hypertensive cardiopathy, and myocardial infarction among others. It is also involved in the coronary arteriopathy associated with advanced arteriosclerosis and transplant arteriopathy (Díez 2000; Chen and Tu 2002). Cardiac surgeries, in particular cardiopulmonary bypass and cardioplegia, have been also reported to trigger myocardial inflammation and apoptosis. During cardiac surgery, many stressful stimuli including ischemia and ischemia–reperfusion, inflammatory response, operative trauma, cardioplegia and oxidative stress have been reported to trigger myocyte death (Anselmi et al. 2004).

Several studies have demonstrated that after an acute myocardial infarction, the level of cardiomyocite apoptosis increases, especially in the peripheral area where the ischemia is less pronounced. Early treatment with thrombolytic drugs or angioplasty restores blood flow and minimizes miocardial necrosis, but the great amount of free radicals released by reperfusion induces apoptosis (Olivetti et al. 1997). In fact, the experimental inhibition of apoptosis has shown that cardiomyocyte apoptosis and the extension of the infarcted area decreased, improving the ventricular function (Moon et al. 2003; Parsa et al. 2003).

Another important area of interest is arterial hypertension, a prevalent disease in a 20–30% of the population. Arterial hypertension induces severe stress in many target organs (heart, blood vessels, brain and kidney) that results in hypertrophy and hyperplasia, an abnormal proliferation needed to maintain organ perfusion. Since apoptosis is the normal physiological mechanism to counteract proliferation, it seems logical that many studies have found elevated indices of apoptosis in different arterial hypertensive states (Díez 2000). Thus, it was proposed that both proliferation and apoptosis contribute to the remodelling of target organs in hypertension (Hamet 1995; Hamet et al. 1995). In the heart, for instance, myocardial remodelling consists of fibrosis, insufficient vascularisation, contractile disturbances, and decrease in the number of cells, because of exaggerated apoptosis. These abnormalities in the myocardial wall may accelerate the development of diastolic and systolic dysfunction, and then an overt heart failure. Among the number of apoptotic factors probably involved, in addition to mechanical stress, some hormones such as angiotensin II and norepinephrine, oxidative stress or inflammatory cytokines play a role in this cardiomyocyte apoptosis. Other changes, such as a downregulation of survival proteins or the activation of death proteins may also play an important role (Salas 2001). The removal of these inducers reduces apoptosis and reverses the loss of contractile function in many cases, indicating the feasibility of the pharmacological application of antioxidants, NOS inhibitors, ACE inhibitors, AII receptor antagonists and adrenergic receptor antagonists (Chen and Tu 2002; Parsa et al. 2003).

7.3.1 Apoptosis in Endothelial Cells

The endothelium lines all vessels on their luminal side, separating the local tissue from the blood compartment. As commented above, a single layer of endothelial cells maintains the vascular tone and the anticoagulant properties of blood vessels. Apoptosis of endothelial cells or endothelial dysfunction is therefore crucial in many pathological states. The endothelium also regulates the apoptosis of VSMCs since products released by endothelial cells promote VSMC survival (Hata et al. 2001). Whereas in the normal vessel wall, there is a very low level of VSMC turnover, and apoptotic and mitotic indices are low (Gordon et al. 1990), in diseased tissue, factors such as inflammatory cytokines alter the balance between cell proliferation and apoptosis.

Apoptosis of vascular cells is an early event in both inflammatory and mechanical injury. In general, the apoptotic mechanism has been associated with higher levels of reactive oxygen species (ROS) than those that support proliferation, such as those produced by macrophages and mechanical injury. Release of cytokines such as TNF-α after sepsis, ischemia-reperfusion, or shock contributes to endothelial apoptosis (Simionescu 2007).

The elevation in oxidative stress can also be the result of excessive formation of NO, which physiologically is an antiapoptotic substance that s-nitrosilates caspases but in excess is converted into peroxynitrite which can initiate apoptosis (Messmer et al. 1994; Tzeng et al. 1997; Kim et al. 1999; Pacher et al. 2007). Physiological levels of NO produced by endothelial eNOS protect against apoptosis through the prevention of caspase 3 activation or by decreasing the non-specific permeability of the inner mitochondrial membrane, thus preventing cytochrome *c* release (Dimmeler et al. 1997; Kim et al. 1999; Brookes et al. 2000). Physiologically, the anti-apoptotic pathway induced by shear-stress involves the phosphorylation of Akt/PKB (protein kinase B), the release of NO and the inhibition of caspase 3 activation (Dimmeler et al. 1998, 1999).

In support of this role of ROS, recent experiments have shown that N-acetylcysteine attenuates TNF-α-induced endothelial apoptosis (Xia et al. 2006). The mechanisms underlying this effect is not completely elucidated, but it seems that ROS regulates caspase 3 activation and that O_2^- production by activated macrophages triggers a Ca^{2+}-dependent, inositol trisphosphate-linked apoptotic cascade in endothelial cells (Madesh and Hajnoczky 2001). The high levels of NO produced by the iNOS in macrophages induce a burst of reactive nitrogen species that promote oxidative damage (Buttery et al. 1996), and can also induce the release of cytochrome *c*, the up-regulation of p53 expression and the activation of the Bcl-2-dependent pathways (Chung et al. 2001). In particular, incubation of endothelial cells with high levels of NO increases apoptosis by decreasing the levels of Bcl-2 and increasing levels of the pro-apoptotic protein Bax (Stoneman and Bennett 2004).

ROS can also promote vascular leakage, which is important in ischemia–reperfusion-induced vascular injury and the induction of inflammatory responses. In addition, the inactivation of NO by O_2^- and increased production of the endothelial hyperpolarizing factor H_2O_2 leads to impaired vasodilation (Papaharalambus and Griendling 2007).

Most types of vascular injury begin with endothelial dysfunction and activation. The induction of adhesion molecules such as E- and P-selectin, vascular cell adhesion molecule-1 and intercellular cell adhesion molecule-1 is an early marker of atherosclerosis. The expression of these adhesion molecules facilitates the transmigration of leukocytes and cytokines, like interleukins, vascular endothelial growth factor and tumor necrosis factor-α, hormones such as AII which potentiate the expression of adhesion molecules in a ROS-dependent fashion. The endothelial apoptosis may be important in the pathophysiology of arteriosclerosis. In fact, apoptotic endothelial cells express more adhesion molecules, which initiate the formation of arteriosclerotic lesion. Oxidized lipoproteins could be the inductors of apoptosis in endothelial cells in this setting (Simionescu 2007). Several reports indicate that endothelial cell apoptotic death is a late event, occurring only in advanced atherosclerosis (Libby 1995; Kockx and Herman 2000). Apoptosis of endothelial cells is assumed to be caused by the local inflammatory mediators or the cytolytic attack of activated killer T cells, cytokines, and oxidized LDL that increase endothelial cell synthesis of matrix metalloproteinases, which degrade components of endothelial cell basal lamina (Rajavashisth et al. 1999) or the oxidative stress (Burlacu et al. 2001).

Endothelial function declines with age, and this is reflected by the attenuation of endothelium-dependent vasodilatory responses, consequence of the alteration in the NO production and increased formation of ROS. Aging is also associated with a reduction in the regenerative capacity of the endothelium and endothelial senescence, which is characterized by an increased rate of endothelial cell apoptosis.

7.3.2 Apoptosis in Vascular Smooth Muscle Cells (VSMC)

Vascular smooth muscle contracts or relaxes, and by changing the resistance to blood flow, modulates the blood pressure and thus the amount of blood that the organs receive according to their needs. Thus the main function of vascular smooth muscle tone is to regulate the diameter of the blood vessels in the body. Excessive vasoconstriction leads to hypertension, while excessive vasodilatation as in shock leads to hypotension. In contrast to endothelial cell death, VSMC apoptosis occurs mainly after mechanical injury. VSMC exhibit a variety of responses after vessel injury, including phenotypic switching, migration, proliferation, protein synthesis, and apoptosis. VSMC apoptosis is evident at short time after balloon injury in rat carotid arteries and is attenuated by treatment with antioxidants (Pollman et al. 1999). The apoptotic response appears to be mediated in part by activation of c-jun N-terminal kinase. Furthermore, *in vitro* studies have shown that H_2O_2 can induce apoptosis through a protein kinase C pathway, which is antagonized by Akt and heme oxygenase-1 (Brunt et al. 2006). NO also plays a role in VSMC apoptosis, in this case by magnifying the effect of other apoptotic inducers; for example, Shichiri et al. (2000) demonstrated that serum deprivation in combination with addition of NO donors increased the rate of apoptosis of rat VSMCs.

The regulation of VSMC proliferation and apoptosis may be an important determinant of the physiological normal structure of the vascular wall (Fortuño et al. 2001). Thus, whereas proliferation is more important in order to facilitate VSMC thickening in arterial hypertension, the migration of VSMC to subintimal space in an arteriosclerotic plaque results in a greater number of apoptotic cells, and a lower capacity to synthesize extracellular matrix, which may be an important factor producing a more fragile fibrous cap, promoting thus plaque rupture.

Atherosclerotic plaques are formed by an accumulation of VSMCs, inflammatory cells (macrophages, T lymphocytes and other cells) on a dysfunctional endothelium. Extracellular lipid, collagen deposition and matrix formation are also important players. Inflammattion triggers proliferation and migration of VSMCs of the vessel wall. For a long period, atherosclerosis is clinically silent, and it is unusual that complications appear before the development of advanced lesions. These consist of a VSMC-rich fibrous cap overlying a lipid- and macrophage-rich necrotic core, and it is the relative proportion of these components that determines the clinical manifestations of the plaque. For example, unstable plaques, which are more likely to break, have more inflammatory cells and lipid, and less VSMCs than stable lesions. Apoptosis of VSMCs in the fibrous cap of advanced plaques is believed to be more usual than rupture to produce fatal heart attacks (Clarke and Bennett 2006).

The mechanisms by which apoptosis is induced in atherosclerotic plaques are not completely known yet (Rossig et al. 2001; Littlewood and Bennett 2003; Stoneman and Bennett 2004; Clarke and Bennett 2006). It seems that VSMC and macrophages located in the intimal part of the plaque show more apoptosis than in normal vessels (Blanco-Colio et al. 2000). Some stimuli, however, have been identified, such as oxidized LDLs that are able to induce apoptosis of endothelial cells of coronary arteries, macrophages and VSMC (Jovinge et al. 1997). It has also been shown that apoptosis-inducing factor plays an important role in oxidized LDL-induced endothelial injury (Zhang et al. 2004). Cytokines released by macrophages (TNF-a, IL-1) and T lymphocytes (INF-g) also induce VSMC apoptosis (Geng et al. 1996). Another important factor relates to the presence of Fas (CD95) and Fas-L, both in endothelial cells and VSMC which indicates that apoptosis may also be induced by these receptors (Geng et al. 1997).

In summary, it seems that normal arteries adapt very well to important cell losses with little change in functional active or passive properties. In contrast, VSMC apoptosis is a critical process which determines plaque stability and thus the most important consequence of atherosclerosis and plaque rupture.

References

Anselmi A, Abbate A, Girola F et al. (2004) Myocardial ischemia, stunning, inflammation, and apoptosis during cardiac surgery: a review of evidence. Eur J Cardiothorac Surg 25:304–311
Baldwin A (2001) The transcription factor NK-6B and human disease. J Clin Invest 107:3–6
Berridge MJ (2008) Smooth muscle cell calcium activation mechanisms. J Physiol 586:5047–5061
Blanco-Colio LM, Hernández-Presa MA, Ortego M et al. (2000) Papel potencial de la apoptosis en la aterosclerosis. Cardiovasc Risk Fact 9:131–137
Brookes PS, Salinas EP, Darley-Usmar K et al. (2000) Concentration-dependent effects of nitric oxide on mitochondrial permeability transition and cytochrome c release. J Biol Chem 275:20474–20479
Brunt KR, Fenrich KK, Kiani G et al. (2006) Protection of human vascular smooth muscle cells from H2O2-induced apoptosis through functional codependence between HO-1 and AKT. Arterioscler Thromb Vasc Biol 26:2027–2034
Buga GM, Gold ME, Fukuto JM et al. (1991) Shear stress-induced release of nitric oxide from endothelial cells grown on beads. Hypertension 17:187–193
Burlacu A, Jinga VV, Gafencu AV et al. (2001) Severity of oxidative stress generates different mechanisms of endothelial cell death. Cell Tissue Res 306:409–416
Buttery LD, Springall DR, Chester AH et al. (1996) Inducible nitric oxide synthase is present within human atherosclerotic lesions and promotes the formation and activity of peroxynitrite. Lab Invest 75:77–85
Celermajer DS (1997) Endothelial dysfunction: does it matter? Is it reversible?. J Am Coll Cardiol 30:325–333
Chen QM, Tu VC (2002) Apoptosis and Heart Failure: Mechanisms and Therapeutic Implications. Am J Cardiovasc Drugs 2:43–57
Chung HT, Pae HO, Choi BM et al. (2001) Nitric oxide as a bioregulator of apoptosis. Biochem Biophys Res Commun 282:1075–1079
Cines DB, Pollak ES, Buck CA et al. (1998) Endothelial cells in physiology and in the pathophysiology of vascular disorders. Blood 91:3527–3561
Clarke M, Bennett M (2006) The emerging role of vascular smooth muscle cell apoptosis in atherosclerosis and plaque stability. Am J Nephrol 26:531–535
Colucci WS (1996) Apoptosis in the heart. N Engl J Med 335:1224–1226
Cooke JP, Tsao PS (1994) Is NO an endogenous antiatherogenic molecule? Arterioscler Thromb 14:753–759
Díez J (2000) Apoptosis en las enfermedades cardiovasculares. Rev Esp Cardiol 53:267–274
Dimmeler S, Assmus B, Hermann C et al. (1998) Fluid shear stress stimulates phosphorylation of Akt in human endothelial cells: involvement in suppression of apoptosis. Circ Res 83:334–341
Dimmeler S, Fleming I, Fisslthaler B et al. (1999) Activation of nitric oxide synthase in endothelial cells by Akt-dependent phosphorylation. Nature 399:601–605

Dimmeler S, Haendeler J, Galle J et al. (1997) Oxidized low-density lipoprotein induces apoptosis of human endothelial cells by activation of CPP32-like proteases. A mechanistic clue to the 'response to injury' hypothesis. Circulation 95:1760–1763
Esper RJ, Nordaby RA, Vilariño JO et al. (2006) Endothelial dysfunction: a comprehensive appraisal. Cardiovasc Diabetol 5:4–18
Feletou M, Vanhoutte PM (1988) Endothelium-dependent hyperpolarization of canine coronary smooth muscle. Br J Pharmacol 93:515–524
Fortuño MA, Ravassa S, Fortuño A et al. (2001) Cardiomyocyte apoptotic cell death in arterial hypertension: mechanisms and potential management. Hypertension 38:1406–1412
Furchgott RF, Zawadzki JV (1980) The obligatory role of endothelial cells in the relaxation of arterial smooth muscle by acetylcholine. Nature 288:373–376
Garland CJ, Plane F, Kemp BK et al. (1995) Endothelium-dependent hyperpolarization: a role in the control of vascular tone. Trends Biol Sci 16:23–30
Geng YJ, Henderson LE, Levesque EB et al. (1997) Fas is expressed in human atherosclerotic intima and promotes apoptosis of cytokine-primed human vascular smooth muscle cells. Arterioscler Thromb Vasc Biol 17:2200–2208
Geng YJ, Wu Q, Muszynski M et al. (1996) Apoptosis of vascular smooth muscle cells induced by in vitro stimulation with interferon-gamma, tumor necrosis factor-alpha, and interleukin-1 beta. Arterioscler Thromb Vasc Biol 16:19–27
Gordon D, Reidy MA, Benditt EP et al. (1990) Cell proliferation in human coronary arteries. Proc Natl Acad Sci USA 87:4600–4604
Hamet P (1995) Proliferation and apoptosis in hypertension. Curr Opin Nephrol Hypertens 4:1–7
Hamet P, Richard L, Dam T-H et al. (1995) Apoptosis in target organs of hypertension. Hypertension 26:642–648
Hata S, Fukuo K, Morimoto S et al. (2001) Vascular smooth muscle maintains the levels of Bcl-2 in endothelial cells. Atherosclerosis 154:309–316
Iñiguez MA, Cacheiro-Llaguno C, Cuesta N et al. (2008) Prostanoid function and cardiovascular disease. Arch Physiol Biochem 114:201–209
James TN (1994) Normal and abnormal consequences of apoptosis in the human heart. From postnatal morphogenesis to paroxysmal arrhythmias. Circulation 90:556–573
James TN, St Martin E, Willis PW et al. (1996) Apoptosis as a possible cause of gradual development of complete heart block and fatal arrthythmias asociated with absence of the AV node, sinus node, and internodal pathways. Circulation 93:1424–1438
Johnston CJ (1992) Renin-angiotensin system: a dual tissue and hormonal system for cardiovascular control. J Cardiovasc Pharmacol 20(Suppl B):S1–S5
Jovinge S, Crisby M, Thyberg J, et al. (1997) DNA fragmentation and ultrastructural changes of degenerating cells in atherosclerotic lesions and smooth muscle cells exposed to oxidized LDL in vitro. Arterioscler Thromb Vasc Biol 17:2225–2231
Kawano H, Do YS, Kawano Y et al. (2000) Angiotensin II has multiple profibrotic effects in human cardiac fibroblast. Circulation 101:1130–1137
Kerr JF, Wyllie AH, Currie AR (1972) Apoptosis: a basic biological phenomenon with wide-ranging implications in tissue kinetics. Br J Cancer 26:239–257
Kim YM, Bombeck CA, Billiar TR (1999) Nitric oxide as a bifunctional regulator of apoptosis. Circ Res 84:253–256
Kockx MM, Herman AG (2000) Apoptosis in atherosclerosis: beneficial or detrimental? Cardiovasc Res 45:736–746
Levin ER (1996) Endothelins as cardiovascular peptides. Am J Nephrol 16:246–251
Libby P (1995) The molecular bases of the acute coronary syndromes. Circulation 91:2844–2850
Littlewood TD, Bennett MR (2003) Apoptotic cell death in atherosclerosis. Curr Opin Lipidol 14:469–475
Loscalzo J, Welch G (1995) Nitric oxide and its role in the cardiovascular system. Prog Cardiovasc Dis 38:87–104
Madesh M, Hajnoczky G (2001) VDAC-dependent permeabilization of the outer mitochondrial membrane by superoxide induces rapid and massive cytochrome c release. J Cell Biol 155:1003–1015

Messmer UK, Ankarcrona M, Nicotera P et al. (1994) p53 expression in nitric oxide-induced apoptosis. FEBS Lett 355:23–26
Moon C, Krawczyk M, Ahn D et al. (2003) Erythropoietin reduces myocardial infarction and left ventricular functional decline after coronary artery ligation in rats. Proc Natl Acad Sci USA 100:11612–11617
Olivetti G, Abbi R, Quaini F et al. (1997) Apoptosis in the failing human heart. N Engl J Med 336:1131–1141
Overgaard CB, Dzavík V (2008) Inotropes and vasopressors: review of physiology and clinical use in cardiovascular disease. Circulation 118:1047–1056
Pacher P, Beckman JS, Liaudet L (2007) Nitric oxide and peroxynitrite in health and disease. Physiol Rev 87:315–424
Papaharalambus CA, Griendling KK (2007) Basic mechanisms of oxidative stress and reactive oxygen species in cardiovascular injury. Trends Cardiovasc Med 17:48–54
Parratt JR (1997) Nitric oxide. A key mediator in sepsis and endotoxaemia? J Physiol Pharmacol 48:493–506
Parsa CJ, Matsumoto A, Kim J et al. (2003) A novel protective effect of erythropoietin in the infarcted heart. J Clin Invest 112:999–1007
Pollman MJ, Hall JL, Gibbons GH (1999) Determinants of vascular smooth muscle cell apoptosis after balloon angioplasty injury. Influence of redox state and cell phenotype. Circ Res 84:113–121
Raffetto JD, Leverkus M, Park H-Y et al. (2001) Synopsis on cellular senescence and apoptosis. J Vasc Surg 34:173–177
Rajavashisth TB, Liao JK, Galis ZS et al. (1999) Inflammatory cytokines and oxidized low density lipoproteins increase endothelial cell expression of membrane type 1-matrix metalloproteinase. J Biol Chem 274:11924–11929
Reed JC (2000) Mechanisms of apoptosis. Am J Pathol 157:1415–1430
Rossig L, Dimmeler S, Zeiher AM (2001) Apoptosis in the vascular wall and atherosclerosis. Basic Res Cardiol 96:11–22
Salas MA (2001) Apoptosis en el sistema cardiovascular: Un camino hacia nuevas estrategias terapéuticas. Rev Fed Arg Cardiol 30:523–532
Sanders KM, Ward SM (1992) Nitric oxide as a mediator of nonadrenergic noncholinergic neurotransmission. Am J Physiol Gastrointest Liver Physiol 262:G379–G392
Shichiri M, Yokokura M, Marumo F et al. (2000) Endothelin-1 inhibits apoptosis of vascular smooth muscle cells induced by nitric oxide and serum deprivation via MAP kinase pathway. Arterioscler Thromb Vasc Biol 20:989–997
Simionescu M (2007) Implications of early structural-functional changes in the endothelium for vascular disease. Arterioscler Thromb Vasc Biol 27:266–274
Snyder SH (1992) Nitric oxide: first in a new class of neurotransmitters. Science 257:494–496
Stamler JS, Singel DJ, Loscalzo J (1992) Biochemistry of nitric oxide and its redox-activated forms. Science 258:1898–1902
Stoneman VEA, Bennett MR (2004) Role of apoptosis in atherosclerosis and its therapeutic implications. Clinical Sci 107:343–354
Tzeng E, Kim YM, Pitt BR et al. (1997) Adenoviral transfer of the inducible nitric oxide synthase gene blocks endothelial cell apoptosis. Surgery 122:255–263
Voetsch B, Jin RC, Loscalzo L (2004) Nitric oxide insufficiency and atherothrombosis. Histochem Cell Biol 122:353–367
Webb RC (2003) Smooth muscle contraction and relaxation. Advan Physiol Edu 27:201–206
Xia Z, Liu M, Wu Y, Sharma V et al. (2006) N-acetylcysteine attenuates TNF-alpha-induced human vascular endothelial cell apoptosis and restores eNOS expression. Eur J Pharmacol 550:134–142
Zhang W, Li D, Mehta JL (2004) Role of AIF in human coronary artery endothelial cell apoptosis. Am J Physiol Heart Circ Physiol 286:H354–H358

Chapter 8
Apoptotic Events in Male Germ Cells and in Mature Mammalian Spermatozoa

J.A. Tapia and F.J. Peña

Abstract The spermatozoon is a highly specialized cell, which main function is to transport the male haploid genome in the female genital tract and to deliver it during fertilization of the oocyte. Several reports have shown that the molecular machinery involved in apoptosis in somatic cells is also present in sperm cells during their generation in the testis as well as in the mature, ejaculated mammalian spermatozoa. The aim of the present chapter is to give an overview of recent research on apoptosis in male germ cells and in ejaculated spermatozoa, pointing out its regulation and potential role in important sperm functions. The potential role that oxidative-stress and mitochondrial dysfunction could play as inductors of apoptosis in both fresh and frozen-thawed spermatozoa will be also discussed.

Keywords Apoptosis · Spermatogenesis · Mammalian Spermatozoa

8.1 Introduction

The mammalian spermatozoa are highly specialized cells generated during a process called spermatogenesis. This process not only consists in increases of the cell number but also is accompanied by profound differentiation changes involving cell shape, looses of some organelles and redistribution of few others, enzymatic activities, and, lastly, acquisition of motility ability (Amann 2008).

Upon ejaculation mammalian spermatozoa are unable to fertilize the oocyte. During the transit throughout the female genital tract, spermatozoa suffer a series of transformations, globally termed as capacitation, which are yet not completely defined but which basically involve profound changes in the membrane fluidity and ionic permeability, concomitantly with increased sperm motility and with activation of intracellular signalling pathways in a cAMP/PKA-dependent manner (Herrero et al. 2000; Gadella et al. 2001; Gadella and Van Gestel 2004; de Laminarde and

J.A. Tapia (✉)
Department of Physiology, Laboratory of Spermatology, University of Extremadura, Avda Universidad s/n, 10071 Cáceres, Spain
e-mail: jatapia@unex.es

G.M. Salido, J.A. Rosado (eds.), *Apoptosis: Involvement of Oxidative Stress and Intracellular Ca^{2+} Homeostasis*, DOI 10.1007/978-1-4020-9873-4_8,

O'Flaherty 2008; Gadella et al. 2008; Gadella 2008). Sperm capacitation can be induced *in vitro* by the incubation of these cells in a chemically defined medium whose composition vary among different mammalian species (Gadella 2008). Capacitated cells are capable to reach the oocyte, bind to *zona pellucida* and suffer a new transformation consisting in the exocytosis of the acrosome content in a process termed acrosome reaction. After this, the proteases released from the acrosome will degrade the glycoproteins surrounding the plasma membrane of the oocyte, allowing the spermatozoon to progress and fuse with the oocyte at fertilization (Gadella et al. 1999; Grunewald et al. 2006; Gadella 2008).

During the last decades an increasing number of reports have evidenced that apoptosis or programmed cell death plays very important roles in the physiological and pathological regulation of both male germ cells and terminally-differentiated ejaculated spermatozoa. Apoptosis in male germ cells is required to accomplish an equilibrated ratio between somatic and germinal component in the testis which will ensure an efficient spermatogenesis (Print and Loveland 2000; Shaha 2007; Sofikitis et al. 2008; Johnson et al. 2008). In other words, a density-dependent downregulation of germ cell population by means of apoptotic mechanisms is required to ensure germ cell homeostasis. This regulation involves factors promoting germ cell apoptosis (Fas, TRAIL, TNFα, Bcl-2, p53) and factors promoting cell survival (Bcl-2, TNFα and Kit/SCF) (Print and Loveland 2000; Shaha 2007; Sofikitis et al. 2008).

Although the role of apoptosis during spermatogenesis and in somatic cells is well established, there is disagreement regarding the importance of apoptotic processes in ejaculated spermatozoa. It is not clear whether the apoptotic markers detected in ejaculated mammalian spermatozoa are residues of an abortive apoptotic process started before ejaculation or whether they result from apoptosis initiated in the post-ejaculation period (Weil et al. 1998; Sakkas et al. 1999a,b; Said et al. 2004; Lachaud et al. 2004; Grunewald et al. 2005; Aquila et al. 2007; Bejarano et al. 2008). Furthermore, independently of the origin of the apoptotic changes that are detected in sperm, the exact relation of these changes with the motility, capacitation, acrosome reaction and other relevant parameters, which globally will determine the fertility ability of these cells, is still a matter of debate. Finally, a correlation among apoptotic changes and impaired sperm function has been also suggested (Paasch et al. 2004c; Said et al. 2005a,b; Barroso et al. 2006).

In this chapter, recent reports providing insight into the induction and regulation of apoptosis in male germ cells as well as in ejaculated mammalian spermatozoa will be reviewed, paying special attention to the physiological and/or detrimental consequences that the apoptotic events could induce in these cells. Furthermore, studies addressing the oxidative stress and mitochondria dysfunction as inductors of apoptosis in both fresh and frozen-thawed spermatozoa will be also discussed.

8.2 Spermatogenesis and Male Germ Cells

The generation of mature spermatozoa in the testis is accomplished by a highly synchronized and regulated process, globally termed as spermatogenesis, which consists in a series of sequential cell divisions and differentiation steps, allowing the transformation of the undifferentiated diploid spermatogonia in the highly differentiated mature haploid spermatozoon (Fig. 8.1) (Franca et al. 2005; Ehmcke and Schlatt 2006; Amann 2008). The process is initiated in the mouse embryo at around day 11 postcoitum, when primordial germ cells (PGC) colonize the genital ridge. Under the influence of Y-chromosome-bearing Sertoli cells, the PGC proliferate. Among these cells some will degenerate – more likely by necrosis rather than apoptosis (Koji 2001) – while the remainder become gonocytes (Wang et al. 1998; Print and Loveland 2000; Koji 2001). The gonocytes remain quiescent until after birth, when they are reactivated and initiate the process of spermatogenesis (Print and Loveland 2000).

The first phase of the spermatogenesis occurs in the seminiferous epithelium of the testis, which consists of Sertoli cells and several types of germ cells and later progress to the epididymis where occur the final stages of maturation of the developing gametes (Amann 2008; Johnson et al. 2008). Epididymis also serves as storage space for sperm cells until ejaculation. In the young animal, development of the initial cohort of gonocytes into sperm is known as the *first wave of spermatogenesis*. It is initiated when gonocytes differentiate into spermatogonia and is characterized by the sequential appearance of progressively more mature germ cell types. In the mouse, this occurs between birth and day 5 post-partum, and in men between birth and 6 months of age (Print and Loveland 2000). While some spermatogonia become self-renewing spermatogonial stem cells, most differentiate into spermatocytes. At approximately day 10 post-partum in mice and at puberty in man, these spermatocytes initiate meiosis which is associated with concomitant increases in gonadotrophin and androgen levels (Print and Loveland 2000). Once initiated, spermatogenesis usually continues uninterrupted until death, although in aged individuals can be discerned a slight decrease in the daily sperm production and other quantitative features of spermatogenesis (Paniagua et al. 1991; Kimura et al. 2003; Amann 2008).

Accordingly with the differentiation status, several types of germ cells can be distinguished during spermatogenesis (Fig. 8.1) (Franca et al. 2005; Ehmcke and Schlatt 2006; Ehmcke et al. 2006a,b; Amann 2008). In humans have been identified A_{dark}-spermatogonia, progenitor A_{pale}-spermatogonia, committed A_{pale}-spermatogonia, B-spermatogonia, preleptotene (also termed young primary spermatocytes), leptotene, zygotene, pachytene, and diplotene spermatocytes; secondary spermatocytes; spermatids; and spermatozoa (Amann 2008). The progenitor A_{dark}-spermatogonia are reserve stem cells on the basis of location and infrequent division, but with the appropriate stimuli, they can produce progenitor A_{pale}-spermatogonia that are renewing stem cells (Ehmcke and Schlatt 2006; Amann 2008). Progenitor and committed A_{pale}-spermatogonia display very similar characteristics, but the latter can further divide to produce B-spermatogonia (Amann 2008). In non-human primates, have been identified four generations of B spermatogonia, namely B_1, B_2, B_3 and B_4 (Ehmcke et al. 2006a; Ehmcke et al. 2006b), whereas the committed A_{pale}-spermatogonia is followed by the intermediate $A_{transition}$-spermatogonia. In mice, rats and pigs (Franca et al. 2005; Ehmcke et al. 2006b; Sofikitis et al. 2008),

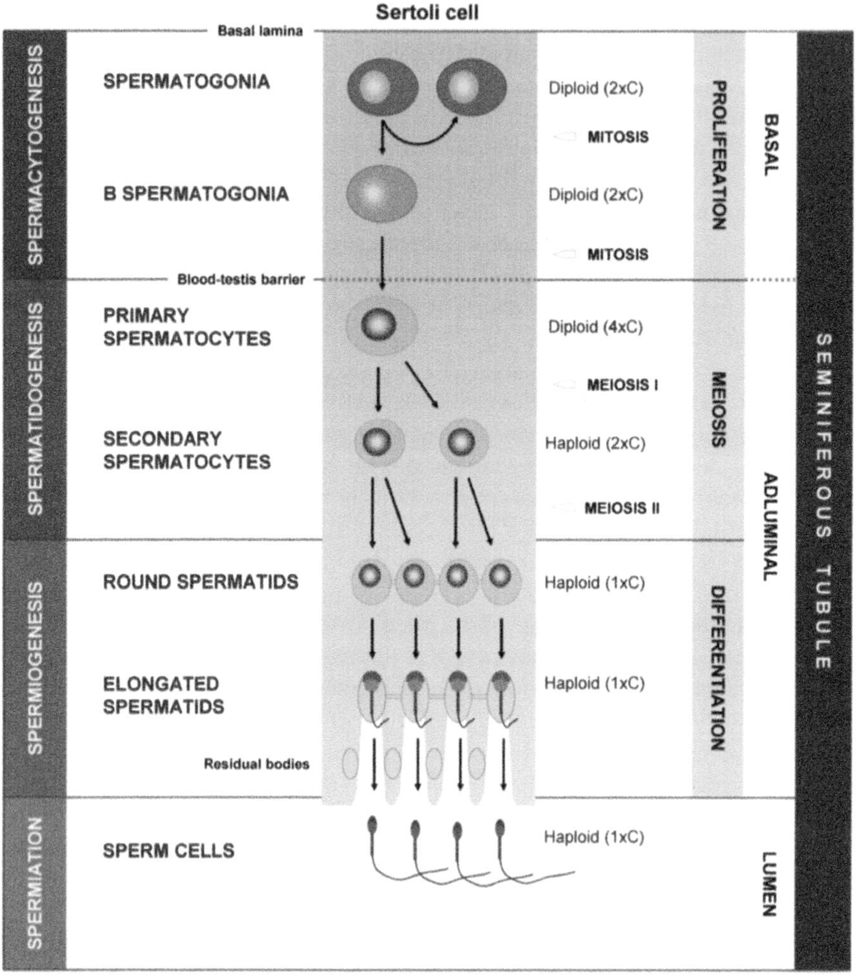

Fig. 8.1 Diagram summarizing the different phases of the spermatogenesis, the characteristics of the germ cells involved in each stage, and the compartments in the seminiferous tubuli. Spermatogenesis is accomplished in the seminiferous tubuli of the testis in four sequential phases, each corresponding with the generation of increasingly differentiated germ cells. The first phase is termed spermacytogenesis, where the spermatogonia, which are the most undifferentiated germ cells, divide by mitosis to renew themselves and to generate B-type spermatogonia. The latter can further divide by mitosis and differentiate forming primary spermatocytes. In the second phase, termed spermatidogenesis, each primary spermatocyte suffer the first meiotic division to generate two secondary spermatocytes which enters second meiotic division to generate four round haploid spermatids. The third phase is termed spermiogenesis and consists in the transformation of a spherical spermatid to a sperm-like mature spermatid. During this transformation the nucleus condenses in size and is stabilized by protamines and most of the cytoplasm of the undifferentiated spermatids is released as residual bodies and phagocyted by the Sertoli cells. The last step, termed spermiation, involves the rupture of the structures and bonds anchoring a mature spermatid to a Sertoli cell, so the spermatozoon is released into the tubule lumen. Accordingly with the cell divisions and the maturation changes, the spermatogenesis show three different phases. The first one is the proliferation phase, involving germ cell division by mitosis and regeneration of cells in

seven types of A spermatogonia (A_{single}, A_{pair}, $A_{aligned}$, A_1, A_2, A_3 and A_4) have been described. Of those, the A_{single} are considered to be the spermatogonial stem cells, whereas the rest are clonally further expanded colonies, either synchronized (A_{pair} and $A_{aligned}$ spermatogonia) or not synchronized (A_1–A_4 spermatogonia) with the seminiferous epithelial cycle. B and Intermediate spermatogonia are morphologically distinct, large interconnected cohorts of spermatogonia which are present at defined spermatogenic stages (Ehmcke et al. 2006b).

In general, distinctive characteristics in the germ cells have been defined for a number of mammalian species, indicating that early differentiation steps in mammalian spermatogenesis are differently developed among species (Franca et al. 2005; Ehmcke et al. 2006b; Amann 2008; Berndtson 2008). As a result, these distinctive characteristics lead to species-specific differences in the organization of the seminiferous epithelium, in the pattern of spermatogonia renewal and proliferation and, ultimately, in the number of sperm cells produced per day (Russell et al. 1990; Franca et al. 2005; Ehmcke et al. 2006b; Amann 2008; Berndtson 2008).

In spite of these differences, from a global point of view spermatogenesis is similarly accomplished in mammals by four interdependent and sequential processes, each corresponding to a particular type of cells (Fig. 8.1) (Franca et al. 2005; Ehmcke and Schlatt 2006; Ehmcke et al. 2006a,b; Amann 2008). The first stage is termed spermatocytogenesis and in humans consists in the spermatogonial proliferation and differentiation, which involve renewal of progenitor A_{dark}- and A_{pale}-spermatogonia, division of committed A_{pale}-spermatogonia to form B-spermatogonia, and their division to form preleptotene spermatocytes. In the second process, termed spermatidogenesis, occurs the meiosis of the spermatocytes. This stage includes the last synthesis of DNA in preleptotene spermatocytes followed by two meiotic divisions to form haploids spermatids. The second meiotic division occurs very rapidly and, as a result, secondary spermatocytes are rarely seen in histological preparations. The third phase is termed spermiogenesis and consists in the transformation of a spherical spermatid to a sperm-like mature spermatid. Finally, the fourth step, termed spermiation, involves the rupture of the structures and bonds anchoring a mature

Fig. 8.1 (continued) the basal compartment; the second is the meiosis phase, in which spermatocytes divide by meiosis to generate haploid cells; the third one is the differentiation phase, in which spermatids do not further divide but suffer profound differentiation changes to generate sperm cells. As indicated in the graph, spermatogenesis develops in an ordered fashion in the seminiferous tubules. Germ cells start to mature on the outside of the tubuli and move towards the lumen with sperm maturation. Sertoli cells are supporting cells that extend from the lumen to the edge of the tubule. They surround and nurture the developing sperm and form a blood-testis barrier which control spermatogenesis and segregate spermatogonia from spermatocytes and more differentiated cells. Accordingly with this barrier two sections can be distinguished in the seminiferous tubuli, the basal area that extend from basal lamina to blood-barrier testis, and the adluminal section that extend from blood-barrier testis to the lumen of the tubuli. Finally, in the graph are also indicate the characteristics of the germ cells in terms of chromosome content (haploid *vs.* diploid) and number of chromatids in each chromosome (1, 2 or 4 chromatids) (See also Plate 10 in the Color Plate Section on page 231)

spermatid to a Sertoli cell, so the spermatozoon is released into the tubule lumen and can be washed out of the seminiferous tubule (Amann 2008). These processes are interdependent, although each has distinct demands for regulatory molecules from nearby Sertoli, peritubular, and Leydig cells, as for the vascular network (Sofikitis et al. 2008; Amann 2008; Johnson et al. 2008).

The Sertoli cells play a major role in regulation of spermatogenesis. Sertoli cell functions include providing structural support and nutrition to developing germ cells, phagocytosis of degenerating germ cells and residual bodies, release of spermatids at spermiation and production of a host of proteins that regulate and/or respond to pituitary hormone release and that influence mitotic activity of spermatogonia (Johnson et al. 2008). Furthermore, a key feature of Sertoli cell structural support for developing germ cells is the blood testis barrier that resides in tight junctions located between adjacent Sertoli cells. This barrier segregates the spermatogonia and early preleptotene primary spermatocytes within the basal compartment. It also permits movement of preleptotene primary spermatocytes into the luminal compartment and spermatocytes and spermatids in the adluminal compartment. This structural arrangement creates an immunologic barrier by isolating the more advanced germ cell types (spermatocytes and spermatids) from the immune system so that their antigens do not stimulate autoimmunity (Johnson et al. 2008).

8.3 Apoptotic Events During the Spermatogenesis

The maintenance of normal architecture of the seminiferous tubuli, thus the sustaining of spermatogenesis at a physiological level, is achieved by a dynamic balance of germ cellular regeneration and elimination (Franca et al. 2005; Johnson et al. 2008). This balance is determined by the limited ability of the Sertoli cells to provide critical factors necessary for the successful progression of spermatogonia into spermatozoa. Therefore, Sertoli cells limit the expansion of the germ cell population because each one is able to nurture a defined number of germ cells (Franca et al. 2005; Sofikitis et al. 2008; Johnson et al. 2008). In support of this concept, experimental reduction of Sertoli cell number in immature rat testes caused a porportinal reduction in the number of round spermatids in the adult animal (Orth et al. 1988).

Although germ cell loss during normal spermatogenesis was clearly established for many years, more recently the concept of apoptosis or programmed cell death in these cells substituted to cell degeneration (Roosen-Runge 1973; Royere et al. 2004). During last years, an increasing number of studies have demonstrated that apoptotic processes may play an important role in the regulation of spermatogenesis in all mammals investigated so far, including human (Kimura et al. 2003; Pareek et al. 2007; Ruwanpura et al. 2008a), pig (Franca et al. 2005), horse (Heninger et al. 2004; Forrest et al. 2006), mouse (De et al. 1993; Wang et al. 1998; Koji 2001; Embree-Ku et al. 2002; Miura et al. 2002), and rat (Yan et al. 2000; Jahnukainen et al. 2004; Tirado et al. 2004). As result, currently it is commonly accepted that germ cells loss by apoptotic mechanisms occurs during normal spermatogenesis in nearly all mammalian species, playing a critical role in determining the size of the sperm output and controlling its final efficiency (Franca et al. 2005; Johnson et al. 2008).

Apoptotic germ cells can be found in all the stages of spermatogenesis, although spermatogonia and spermatocytes are likely the main targets of apoptosis both under physiological and pathological situations (Koji 2001; Franca et al. 2005; Shaha 2007; Sofikitis et al. 2008). Apoptosis in germ cells is also frequent during meiosis, thus in the spermatidogenesis phase, and is probably related to chromosomal damage (Franca et al. 2005; Shaha 2007; Sofikitis et al. 2008). In addition, another loss of germ cells also occurs during spermiogenesis. However, round spermatids behave rather differently than spermatogonia or spermatocytes since the loss of these cells usually occurs *en masse* by forming multinucleated symplasts (Kierszenbaum 2001).

Interestingly, germ cell loss of up to 15% takes place during spermiogenesis in boars from newly-formed to elongating spermatids (Okwun et al. 1996; Franca et al. 2005), which is likely dependent of a regulatory mechanism distinctive of this specie that is still unknown (Franca et al. 2005). For all species, it is estimated that up to 75% of type A spermatogonium plus up to 25% of cells that pass early divisions (namely spermatocytes and spermatids) degenerate by apoptotic mechanism in the testes of adult animals, indicating that normal spermatogenesis occurs at the cost of substantial germ cell wastage (Levy and Seifer-Aknin 2001; Franca et al. 2005).

8.3.1 Apoptotic Pathways in Male Germ Cells

As has been previously reviewed (Chapter 2), the signalling events leading to apoptosis can be divided into two major pathways, involving either mitochondria (intrinsic pathway) or death receptors (extrinsic pathway). Typically, the death receptor pathway is mediated by cell surface receptors that are activated by soluble ligands leading to intracellular effects through adaptor proteins. It is worth mentioning that the cohort of intracellular adapter proteins associated with a given death receptor mediate its effects and ultimately can determine the cell fate. The mitochondrial pathway, on the other hand, is activated by intrinsic mechanisms and primarily do not involve cell death receptors. Both pathways are interconnected at different levels, since the death receptor pathway can use the mitochondrial pathway to amplify apoptotic signals upon receptor activation. Finally, all pathways to apoptosis converge on the activation of caspases, which are cysteinyl aspartate proteases that coordinate the efficient dismantling and engulfment of doomed cells (Said et al. 2004; Jin and El-Deiry 2005; Falschlehner et al. 2007; Youle and Strasser 2008; Schutze et al. 2008).

There are several pathways by which germ cells can undergo cell death, including pathways mediated by members of the Bcl-2 family proteins; by Fas ligand, tumour necrosis factor (TNF), and tumour necrosis factor alpha-related apoptosis-inducing ligand (TRAIL) which activate their counterpart receptors belonging to the tumour necrosis factor receptor superfamily (TNF-R); and by tumour suppressors such as p53 (For a complete description of these pathways see Chapter 2).

8.3.2 Bcl-2 Family Proteins in Spermatogenesis

The role of Bcl-2 family proteins in regulating the primordial germ cells (PGC) fate during embryogenesis has been inferred from studies involving transgenic mice and models misexpresing Bcl-2 genes. Since male *bax* and *bcl-w* knockout mice possess germ cells (Knudson et al. 1995; Print et al. 1998; Ross et al. 1998; Print and Loveland 2000; Meehan et al. 2001) apparently they are not directly required for germ cell survival in the embryo. In addition, as will be discussed later, Bcl-w appears to be also dispensable during the first two weeks after birth (Print et al. 1998; Meehan et al. 2001). However, Bcl-xL potentially promotes germ cell survival during embryogenesis, since the survival of cultured PGC was dramatically increased by ectopic Bcl-xL expression (Watanabe et al. 1997). Nevertheless, since Bcl-xL knockout mice die during embryogenesis, undisputable evidence showing the role of Bcl-xL in the fate of germ cells during embryo and foetal development is still pending.

More evidences exist, however, supporting the role of the Bcl-2 family in the regulation of germ cell fate during early postnatal development. The first wave of spermatogenesis is accompanied by extensive germ cell apoptosis, which in the mouse peaks at approximately 14 days after birth (Rodriguez et al. 1997; Wang et al. 1998; Print and Loveland 2000) and in the rats at approximately 16–18 days after birth (Yan et al. 2000; Zheng et al. 2006). Bax and Bcl-xL and perhaps other Bcl-2 family members are likely involved in the regulation of this early apoptotic wave. Bax is abundantly present in mouse testis between one and three weeks after birth (Rodriguez et al. 1997), and appears to be localized to those germ cells which are undergoing apoptosis during the first spermatogenic wave (Rodriguez et al. 1997). In addition, upregulation of Bax expression is a feature of germ cell apoptosis *in vitro*, since increased levels of Bax expression accompany cytotoxic drug-induced apoptosis of a human germ cell line (Boersma et al. 1997; Print and Loveland 2000).

Studies conducted in genetically modified mice further probe the involvement of Bax in the first spermatogenic wave, since testes of adult *bax* knockout mice contained excessive numbers of spermatogonia and pre-leptotene spermatocytes, findings which are consistent with failed apoptosis during the first wave of spermatogenesis (Knudson et al. 1995; Print and Loveland 2000; Russell et al. 2002). The excess in the number of germ cells were subsequently removed, presumably by Bax-independent apoptosis, although normal adult spermatogenesis was never established and, consequently, Bax knockout mice are sterile as the result of the lack of apoptotic loss of germ cells in the neonatal stage (Knudson et al. 1995; Russell et al. 2002). More recently, Jahnukainen et al. (2004) have reported an increase in the expression of Bax and cleaved caspase 3 in the rat testis at 18 day post-partum, a time when peak germ cell apoptosis was observed (Yan et al. 2000; Zheng et al. 2006). Bax was detected in pachytene spermatocytes, the same cells that are undergoing apoptosis, from days 18 to 26 post-partum, suggesting that Bax regulates apoptosis in midpachytene spermatocytes during the first phase of spermatogenesis in rats (Jahnukainen et al. 2004). In view of the above studies in murine models, it appears that Bax is playing an essential role in regulating apoptosis during the first wave of spermatogenesis.

Similarly to Bax, Bcl-xL may also regulate germ cell survival during the first wave of spermatogenesis, since it is expressed at high levels in testis at this time (Rodriguez et al. 1997). In addition, overexpression of a Bcl-xL in germ cells disrupted the first wave of spermatogenesis by reducing the incidence of apoptosis, and produced a pathology in young adult mice which appears similar to that found in *bax* knockout animals, characterized by an increase in spermatogonia density in seminiferous tubules and few apoptotic cells at day 21 (Knudson et al. 1995; Rodriguez et al. 1997; Russell et al. 2002). A similar phenotype is produced when Bcl-2 is overexpressed in spermatogonia (Furuchi et al. 1996), indicating that the prosurvival functions of Bcl-xL and Bcl-2, when expressed ectopically in the testis, are similar (Print and Loveland 2000). However, Bcl-2 expression in normal testes has not been detected (Hockenbery et al. 1991) and Bcl-2-deficient mice display normal spermatogenesis (Veis et al. 1993), indicating that Bcl-2 is not primarily involved in the regulation of germ cells apoptosis in adult testis. In support of this, Sugiyama et al. (2001) have shown that spermatogonia overexpressing Bcl-2 transplanted into the seminiferous tubules of sterile mice survived longer in their new environment, compared to spermatogonia from wild-type mice, but most of them failed to advance in their development. Thus, spermatogonia overexpressing Bcl-2 remained in a quiescent stage and only reassumed cell differentiation after Bcl-2 gene expression became silent, which occur after a lag period of 5 months following transplantation (Kierszenbaum 2001; Sugiyama et al. 2001). In view of the above reports, it appears that early apoptotic wave in mice, which is required for the formation of mature sperm, is dependent on the proper balance of anti-apoptotic genes, such as *Bcl-xL*, and pro-apoptotic genes such as *bax*, being specially critical the spatial and the temporal expression of these genes in order to correctly set up the spermatogenesis.

Apparently the pro-survival protein Bcl-w is not primarily involved in the regulation of the first wave of spermatogenesis, since the first wave of spermatogenesis progress at a normal rate in the *Bcl-w* knockout mice. In these animals, however, the incidence of apoptosis becomes dramatically elevated between two and four weeks of age resulting in a significant reduction of post-meiotic germ cells in young adult males (Print et al. 1998; Ross et al. 1998; Print and Loveland 2000; Meehan et al. 2001). These results are likely indicating that Bcl-w display a major role in the regulation of germ cell apoptosis upon the first wave of spermatogenesis is accomplished in mice.

The apparently normal phenotype of the *bcl-w* knockout mouse testis, prior to two weeks of age, may be due to the compensatory role of other anti-apoptotic proteins such as Bcl-xL, which appears to be abundant in spermatogonia and pre-leptotene spermatocytes until two weeks of age (Ross et al. 1998; Print and Loveland 2000). However, the possible compensatory role of the anti-apoptotic protein Bcl-xL in transgenic mice deprived of Bcl-w is apparently transient, since the testes of six week old *bcl-w* knockout mice contained numerous apoptotic cells, many of which remained linked as multinucleate symplasts. By six months of age, many tubules of Bcl-w-deficient mice were completely devoid of germ cells (Print et al. 1998), indicating that this pro-survival protein is absolutely required for the normal progression of spermatogenesis in adult testis. Interestingly, transgenic mice overexpressing Bcl-w are also infertile, showing disrupted spermatogenesis with various severities ranging from thin seminiferous epithelium containing less germ cells to Sertoli cell-only appearance within the adult testes of the transgenic mice overexpressing Bcl-w (Yan et al. 2003). These results suggest that regulated spatial and temporal expression of Bcl-w is also required for normal testicular development and spermatogenesis, and that overexpression of Bcl-w inhibits germ cell cycle entry and/or cell cycle progression leading to disrupted spermatogenesis (Yan et al. 2003). Further supporting this are the dynamic changes in the expression profiles of Bcl-2 family proteins during the testicular development in the rat that are consistent with a model in which germ cells are primed for apoptosis during the first cycle of spermatogenesis by *de novo* expression of the death effectors Bax and Bad. These proteins can not initiate further apoptosis until the early spermatogenic wave has been set up by anti-apoptotic Bcl-2 family proteins Bcl-xL and Bcl-w (Yan et al. 2000).

In addition to the murine studies discussed so far, the expression pattern of Bcl-2 proteins has also been reported in normal human testicular tissue by Oldereid et al. (2001), showing that these proteins are distributed preferentially within distinct germ cell compartments. Bcl-x is preferentially expressed in human spermatogonia, whereas Bcl-2 and Bak are preferentially expressed in the compartments of human spermatocytes and differentiating human spermatids. Bax demonstrates a preferential expression in the nuclei of human round spermatids, whereas Bad can be detected in the acrosome region of various stages of human spermatids. Contrarily to the Bcl-2 family members indicated above, Mcl-1 staining does not demonstrate a particular distribution pattern while Bcl-w, p53 and p21 were not detected by immunochemistry in the human testis in this study (Oldereid et al. 2001). The distribution pattern of the Bcl-2 proteins in the human germ cells could be reflecting a differential regulation of the apoptotic pathways to accurately respond to specific signals or injuries accordingly with the differentiation status. For example, since Bax is an apoptotic promoter, the Bax preferential expression in human round spermatids may suggest that round spermatids may be particularly predisposed to apoptosis when DNA is damaged (Sofikitis et al. 2008).

More recently, Kimura et al. (2003) have demonstrated that the apoptotic rate of primary spermatocytes in aged men is significantly elevated compared with younger controls, resulting in a decrease of the number of human primary spermatocytes per Sertoli cell. This germ cell loss directly correlates with a decrease of Bcl-xL in the primary spermatocytes of aged men, indicating that the expression of Bcl-xL is inversely related with the apoptotic rate in human primary spermatocytes and suggesting that Bcl-xL may contribute to the regulation of human primary spermatocyte apoptosis. Contrarily, the apoptotic rate in spermatogonia is significantly lower in aged men compared with younger controls. Despite this lower rate, the balance of spermatogonial proliferation and apoptosis showed no significant difference between the group of aged men and the control group, which could explain why the overall number of spermatogonia in aged men is similar to that described in younger controls (Kimura et al. 2003). These results are indicating that, similarly to the changes described during embryogenesis and first spermatogenic wave in the murine models, the control of germ cell apoptosis in aged individuals is also more than likely regulated by the dynamic changes in the expression profiles of Bcl-2 family proteins.

8.3.3 Death Receptor Pathway: The Fas/FasL System in Spermatogenesis

Fas ligand (FasL) is a transmembrane protein belonging to the large TNF family of proteins which binds and activates the Fas receptor (CD95/APO1) member of the TNF receptor superfamily (Schutze et al. 2008). FasL requires its proteolytic processing in the membrane, into an active soluble form to become accessible to Fas receptor. Upon its activation by FasL, Fas receptor recruits adaptor proteins to its intracellular domains forming a multi-molecular signalling complex that initiates a pro-apoptotic death signal in the bearing cells (Schutze et al. 2008). The existence of both Fas and FasL in normal male germ cells has been reported. However, discrepancies exist regarding the localization and role of the Fas/FasL system. Some studies have reported that Sertoli cells express Fas ligand on their surfaces, whereas Fas receptor is present on the surface of associated spermatogenic cells (French et al. 1996; Lee et al. 1997; Koji et al. 2001). Contrarily, other reports have shown that both FasL and Fas are concomitantly localized to different generations of germ cells (D'Alessio et al. 2001; Nair and Shaha 2003; Shaha 2007). Spermatocytes and round spermatids expressed FasL at the mRNA level, while only elongated spermatids and spermatozoa expressed FasL at the protein level (Riccioli et al. 2003). Nevertheless, as will be discussed later, the expression of the Fas/FasL system in ejaculated mammalian spermatozoa is also controversial.

Independently of the cell type bearing the activator component (FasL), the role of Fas/FasL system in germ cell apoptosis is a matter of debate. A number of studies could not obtain solid evidence on the relationship between the Fas/FasL system and germ cell apoptosis in fetal, neonatal, and adult testes of normal mice (Lee et al. 1997; Wang et al. 1998; Koji 2001; Koji et al. 2001). Furthermore, Fas-deficient *lpr/lpr* mice, are fertile with the apoptosis of germ cells occurring at a rate similar to that obtained in normal mice testis (Adachi et al. 1995; Lee et al. 1997; Koji 2001; Shaha 2007). However, it is suggested that these Fas-deficient mice suffer a spontaneous restoration of Fas levels in the testis by an unknown mechanism. This assumption is supported by the fact that similar amount of Fas protein is found in the testes of both *lpr/lpr* and control mice, whereas the level of Fas protein is significantly lower in all other organs tested in the *lpr/lpr* mice compared to control (Lee et al. 1997). Similarly, the *gld/gld* mice lacking functional FasL are also fertile (Koji 2001), although display a small increase in the number of early stages spermatids compared to control mice (Shaha 2007).

These studies did not definitely support a physiological role for Fas/FasL system during the development of the male germ cells. However, they did not exclude some immunoprotective role for the Fas/FasL system both *in situ* during spermatogenesis (Koji 2001) or in the female genital tract upon ejaculation (Riccioli et al. 2003). In this regard it should be noted that the testis is an immune-privileged organ where FasL may be involved in the killing of infiltrating activated T-cells that express Fas (Koji 2001). Further research work is required to determine whether the Fas/FasL system is effectively involved in the regulation of normal spermatogenesis or, by contrast, is only playing an immunoregulative role in the testis.

Despite its modest role in regulating normal germ cell apoptosis, the Fas/FasL system appears to play an important role during the induction of germ cell apoptosis by a number of testicular stresses and disorders, including ionizing radiations (Embree-Ku et al. 2002), ischemia-reperfusion (Koji 2001; Koji et al. 2001), elevated testicular temperature (Miura et al. 2002), cryptorchidism (Yin et al. 2002), orchities (Theas et al. 2003, 2006), androgen deprivation (Sofikitis et al. 2008), and estrogen stimulation (Koji 2001; Nair and Shaha 2003; Shaha 2007), all of which can disrupt normal spermatogenesis.

The requirement of testosterone and follicle stimulating hormone (FSH) in promoting germ cell survival has been evidenced by experiments using approaches such as hypophysectomy, GnRH antagonists, hormonal supplementation, or hormone immunoneutralization with specific antibodies, leading to the conclusion that both hormones act as germ cell survival factors (Tapanainen et al. 1993; Sinha Hikim et al. 1995; Marshall et al. 1995; Tesarik et al. 2002; Royere et al. 2004; Pareek et al. 2007; Sofikitis et al. 2008; Ruwanpura et al. 2008a,b). However, testosterone shows a dual action on germ cell apoptosis, depending on the experimental conditions and the spermatogenetic stages (Troiano et al. 1994; Royere et al. 2004). In this regard, it has been reported that the excess of testosterone increases the apoptosis of male germ cells with a concomitant expression of the Fas/FasL system in the testis (Zhou et al. 2001). In contrast, toxins that destroy Leydig cell resulting in a rapid decrease of testosterone concentrations stimulate an increase in testicular Fas with a concomitant elevation of germ cell apoptosis (Nandi et al. 1999). These reports, therefore, are pointing out that both the excess of testosterone as well as the deprivation of this hormone likely induce the apoptosis of spermatogonia in a Fas-dependent manner. Nevertheless, in addition to the role that the Fas/FasL plays, the germ cell apoptosis following decrease in FSH and testosterone levels is also influenced by the intrinsic apoptotic pathway (Pareek et al. 2007). In fact, it has been described that spermatogonial apoptosis in men and rats induced by deprivation of testosterone or FSH is predominantly influenced *in vivo* by the intrinsic pathway rather than by the cell death receptor pathway (Ruwanpura et al. 2008a,b), further reinforcing the idea that both pathways are likely involved in germ cell apoptosis after the removal of these survival factors. Finally, the cell primarily targeted by testosterone withdrawal is not clear, because it has been described that testosterone withdrawal stimulates caspase activity and produces DNA fragmentation in Sertoli cell, with only a weak effect on DNA fragmentation and caspase activity in germ cells. These results may suggest that the apoptosis of germ cells induced by testosterone withdrawal is controlled by Sertoli cells via a yet undefined mechanism (Tesarik et al. 2002; Shaha 2007).

In addition to testosterone and FSH, other studies have revealed that estrogens can also induce germ cells apoptosis by means of a Fas-mediated mechanism, further reinforcing the idea that the Fas/FasL system may play a role in the hormonal regulation of spermatogenesis. Nair and Shaha (2003) demonstrated that haploid cells undergo apoptosis in response to diethylstilbestrol (DES), a compound that mimic the estrogen action. In this study the apoptotic pathways involved were also characterized, showing that Fas and FasL were overexpressed in spermatids after DES exposure and that this overexpression was accompanied by the activation of caspase 8. Interestingly, the expression of Fas and FasL proteins were most prominent in germ cells belonging to spermatogenesis stages with a higher ratio of apoptotic processes, reinforcing the idea that these two proteins are major modulators of cell germ death and also indicating that germ cells could modify the expression of death proteins to orchestrate their own death (Nair and Shaha 2003; Shaha 2007). Finally, the estrogen-mediated apoptosis of germ cells is likely amplified by the intrinsic pathway, since there is a DES-induced translocation of Bax from the cytosol to the mitochondria causing the release of cytochrome c and a concomitant reduction of the mitochondrial membrane potential (Mishra and Shaha 2005; Shaha 2007).

Similar results were reported by Koji T. (2001), showing that long-term treatment of adult male mice with estradiol-3-benzoate (EB) caused a marked increase in the apoptosis in germ cells at stages later than type B spermatogonia. Interestingly, even 90 days after treatment with EB, apoptosis-positive cells persisted, indicating that spermatogonial stem cells, which are resistant to estrogen, were still actively proliferating. The involvement of the Fas/FasL system in the induction of germ cell apoptosis by EB was assessed by immunohistochemistry studies, showing a close topographical association among TUNEL-positive germ cells, Fas-positive germ cells, and FasL-positive Sertoli cells (Koji 2001). In view of the above data, it could be inferred that normal testicular homeostasis may involve the Fas/FasL system to maintain proper spermatogenic cell number, which may be under estrogen regulation. The localization of estrogen receptor beta (ERβ) and P450 aromatase in pachytene spermatocytes as well as in other germ and/or somatic cells of the testis (Saunders et al. 1998) suggests a role for estrogens in this step of spermatogenesis, which could be regulating normal spermatogenesis (O'Donnell et al. 2001) as well as germ cell losses induced by testicular toxicants or long-term hormonal treatment (Nair and Shaha 2003; Tirado et al. 2004; Mishra and Shaha 2005; Chaki et al. 2006; Gautam et al. 2007).

On the other hand, the Fas/FasL system could also be playing a role in the germ cell apoptosis induced by cryptorchidism. This is a frequent sexual disorder characterized by the misplacement of the gonads during development by which testis is exposed to body temperature rather than to scrotal temperature. At the spermatogenesis level this is detrimental, since it is known that germ cell apoptosis occurs following their exposure to elevated temperatures (Vera et al. 2004; Shaha 2007). In experimental cryptorchidism in a p53-deficient mice model, a delay of 2–3 days in the spermatogenesis has been demonstrated, indicating that p53 could be involved in the regulation of apoptosis in the early spermatogenesis stages in the cryptorchid testis (Yin et al. 1998). Furthermore, in a double mutant mice bearing Fas- and p53-defective genes, germ cell apoptosis were delayed by an additional 3 days. In addition, the Fas production increased in the time frame of p53-independent apoptosis in the experimental cryptorchid testis of wild-type mice (Yin et al. 2002), likely indicating that the Fas/FasL system is playing a role in this process. However, this double mutant still showed apoptosis of germ cells during experimental cryptorchidism, indicative of that alternative apoptosis pathways must be activated during spermatogenesis in the cryptorchid testis (Yin et al. 2002). In this regard, it has been reported that heat-induced germ cell apoptosis is not impaired in transgenic mice bearing lost-of-function mutations of neither *fasL* nor *fas* genes (Sinha Hikim et al. 2003; Vera et al. 2004; Shaha 2007). In contrast, other study showed that the expression level of *Fas* and *p53* increased significantly from 1 to 3 days after heat exposure, suggesting that germ cell apoptosis induced by heat exposure is possibly mediated by the Fas/FasL system (Miura et al. 2002). In view of these results, it seems likely that additional studies are required to completely understand the regulation of the apoptotic processes in the heat-exposed germ cells, including that occurring in cryptorchid individuals, and particularly the role that the Fas/FasL system could be playing in such regulation.

8.3.4 Other Death Receptor Signalling in Spermatogenesis

In addition to FasL, other ligands acting through the TNF-R superfamily could be also involved in the regulation of apoptosis in male germ cells, including tumor necrosis factor alpha (TNFα) and the tumor necrosis factor alpha-related apoptosis-inducing ligand (TRAIL). TNFα can bind two different receptors (TNFR1 and TNFR2) (Schutze et al. 2008), whereas TRAIL can bind two apoptosis-inducing receptors, TRAIL-R1 (DR4) and TRAIL-R2 (DR5); two additional cell-bound receptors incapable of transmitting an apoptotic signal, TRAIL-R3 (LIT, DcR1) and TRAIL-R4 (TRUNDD, DcR2), sometimes also called *decoy receptors*; and, finally, a soluble receptor called osteoprotegerin (OPG) (Falschlehner et al. 2007). Similarities between TNF and TRAIL receptors were previously underlined about trimerization and death domains. Although the recruited proteins may differ between both systems (FADD *vs.* TRADD) the signal transduction leads to procaspase 8 activation and cell death. However, TNFα may mobilize an alternate anti-apoptotic signal transduction pathway that involves association of TNF-R with another protein (TNFR associated factor 2) able to recruit proteins that ultimately activate the transcription of inhibitors of apoptosis. Thus, depending on the situation and the cell compartment, TNFα may behave as a pro- or anti-apoptotic factor (Falschlehner et al. 2007; Schutze et al. 2008).

Grataroli et al. (2002) studied the ontogenesis of the TRAIL and its receptors in rats. In this study TRAIL was detected both at the mRNA and protein level in foetal and adult testis, within Leydig and germ cells. Additionally, DR5 receptors were observed in Leydig cells whereas DR4 receptors were expressed in post-meiotic germ cells. In a later study, Grataroli et al. (2004) demonstrated by using immunohistochemistry that TRAIL and its receptors were expressed in different human testicular germ cell types. TRAIL and receptor DR5 as well as decoy receptors were localized in Leydig cells, whereas DR4 receptor was detected in human peritubular and Sertoli cells.

McKee et al. (2006) have recently reported that recombinant TRAIL caused a three-fold increase in germ cell apoptosis in primary testis explants, demonstrating that testicular germ cells can undergo TRAIL-mediated apoptosis. Histological observation of the germ cell sub-types undergoing apoptosis shows that it was the post-mitotic spermatocytes that undergo apoptosis following exposure to exogenously added TRAIL. However, mice with a gene deficiency in either DR5 or TRAIL failed to show any morphological differences in the spermatogenesis of adult testis as compared to the wild type strain and no changes were observed in the fecundity or fertility of these mice as compared to control (McKee et al. 2006).

All the above studies (Grataroli et al. 2002, 2004; McKee et al. 2006) may suggest a physiological role for TRAIL in spermatogonia apoptosis, especially in the later stages, although further research is likely required to completely elucidate this potential role.

On the other hand, TNFR was detected in Sertoli and Leydig cells on adult human testis biopsies whereas its ligand, namely TNFα, was detected in pachytene spermatocytes and round spermatids (Pentikainen et al. 2001). In the mouse testis, TNFα is produced in the pachytene spermatocytes and round spermatids and the receptor TNFR1 has been identified in Sertoli cells (De et al. 1993). The localization of the components of the TNFR/TNFα system on germ cells likely indicates that the effect of TNFα is paracrine rather than direct. A role of TNFα in regulating sperm germ apoptosis during spermatogenesis was inferred from various studies. In rats, TNFα promotes cell survival in the seminiferous epithelium, and this effect can be blocked by infliximab, a TNFα antagonist. The prosurvival effect of TNFα in this study might be at least partly mediated by modulating the expression and subcellular localization of Bcl-2 family proteins (Suominen et al. 2004). One additional study showed that TNFα dose-dependently inhibited germ cell apoptosis induced by hormonal or serum deprivation during *in vitro* culture of segments of seminiferous tubules (Pentikainen et al. 2001). The system TNFR/TNFα, however, is not always acting in the germ cells as anti-apoptotic mediators. In agreement with its dual role, it has been shown that soluble factors released from testicular macrophages during experimental autoimmune orchitis in rats induce apoptosis of germ cells and that TNFα is a relevant cytokine involved in testicular damage during severe orchitis (Suescun et al. 2003; Theas et al. 2006, 2008), indicating that TNF-R can also promote germ cell apoptosis in experimentally damaged adult testes.

Interestingly, several studies suggest that the TNFR/TNFα system can regulate germ cell apoptosis by modulating the level of expression of the Fas/FasL system, although the precise interrelation between both signalling pathways is still not completely understood. Pentikainen et al. (2001) have reported that TNFα induced a partial decrease in FasL expression during *in vitro* culture of segments of seminiferous tubules deprived of hormone or serum, which was concomitant with its anti-apoptotic effect. In contrast, in cultured mouse Sertoli cells it has been reported that TNFα (and IFNγ) stimulation markedly increase the expression of both soluble and membrane-bound Fas in a dose-dependent manner (Riccioli et al. 2000). Furthermore, recently has been proposed a mechanism by which germ cell apoptosis is triggered by the expression of FasL in Sertoli cells in a TNFα-dependent manner (Yao et al. 2007). In this model, a toxic injury to the Seroli cells increase FasL transcription and FasL protein levels on the Sertoli cell plasma membrane, which can then interact with Fas receptor expressed on adjacent germ cells to trigger them to undergo apoptosis. In addition, toxic-induced Sertoli cell injury also decreases physical, hormonal, and nutritive support for germ cells leading to an increased production and release of the soluble form of TNFα released from these cells. Germ cell-produced TNFα can then bind to TNFR1 that is present on the Sertoli cell resulting in additional in creases in the FasL gene transcription. This model points to a potential feed-forward signalling mechanism by which germ cells prompt Sertoli cells to trigger their own apoptotic elimination, more than likely with the aim of demise the number of germ cells in the injured testis (Yao et al. 2007).

8.3.5 Other Modulators of Male Germ Cell Apoptosis

The cell cycle regulator p53 appears to be involved in the apoptosis of sperm germ cells, since its expression was reported to increase in the first 4 weeks of life in mice (Rossi et al. 2000) and p53 knockout mice exhibited a doubling of undifferentiated spermatogonia compared with wild type mice (Beumer et al. 1998). In addition, overexpression of p53 in post-meiotic cells was reported to generate a phenotype varying from a normal fertility to a severely altered spermatogenesis, depending on the level of overexpression (Allemand et al. 1999), suggesting an important concentration-dependent effect of this tumour suppressor on spermatogenesis. Furthermore, the apoptosis induced by p53 overexpression in meiotic and post-meiotic cells appears to be caspase independent (Coureuil et al. 2006). P53 is also required for radiation-induced apoptosis of spermatogonia (Hasegawa et al. 1998; Beumer et al. 1998) and is also involved in the regulation of sperm germ apoptosis during heat stress of the testis (Yin et al. 2002). In view of these reports, p53 gene could account for the selective apoptosis of damaged germ cells in order to prevent the transmission of genetic abnormalities to offspring (Print and Loveland 2000), accordingly with the p53 role in conserving genomic stability. In addition, other transcription factors such as c-Myc, E2F-1 or NF-κB may provide additional fate-determining signals during spermatogenesis (Kodaira et al. 1996;

Holmberg et al. 1998; Teng and Vilagrasa 1998; Pentikainen et al. 2002; Suominen et al. 2004; El-Darwish et al. 2006; Yao et al. 2007).

Programmed cell death can also be modulated in somatic cells by a large cohort of kinases and phosphatases. By using different experimental approaches, many of these components have also been detected in germ cells, although their hypothetical role in regulating germ cell apoptosis is, to date, largely unknown. It is worth mentioning that PI3-K (Blume-Jensen et al. 2000; Lee et al. 2007), Akt (Hixon and Boekelheide 2003; Stabile et al. 2006; Lee et al. 2007), PKC (Um et al. 1995; Niino et al. 2001; Feng et al. 2005), tyrosine phosphatases (Kaneko et al. 1993), and cyclins (Liu et al. 2000) among other signalling proteins, all of which are well known regulators of apoptosis in somatic cells, have been found in mammalian germ cells. Furthermore, other signalling pathways also converge to apoptosis in somatic cells to which attention should be directed in order to completely understand the regulation of apoptosis during spermatogenesis, such as the role integrins in maintaining the cohesion of the seminiferous epithelium (Beardsley and O'Donnell 2003; Siu et al. 2005; Beardsley et al. 2006).

8.3.6 Role of Oxidative Stress and Ca^{2+} Signalling in Germ Cell Apoptosis

Many of the chemical and physical treatments capable of inducing apoptosis are known to evoke oxidative stress. In recent years, a large body of evidence has accumulated to indicate that oxidative stress plays a fundamental role in the induction of germ cell apoptosis under conditions of stress to testis. Elevation of oxidative stress in testes was reported under conditions like heat (Ikeda et al. 1999), ischemia (Chaki et al. 2003), or toxicant damage (Mishra et al. 2006; Shaha 2007).

Alteration of normal hormonal milieu by estrogen administration has been also reported to induce oxidative stress and apoptotic changes during spermatogenesis (Mishra and Shaha 2005; Chaki et al. 2006; Gautam et al. 2007). Gautam et al. (2007) have reported a significantly higher H_2O_2 level in the tubular cells in rats following a long–term treatment with human chorionic gonadotropin (hCG) during which lipid peroxidation in the testis was increased and activities of enzymatic antioxidants such as superoxide dismutase, catalase and glutathione-s-transferase were significantly down-regulated. These data suggest that the pathogenesis of germ cell apoptosis following chronic hCG treatment was due to increased H_2O_2 levels and oxidative stress in the testis. The apoptosis was mediated through the activation of pro-apoptotic factors such as Fas protein expression along with a multifold increase in caspase 3 activity in the testis. Impairment of spermatogenesis was consistent with the decrease in the number of maturing germ cells in the seminiferous epithelium of the treated rats (Gautam et al. 2007).

Furthermore, Mishra and Shaha (2005) showed that estradiol causes increases in FasL expression and a subsequent activation of the Fas/FasL system in the rat testis. Interestingly, this system induces a concomitant increase of NO formation through the activation of inducible nitric oxide synthase. The peroxynitrites formed as reactive products of superoxide and NO lead to the induction of a transient mitochondrial hyperpolarization resulting in the release of cytochrome c from the organelle. Scavenging of the nitrites with an intracellular peroxynitrite scavenger reduced hyperpolarization and cell death. Interestingly, prevention of NO generation to reduce nitrite formation leads to the inhibition of hyperpoarization as well as caspase 3 activation, however, inhibition of caspase 8 cleavage was unable to prevent hyperpolarization but reduced caspase 3 activity indicating the existence of a dual pathway (Mishra and Shaha 2005; Shaha 2007).

On the other hand, recent studies have also demonstrated a crucial role for Ca^{2+} homeostasis in the regulation of the apoptotic processes in male germ cells. For example, by using an *in vitro* spermatogenic cell apoptosis model, Mishra et al. (2006) have shown that toxin-induced apoptosis caused a significant increase in reactive oxygen species (ROS) followed by an enhancement of intracellular Ca^{2+} through the T-type Ca^{2+} channels, which correlated with concomitant changes in the ratio of pro-apoptotic and pro-survival Bcl-2 proteins. Impediment of Ca^{2+} influx by blocking the T-type Ca^{2+} channel, restored the normal ratio of the Bcl-2 proteins and caused prevention of mitochondrial potential loss, reduction of caspase 3 activity, inhibition of DNA fragmentation, and increase in cell survival (Mishra et al. 2006).

8.4 The Ejaculated Mammalian Spermatozoa

From a morphological point of view, in the spermatozoa can be distinguished two main structures: the head and the tail or flagellum (Fig. 8.2) (Mortimer 1997; Flesch and Gadella 2000). The sperm head contains, besides a very low amount of cytosol, the nucleus and the acrosome. Accordingly with the extension of the latter, in sperm head plasma membrane of most mammals are defined two principal regions termed acrosomal region and postacrosomal region. The acrosomal region, often also referred as acrosomal cap, is positioned in the anterior part of the head and contains a posterior region termed equatorial segment which is a domain located immediately posterior to the acrosome. The size and shape of the acrosomal region show profound differences among different mammalian species. The postacrosomal region includes the plasma membrane between the posterior margin of the acrosome and the posterior ring or connecting piece at the flagellum (Mortimer 1997; Flesch and Gadella 2000).

The flagellum is structurally divided into four major parts: the connecting piece, the midpiece, the principal piece, and the end piece (Mortimer 1997; Inaba 2003). In the flagellum is located the axoneme, structure absolutely required for the flagellum to perform motility. This structure is highly conserved in all ciliated and flagellated eukaryotic cells. However, only mammalian sperm flagella contain three additional accessory structures: the mitochondrial sheath, outer dense fibres, and fibrous sheath, all of which can be distinguished within different extensions of the sperm flagellum (Mortimer 1997; Inaba 2003).

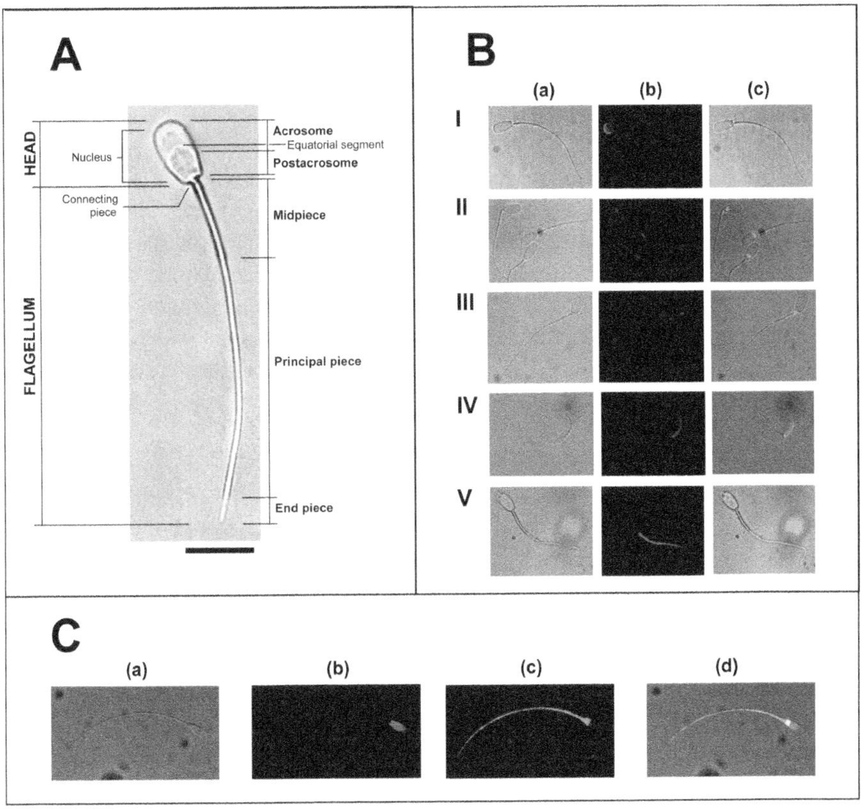

Fig. 8.2 Subcellular structures in a mammalian spermatozoon. *A*. Microphotograph of a boar spermatozoon showing the distribution of the different subcellular structures that can be distinguished in the ejaculated mammalian spermatozoa. The image was obtained with transmission light in an invert microscope (×100). The *black* calibration bar in the *bottom* represents 10 μm. *B*. Fluorescent images of boar spermatozoa labelled with antibodies which specifically recognize proteins located in the acrosome (I), postacrosome (II), connecting piece (III), midpiece (IV), and rest of the tail (V). All images were obtained with a Bio-Rad MRC1024 confocal microscope with a ×60 objective in oil immersion. In each column, panel (**a**) represent transmission images; (**b**) is the immunofluorescence obtained in the samples excited at 488 nm with an argon laser and recorded with a 515-nm longpass emission filter; and (**c**) is the superposition of a and b. *C*. Fluorescent images of a stallion spermatozoon labelled with a nuclear marker and a fluorogenic marker for active caspases. The images were obtained at ×60 magnification in a Bio-Rad MRC1024 confocal microscope equipped with an argon laser. All samples were excited at 488 nm. **Panel (a)** represents a transmission image. **Panel (b)** shows the nuclear fluorescence recorded with a 680/32 nm emission filter. **Panel (c)** is the fluorescence recorded with a 540/30 nm emission filter which corresponds with active caspases. **Panel (d)** is the merge of a + b + c. (See also Plate 11 in the Color Plate Section on page 232)

In relation to these ultrastructural characteristics, it is worth mentioning that spermatozoa are highly compartmentalized cells. In the head is stored the genome in an extremely compact nucleus, which is covered by the acrosome that contains hydrolytic enzymes. Mitochondria are located at the junction between the head and the flagellum, forming a tight helical structure around the fibrous sheath in the middle-piece. The flagellum, as has been indicated above, contains the motile apparatus, the axoneme, which is surrounded in mammalian spermatozoa by nine outer dense fibres and a thick fibrous sheath that run for most, but not all, the flagellum length (Mortimer 1997; Abou-Haila and Tulsiani 2000; Inaba 2003; de Laminarde and O'Flaherty 2008). Accordingly with this compartmentalization, three highly specialized regions in the mature sperm cell can be distinguished (Fig. 8.2): (i) the sperm head, involved in sperm-oocyte interaction; (ii) the midpiece with mitochondria, involved in energy production; (iii) the flagellum, involved in motility (Flesch and Gadella 2000).

Mammalian mature spermatozoa have only two major functions. In first place, they carry the male genome to the oocyte and, in second place, spermatozoa must then bind and fertilize oocytes (Mortimer 1997; de Laminarde and O'Flaherty 2008). However, during its formation in the testes, testicular spermatozoa are non-motile and are unable to bind to the oocyte extracellular coat, the *zona pellucida* (ZP), thus they do not possess fertilizing ability. On contrary, the mammalian spermatozoon must undergo a series of profound biochemical and functional changes in order to acquire fertilizing ability. These changes occur after spermiation during the epididymal transit, in a final maturational step termed as epididymal maturation, and in the female genital tract, in a process collectively termed as capacitation (de Laminarde et al. 1997; Abou-Haila and Tulsiani 2000; Gadella and Van Gestel 2004; de Laminarde and O'Flaherty 2008; Gadella et al. 2008; Gadella 2008).

The net change during capacitation is a combined effect of multiple molecular modifications in sperm plasma membrane proteins/glycoproteins and lipid components that modify the ion channels in the plasmalemma of spermatozoa (Abou-Haila and Tulsiani 2000; Flesch and Gadella 2000; Gadella et al. 2008; Gadella 2008). The preparatory modifications include removal of seminal plasma proteins/glycoproteins adsorbed to the surface of ejaculated spermatozoa and modifications/reorganization of sperm surface molecules. These changes induce an increase in membrane fluidity, based on cholesterol efflux, and changes in ion permeability resulting in alteration of sperm membrane potential. During capacitation is also detected a striking increment in the tyrosine phosphorylation of diverse proteins (Abou-Haila and Tulsiani 2000; Flesch and Gadella 2000; Gadella et al. 2008; Gadella 2008). Another characteristic of spermatozoa also associated to capacitation is the initiation of an exacerbated motility pattern, termed as hyperactivation (Mortimer 1997; Ho and Suarez 2001; de Laminarde and O'Flaherty 2008). These changes will allow the spermatozoa to bind to ZP in a carbohydrate-mediated receptor-ligand process, which ultimately will lead to the acrosome reaction (see below) (Abou-Haila and Tulsiani 2000; Flesch and Gadella 2000). Although capacitation of a sperm cell is required before fertilization virtually in every mammalian species studied, the molecular mechanisms and signal transduction pathways involved in this process are not clearly understood.

The head region of a capacitated spermatozoon expresses specific receptors that enable it to bind to the ZP of an oocyte in a species-specific manner. Binding to the ZP stimulates the spermatozoon to undergo the acrosome reaction, in which the outer acrosomal membrane fuses with the overlying plasma membrane (Abou-Haila and Tulsiani 2000; Flesch and Gadella 2000; Gadella 2008). This exocytotic event results in the release of hydrolytic enzymes, principally the trypsin-like acrosin, and in the exposure of new membrane domains, both of which are essential for fertilization. The ZP-induced acrosome reaction enables the hyperactive motile sperm to penetrate through the ZP. Once one sperm reaches the perivitelline space, its destiny is to bind to and fuse with the oolemma, thus initiating the activation of the oocyte. In contrast to the ZP-binding event, this event is far less species-specific (Flesch and Gadella 2000; Gadella 2008).

The regulation of signalling pathways that are activated during acrosome reaction are poorly understood. However, one of the very early responses generated upon interaction with ZP in spermatozoa is the activation of mechanisms leading to Ca^{2+} influx (Roldan and Shi 2007). This ion is essential for acrosomal exocytosis (Roldan and Shi 2007) since it is necessary for the activation of intracellular enzymes and for the actual fusion of membranes. Accordingly, various Ca^{2+}-dependent steps have been identified in the sequence underlying acrosomal exocytosis (Roldan and Shi 2007). In addition to Ca^{2+}, protein tyrosine phosphorylation seems to be also important during the early steps of the acrosome reaction (de Laminarde et al. 1998; Lefievre et al. 2002; de Laminarde and O'Flaherty 2008).

8.5 Apoptotic Events in Ejaculated Mammalian Spermatozoa

Features typically associated with apoptosis in somatic cells have been identified in ejaculated human, boar, horse, bull, rat, hamster and mouse spermatozoa, including DNA fragmentation, plasma membrane blebbing, chromatin condensation, caspase activation, loss of mitochondrial membrane potential (MMP or $\Delta\Psi_m$), decrease in plasma membrane integrity, and externalization of phosphatidylserine on the plasma membrane (PSE) (Paasch et al. 2004a; Martin et al. 2005; Said et al. 2005a,b, 2006; Grunewald et al. 2006; Barroso et al. 2006; Cisternas and Moreno 2006; Nunez-Martinez et al. 2007; Aziz et al. 2007; Martin et al. 2007; Moran et al. 2008; Ortega-Ferrusola et al. 2008; Said et al. 2008; Brum et al. 2008).

Although the role of apoptosis in somatic cells and during spermatogenesis is well established, the presence of apoptotic markers in mature sperm cells has been more controversial and there is disagreement regarding the importance of these markers in these cells (Brum et al. 2008). There are two main theories regarding why ejaculated spermatozoa exhibit certain characteristics of apoptosis. The first is that during spermatogenesis defective germ cells are marked for elimination and begin undergoing apoptosis, but some of these cells escape elimination in the testis and therefore appear in the ejaculate as apoptotic spermatozoa. This first theory has been termed abortive apoptosis (Sakkas et al. 1999a,b, 2003, 2004). The second theory is that an active apoptotic process occurs in mature spermatozoa.

In the abortive apoptosis scheme, sperm production is mediated by Sertoli cell support and the number of Sertoli cells within the testis defines an upper limit for sperm production. In a number of mammalian species, Sertoli cells detect damage in maturing sperm cells and eliminate damaged germ cells through apoptosis (Sakkas et al. 1999a,b, 2004). Interruptions in this process may lead to increased numbers of ejaculated spermatozoa exhibiting apoptotic-like characteristics and result in decreased fertility. Previous studies have found that immature and abnormal spermatozoa had higher levels of apoptotic markers, including active caspases, Fas, Bcl-xL, p53, and PSE (Sakkas et al. 1999a,b; Barroso et al. 2000; Muratori et al. 2000; Sakkas et al. 2003; Lachaud et al. 2004; Sakkas et al. 2004). The apoptotic appearing spermatozoa observed in a number of studies may have arisen from the abortive apoptosis process during spermatogenesis in which germ cells expressed various apoptotic markers, yet they escaped cell death due to the failure of the Sertoli cells to eliminate these defective germ cells, leading to the appearance of apoptotic-like spermatozoa cells in the ejaculate (Sakkas et al. 1999a,b, 2003, 2004).

The second theory suggests that mature, ejaculated spermatozoa are autonomously capable of undergoing apoptosis or an apoptotic-like process (Brum et al. 2008). It had been previously suggested that mature spermatozoa do not have efficient operative mechanisms for protein synthesis or DNA degradation, thus they could not complete the apoptosis process in their mature state. However, it has been recently reported that ejaculated spermatozoa are capable of triggering a nuclear matrix-associated topoisomerase IIB that, concomitantly with a nuclease, can cleave the entire sperm DNA into small fragments (Shaman et al. 2006, 2007). Furthermore, current data demonstrates that apoptosis, or at least an apoptotic-like process, can be induced in ejaculated spermatozoa by a variety of stimuli and stresses (Ball 2008; Brum et al. 2008). Specifically, heat stress, cryopreservation, androgen receptor activation, progesterone, staurosporine, betulinic acid, active lipids, oxidative stress and TNFα, have recently been shown to cause ejaculated spermatozoa to display signs of apoptosis, including caspase activation, loss of MMP, PSE, and DNA fragmentation (Taylor et al. 2004; Paasch et al. 2004c; Grunewald et al. 2005; Aquila et al. 2007; Bejarano et al. 2008; Perdichizzi et al. 2007; Choi et al. 2008; Brum et al. 2008), and also to cause the appearance of modulating factors of both the intrinsic and the extrinsic apoptosis pathways, such as Bcl-2 proteins or inhibitors of apoptosis (Taylor et al. 2004; Paasch et al. 2004c; Aquila et al. 2007; Choi et al. 2008), likely indicating that both pathways can be autonomously initiated in ejaculated spermatozoa.

Nevertheless, independently of whether the apoptotic changes detected in ejaculated spermatozoa are triggered by the abortive apoptosis or are active apoptotic processes, remains to be determined if these changes are detrimental or, indeed, are required, for the spermatozoa to complete their maturation or to gain fertilizing ability. A number or reports likely indicate that apoptotic changes are detrimental for the spermatozoa and that the selection of the non-apoptotic spermatozoa seems to be an advantage in order to obtain the sperm subpopulation with the highest fertilizing potential (Paasch et al. 2004a; Said et al. 2005a,b, 2006; Grunewald et al. 2006; Aziz et al. 2007; Said et al. 2008).

Although apoptosis is a characteristic form of cell death that usually serves to remove unwanted and potentially dangerous cells, there are some examples where apoptosis-like events do not lead to death but rather are involved in the terminal differentiation of certain cell types (Said et al. 2004). A good example of this is the terminal differentiation of sperm, which shares many morphological and biochemical features with apoptosis. However, rather than causing the death of the entire cell, apoptotic molecular machinery is used to specifically eliminate cytoplasmic components, thereby producing a highly specialized living cell (Said et al. 2004). This maturation, physiological process is later reflected in the cytoplasm collections in the residual bodies detected in mouse and human mature spermatozoa, which display several features of apoptosis (Marchiani et al. 2007a,b). On the other hand, since the overall survival time of the spermatozoa (both fresh and frozen-thawed) from different species could not be prolonged by inhibitors of caspases (Weil et al. 1998; Heninger et al. 2004; Peter et al. 2005), the mere presence of active caspases does not seem to be a crucial lethal factor in the ejaculated spermatozoa. Thus, it can be speculated that active caspases potentially display additional physiological roles in mature spermatozoa rather than being only involved in detrimental changes.

In support of this view it is worth mentioning that an early activation of caspase 3 has been described upon stimulation of intact terminally differentiated cells with physiological concentrations of agonists. Early caspase 3 activation under these circumstances in these cells (platelets and pancreatic acinar cells) is independent on caspase 9 or mitochondrial cytochrome c release, likely indicating that this is a non-apoptotic event required for cellular function rather than for cell death (Rosado et al. 2006). A similar view can be also applied for additional markers of apoptosis, such as PSE which has been related with the initiation of capacitation and acrosome reaction in ejaculated spermatozoa (Pena et al. 2003; Martin et al. 2005; Moran et al. 2008; Gadella et al. 2008).

Furthermore, in a recent report in boar spermatozoa, Choi et al. (2008) have described that the cytochrome *c* expression is increased during capacitation, whereas apoptotic executors, such as caspase 3, were significantly decreased. Therefore, it is proposed that cytochrome *c* upregulation in boar spermatozoa is involved in the activation of signalling pathways leading to tyrosine phosphorylation and capacitation, rather than to be related with apoptotic changes (Choi et al. 2008). Finally, in boar spermatozoa it has been also reported that botulinic acid induces apoptotic-like changes without caspase activation, indicating that both processes can be independently activated under certain circumstances in boar spermatozoa (Moran et al. 2008).

8.5.1 Apoptotic Pathways in Ejaculated Mammalian Spermatozoa

The mechanisms triggering apoptosis in somatic and male germ cells are well characterized (although not yet completely understood). However there is comparatively less information in ejaculated spermatozoa, especially in regard to the molecular machinery controlling both the intrinsic and the extrinsic pathways of apoptosis.

In general, the members of the anti-apoptotic Bcl-2 family proteins display a lower expression in ejaculated spermatozoa compared with germ cells, as a reflect of a different expression pattern of these proteins through the spermatogenic development (Oldereid et al. 2001). For example in adult testis, in situ hybridisation detected abundant *bcl-w* mRNA in spermatogonia, and lower levels in Sertoli cells, spermatocytes and round spermatids, but none was detected in elongating spermatids (Print et al. 1998; Print and Loveland 2000; Meehan et al. 2001). Furthermore, Bax is present in bovine spermatozoa, whereas Bcl-2 is not detected (Martin et al. 2007). Rat, mouse, and hamster spermatozoa express the proapoptotic protein Bid (Heninger et al. 2004), whereas anti-apoptotic proteins were not found (Hockenbery et al. 1991). Although conflicting reports exist (Hockenbery et al. 1991), it seems likely that ejaculated human spermatozoa express a modest amount of Bcl-2 which is increased during capacitation (Aquila et al. 2007). However, this study has also revealed that, regardless of its modest level of expression, this anti-apoptotic protein could be playing an important role in maintaining sperm vitality, since the stimulation of spermatozoa with dihydrotestosterone leads to the phosphorylation of Bcl-2 concomitantly with an increase in apoptotic changes, all of which are apparently downstream on the PI3K/AKT pathway (Aquila et al. 2007).

On the other hand, the expression and regulation of the components involved in the activation of the death receptor pathway, both the ligands and the receptors, has rendered conflicting data in ejaculated spermatozoa. It has been reported that in normal human males less than 10% of spermatozoa express Fas. This number, however, is much higher in males with defects in sperm number, motility and morphology (Sakkas et al. 1999b). In spite of this, in a recent study it has been described that human ejaculated sperm from both normozoospermic and non-normozoospermic men do not express any detectable quantity of Fas (Perticarari et al. 2008). Furthermore, the incubation of human ejaculated spermatozoa with activating anti-Fas antibody was unable to induce apoptosis in these cells (Grunewald et al. 2005; Perticarari et al. 2008). These results suggest that Fas has no functional relevance in mediating caspase activation in human ejaculated spermatozoa. On the other hand, TNFα displays a positive role in the induction of apoptotic changes in ejaculated spermatozoa. TNFα caused in a concentration- and time-dependent manner a reduction of motility, decreased MMP and increased PSE leading to sperm chromatin and DNA damage in ejaculated human spermatozoa from normozoospermic donors (Perdichizzi et al. 2007). Finally, exogenously added TRAIL did not show any effect in the viability of mouse spermatozoa, despite the presence of DR4 and DR5 receptors in these cells (Heninger et al. 2004). DR4 and DR5 receptors were also found in rat spermatozoa, but not in hamster spermatozoa (Heninger et al. 2004).

8.5.2 Apoptotic Changes in Ejaculated Spermatozoa and Their Relation with Sperm Function

As has indicated earlier in this chapter, features typically associated with apoptosis in somatic cells have been identified in ejaculated human, boar, horse, bull, rat, hamster and mouse spermatozoa, including DNA fragmentation, plasma membrane blebbing, chromatin condensation, caspase activation, loss of MMP, decrease in plasma membrane integrity, and externalization of phosphatidylserine on the plasma membrane (PSE) (Paasch et al. 2004a; Martin et al. 2005; Said et al. 2005a,b, 2006; Grunewald et al. 2006; Barroso et al. 2006; Cisternas and Moreno 2006; Nunez-Martinez et al. 2007; Aziz et al. 2007; Martin et al. 2007; Moran et al. 2008; Ortega-Ferrusola et al. 2008; Said et al. 2008; Brum et al. 2008). In a number of studies these changes seem to be related with detrimental changes in spermatozoa, whereas other studies attributed a physiological-like role to some of the apoptotic features. Therefore the precise significance of apoptosis in ejaculated mammalian spermatozoa is still matter of debate. In addition, increases in the indicators of the apoptotic changes are commonly described during cryopreservation, suggesting that a correlation among apoptosis and cellular damage in frozen-thawed spermatozoa likely exists. In the following lines some of these apoptotic features detected in ejaculated spermatozoa will be discussed in relation with their potential physiological relevance.

8.5.2.1 Caspase Activation

In human ejaculated spermatozoa, active caspases have been observed predominantly in the postacrosomal region (caspases 8, 1 and 3) and caspase 9 has been particularly localized in the midpiece, consistent with the localization of mitochondria in this structure (Paasch et al. 2004a). Some of these caspases, specially caspases 3, 8, and 9, appears be activated in frozen-thawed semen from patients and donors (Paasch et al. 2004a,c; Wundrich et al. 2006). The types of caspases found indicate that apoptosis might not only act via the receptor–death-inducing signalling complex–caspase 8 pathway in human sperm. The presence of caspase 9 emphasizes the important role of mitochondria in the apoptosis signalling in these cells. These findings are supported by the prominence of activated caspase 3, acting as a common effector caspase of both pathways (Paasch et al. 2004a). However, the diversity and subcellular distribution of caspases within the human spermatozoa does not seem to be identical in all mammalian species. In equine spermatozoa we have described that caspase 3 is mainly localized to the acrosome and midpiece whereas caspase 9 is found along the tail with little or no signal in the head. Interestingly, caspase 7, which primarily functions as disruptor of cellular components involved in DNA repair, is mainly located in the midpiece and shows a weaker signal in the head and rest of the tail (Ortega-Ferrusola et al. 2008). The fact that a large proportion of ejaculated stallion spermatozoa display caspase activity may be indicative of an apoptotic mechanism involved in sperm selection within the mare genital tract

(Ortega-Ferrusola et al. 2008). Bovine spermatozoa possess caspase 9 but are deprived of caspase 3 and caspase 8, likely indicating that intrinsic-pathway is primarily involved in the activation of caspases in this specie (Martin et al. 2007). Active caspase 3 and 7 has been described in ram spermatozoa (Marti et al. 2006, 2008), allowing sperm to be allocated in different subtypes depending on caspase localization at apical, equatorial, post-acrosomal regions on the head and in the tail (Marti et al. 2008). Finally, Cisternas and Moreno (2006), have shown that, in conjunction with other proteins involved in apoptosis, rat, mouse, and hamster spermatozoa have both active and inactive caspase 8 and 3, whereas active caspase 9 was not found.

It is clear from these studies that at least low levels of active caspases, especially caspase 3, were present in mammalian spermatozoa. Interestingly, these levels were significantly higher in the subpopulations known to contain morphologically abnormal and immature forms of spermatozoa (Said et al. 2004), indicating that the caspase content may be directly related with a lower spermatozoa vitality and lower sperm function. Moreover, a wide spectrum of cell cytoskeletal proteins and membrane components are also targets of caspase 3 in mammalian spermatozoa, therefore its role in decreased fertilization capacity is presumed (Said et al. 2004; Paasch et al. 2004a,c). In this regard, it has been reported that human sperm motility has been negatively correlated with caspases activation (Weng et al. 2002; Taylor et al. 2004; Paasch et al. 2004a). Specifically, sperm fractions with low motility exhibit more active caspases positive cells than the high-motility fractions in donors and patients. In support of this, higher levels of both the active caspase 3 and the inactive procaspase 3 were present in the low-motility fraction compared with the high-motility fractions from donors and patients (Weng et al. 2002). Furthermore, there are lower and more variable levels of procaspase 3 in high-motility fractions of patients and donors and a virtual absence of active caspase 3 (Weng et al. 2002). However, it is worth mentioning that the addition of a pan caspase inhibitor did not lead to any improvement in sperm post-thaw motility (Peter et al. 2005). The activation of caspases in low-motility sperm samples may be attributed to the role played by cytochrome c–Apaf-1 complex, which can activate caspase 9 and is followed by activation of downstream death effectors such as caspase 3, 6 and 7 (Said et al. 2004).

The increment of caspase activity is also a common finding during cryopreservation, being described such increment in frozen-thawed spermatozoa from man, horse, ram, and bull (Martin et al. 2004; Paasch et al. 2004a,c; Wundrich et al. 2006; Ortega-Ferrusola et al. 2008; Brum et al. 2008; Marti et al. 2008). The increment of caspase activation in frozen-thawed semen is usually accompanied by a significantly greater population of spermatozoa that displayed further increments in other apoptotic features, specially an increment of membrane permeability, poor motility and lower vitality. As apoptosis is an active process, it seems likely that in the later stages active caspases would be seen along with increased membrane permeability and breakdown of the nucleus. However, it has been reported that the addition of caspase inhibitors to the cryopreservation medium failed to improve the acrosome and plasma membrane integrity of frozen-thawed ram, dog, and stallion spermatozoa (Peter and Linde-Forsberg 2003; Peter et al. 2005; Marti et al. 2008), likely indicating that caspases are not directly involved in the stress of cryopreservation.

Finally, the role of caspases during capacitation of human sperm has also been investigated. Media containing bicarbonate was used to induce protein kinase-mediated alterations in the phospholipid bilayer, which correlate with capacitation markers. Caspase inhibitors failed to block the capacitation-related modifications in the phospholipid bilayer, and caspase 3 was not detected in the mature sperm. Therefore, it appears that there is no current evidence implicating caspase 3 in mammalian capacitation (de Vries et al. 2003; Said et al. 2004).

8.5.2.2 Changes in Membrane Permeability and Structure: Phosphatidylserine Externalization (PSE)

The combination of fluorescent probes showing different degree of membrane permeability has been used as an indicator of apoptotic-like changes in spermatozoa based on membrane permeability (Silva and Gadella 2006; Ball 2008; Brum et al. 2008; Gadella 2008). Probes that fluoresced in cells with increased plasma membrane permeability in combination with probes that fluoresced in cells with increased nuclear permeability (dead cells) have allowed to differentiate cells that are apoptotic from dead cells (Silva and Gadella 2006; Gadella 2008). By using these approaches it has been described that in cells undergoing the later stages of apoptosis, there was increased membrane permeability prior to increased nuclear permeability (Jin and El-Deiry 2005), which is a common finding in a number of reports in fresh spermatozoa from different species (Pena et al. 2003; Martin et al. 2004; Silva and Gadella 2006; Ortega-Ferrusola et al. 2008; Brum et al. 2008). Furthermore, the permeability of the membrane in conjunction with other apoptotic markers is also increased in frozen-thawed semen from different species (Pena et al. 2003; Martin et al. 2004; Ortega-Ferrusola et al. 2008; Macias-Garcia et al. 2008; Marti et al. 2008).

During apoptosis in somatic cells, phosphatidylserine externalization (PSE) marked the apoptotic cells for destruction by phagocytes and occurred prior to loss of plasma membrane integrity (Jin and El-Deiry 2005). However, controversy exists on how well PSE correlated to other apoptotic markers in spermatozoa. (Pena et al. 2003; Paasch et al. 2004c; Barroso et al. 2006; Ball 2008; Brum et al. 2008). For instance, after the separation of fresh equine spermatozoa by density-gradient centrifugation the percent of cells showing PSE in high- and low-density spermatozoa were quite similar, whereas other apoptotic-like changes were markedly different in both subpopulations (Brum et al. 2008). These results are likely indicating that PSE is not primarily related with other concomitant apoptotic-like changes in equine spermatozoa. Contrarily, PSE correlates very well with other features of apoptosis in fresh and frozen-thawed spermatozoa from boar (Pena et al. 2003) or men (Said et al. 2005a,b; Aziz et al. 2007; Said et al. 2008).

Furthermore, it is also unknown whether PSE is, indeed, consequence of the changes that mammalian spermatozoa must undergo to gain fertilizing competence, such as capacitation and acrosome reaction. In this regard, Martin et al. (2005) have shown a significant increase of PSE to the outer leaflet of the plasma membrane in living human sperm after the induction of the acrosome reaction by their incubation with a Ca^{2+} ionophore, whereas the proportion of apoptotic sperm cells with low MMP, activated caspases, high membrane permeability and DNA fragmentation, did not show any significant increase after such treatment. Thus, it was proposed that the induced PSE could not be related simply to apoptosis but rather be linked to physiological changes needed for the acrosome reaction to occur (Martin et al. 2005). In line with this, PSE also occurred during sperm capacitation, due to bicarbonate dependent changes in order for the spermatozoa to undergo the acrosome reaction and fertilize an oocyte (de Vries et al. 2003; Pena et al. 2003; Harrison and Gadella 2005; Moran et al. 2008; Gadella et al. 2008), although it is yet largely unknown if the apoptotic-like changes observed in spermatozoa are unequivocally related to capacitation (Ball 2008; Brum et al. 2008).

Recently, in a number of studies the properties of spermatozoa showing PSE have been characterized by using an approach consisting in sperm labelling with paramagnetic annexin V-conjugated microbeads followed by magnetic cell sorting (MACS) with the aim of separate two sperm populations on the basis of PSE (Paasch et al. 2004a; Said et al. 2005a,b, 2006; Grunewald et al. 2006; Aziz et al. 2007; Said et al. 2008). In these studies it does appear that magnetic separation of human spermatozoa based on annexin-V labelling leads to decreased caspase activation, increased ability to capacitate and undergo the acrosome reaction, increased motility of spermatozoa, increased mitochondria integrity, and smaller extent of DNA fragmentation in the annexin-V negative population (Paasch et al. 2004a; Said et al. 2005a,b, 2006; Grunewald et al. 2006; Aziz et al. 2007; Said et al. 2008).

Overall, these results likely indicate that PSE (in conjunction with other apoptotic changes) is detrimental for the spermatozoa and that the selection of the PSE-negative spermatozoa should be an advantage in order to obtain the sperm subpopulation with the highest fertilizing potential. Specifically, the MACS separation according to annexin-V labelling resulted in a significant depletion of sperm having activated caspases 8, 9, 1, or 3 within the negative PSE fraction and a simultaneous enrichment of sperm bearing active caspases 8, 9, 1, or 3 into the sperm fraction showing PSE, indicating that a relationship between caspase activation and PSE likely occurs in human spermatozoa (Paasch et al. 2004a). Caspase activation was also increased in frozen-thawed human spermatozoa and this effect correlated with annexin-V labelling (Paasch et al. 2004a). Furthermore, annexin V-negative sperm demonstrated higher oocyte penetration capacity, however both sperm subpopulations, i.e. annexin V-negative and annexin V-positive, showed comparable sperm chromatin decondensation (SCD) following oocyte-intracytoplasmic sperm injection, indicating that PSE appears to impact sperm-oocyte penetration rate but it does not seem to affect early stages of fertilization such as SCD in spermatozoa of healthy donors (Said et al. 2006).

In addition, a correlation between PSE and morphological abnormalities related to impaired sperm function has also been described in human spermatozoa. Aziz et al. (2007), showed that the PSE-negative sperm subpopulations had an improved sperm morphology profile as demonstrated by significantly higher proportions of sperm with normal morphology and significantly lower sperm deformity index scores and percentages of sperm with acrosomal defects, midpiece defects, cytoplasmic droplet and tail defects. Moreover, Aziz et al. (2007) showed a significant correlation between the sperm morphology attributes studied and the expressed apoptotic markers, such as caspase 3 activation and MMP integrity, concluding that the non-apoptotic sperm fractions have morphologically superior quality sperm. Thus, this study may support abortive apoptosis, where the apoptotic mechanism of sperm is already triggered prior to ejaculation and is not susceptible of change after release from the seminiferous tubuli. As a result, persistent active apoptotic signals in ejaculated sperm are likely to develop during spermatogenesis in defective spermatozoa to mark them for removal before entering the epididymis (Aziz et al. 2007). This hypothesis is further supported by the evidence that spermatozoa that are healthy after ejaculation are incapable of becoming apoptotic spontaneously, as inferred from one study demonstrating that the overall increase in the percentage of immotile and non-viable spermatozoa after its storage for 24 h was not accompanied by any detectable increase in the population of healthy human spermatozoa showing PSE and DNA fragmentation and their demise likely occur by necrosis rather than apoptosis (Lachaud et al. 2004). Other reports, however, have described that PSE as well as other apoptotic-like changes are largely unrelated with morphological abnormalities in human, porcine, bovine or equine spermatozoa (Muratori et al. 2000, 2003; Franca et al. 2005; Almeida et al. 2005; Said et al. 2005a; Brum et al. 2008).

On the other hand, as indicated above, the percent of human spermatozoa showing PSE was increased after the cryopreservation of these cells (Paasch et al. 2004a). Interestingly, PSE was not increased in bull and equine frozen-thawed spermatozoa (Martin et al. 2004; Brum et al. 2008). In these two species, PSE occurred at very low level in fresh spermatozoa and no significant difference exists after cryopreservation in regard to PSE. Based on these results, it could be speculated that PSE is not associated with apoptotic-like changes in spermatozoa from both species (Brum et al. 2008).

8.5.2.3 Fragmentation of Nuclear DNA

DNA fragmentation has been negatively correlated to *in vitro* fertilization and pregnancy success rates (Sun et al. 1997; Host et al. 2000a,b; Fatehi et al. 2006; Erenpreiss et al. 2006; Shamsi et al. 2008), but there is still much debate as to whether this is due to an active apoptotic-like process in mature spermatozoa (Sakkas et al. 1999a,b; Said et al. 2004; Brum et al. 2008).

DNA-fragmented sperm could be interpreted as a fraction of cells that have failed to complete maturation and, in particular, to complete the packaging of chromatin during spermiogenesis. Indeed, in the elongating phase, protamines substitute histones and associate with DNA to form a very condensed complex that is stabilized by disulfide bonds among the same protamines (Muratori et al. 2000; Shamsi et al. 2008). Because of this unique packaging mechanism, sperm DNA develops an extreme stability and resistance to standard lysis agents and enzymatic digestion

(Shamsi et al. 2008). During spermiogenesis, McPherson and Longo (1992) observed the presence of DNA strand breaks and proposed that they may have a role in favouring the histone-to-protamine substitution. Later on in the maturation process, DNA strand breaks are likely to be ligated by the action of nuclear enzymes, such as topoisomerase II (McPherson and Longo 1992). Sperm that fail to complete ligation and maturation may retain DNA strand breaks.

The presence of DNA damage (detected by terminal deoxynucleotidyl transferase-mediated dUDP nick-end labelling or TUNEL) in subpopulations of human, equine or bovine spermatozoa concomitantly with other markers of apoptosis (Barroso et al. 2000; Muratori et al. 2000; Lachaud et al. 2004; Fatehi et al. 2006; Brum et al. 2008), has raised the idea that apoptotic-like changes occurring in ejaculated spermatozoa are leading to the degradation of DNA as a distal step of the process. However, in a number of reports data presented have dissociated both DNA-fragmentation and apoptosis-like changes. For example, both fresh and frozen-thawed equine spermatozoa show a number of apoptotic-like changes, including increased active caspases, low MMP and altered plasma membrane permeability, all of which were unrelated to DNA fragmentation (Brum et al. 2008). Moreover, in normal human spermatozoa only a negative correlation was found between DNA breakage and progressive motility, whereas was not observed any significant correlation between DNA breakage and the characteristics resembling those of somatic apoptosis (Muratori et al. 2000). Similar results were observed when ejaculates from infertile men were evaluated following gradient centrifugation, showing that a direct correlation exist between DNA fragmentation and sperm motility of both high and low sperm motility fraction, whereas no correlation was observed among DNA fragmentation and apoptotic-like changes such as phosphatidilserine externalization (Barroso et al. 2000).

In view of these results Muratori et al. (2000) have proposed that DNA fragmentation in ejaculated sperm should be considered as a marker of poor functional activity rather than an index of apoptosis. In other words, DNA-fragmented sperm are not cells that are committed to death, but rather, they retain several abnormalities that are compatible with a lower degree of maturation. DNA-fragmented sperm are in fact less motile, more immature, and even less susceptible to hypo-osmotic swelling, which indicates a lower functional integrity of the sperm membrane (Muratori et al. 2000).

However, the reports lacking a correlation among DNA fragmentation (based on the TUNEL assay) and other apoptotic changes, both in fresh as well as in frozen-thawed spermatozoa, could be explained in two circumstances. First, it could be reflecting methodological deficiencies because the TUNEL method may not be sensitive enough to detect the DNA fragmentation that occurred in spermatozoa (Brum et al. 2008; Shamsi et al. 2008). TUNEL is a commonly used method for detecting DNA fragmentation that results from apoptotic signalling cascades. The assay relies on the presence of nicks in the DNA which can be identified by terminal deoxynucleotidyl transferase, an enzyme that will catalyze the addition of dUTPs that are secondarily labeled. This method is often used as a specific marker for apoptosis due to its ability to detect double-strand DNA nicks without need for denaturation;

however, unless DNA damage is very extensive the TUNEL method will not be able to detect it (Brum et al. 2008). The second possibility is that the DNA fragmentation index obtained in frozen-thawed spermatozoa could be underestimated due to partial loss of the starting sperm subpopulation showing DNA fragmentation prior to cryopreservation. This possibility was evidenced by Anzar et al. (2002), which demonstrated that the subpopulation of dead spermatozoa showing DNA fragmentation can undergo cellular fragmentation during the freezing and thawing process. The partial or complete loss of this subpopulation could be masking a true increment in the amount of spermatozoa showing DNA fragmentation after thaw (Brum et al. 2008).

8.5.3 Mitochondria Dysfunction and Apoptotic Processes in Ejaculated Spermatozoa

As previously mentioned, the transformation of a round spermatid to a sperm-like mature spermatid occurs during spermiogenesis. This process involves the loss of most of the cellular organelles, with the exception of the acrosome, derived from the Golgi vesicle, and a certain sub-set of mitochondria that are allocated in the mid-piece of the sperm tail in the mature spermatozoon (Otani et al. 1988; Mortimer 1997; Franca et al. 2005). In addition to these early changes, recent research indicate that critical steps of sperm maturation in the epididymis also involve changes in the mitochondria (Aitken and Baker 2004; Aitken et al. 2007). For example, tyrosine phosphorylation, first in the mid piece and later in the tail, is a major step in sperm maturation in the epididymis. This change is associated with the phosphorylation of several mitochondrial proteins, creation of MMP and activation of mitochondrial free-radical production (Aitken and Baker 2006; Aitken et al. 2007).

Upon ejaculation, three major functions have been attributed to mitochondria in sperm cells. The first one is to generate energy for sperm motility through oxidative phosphorylation. This concept, however, is currently under revision and debate. Mitochondria are found only in the mid piece, thus oxidative phosphorylation occurs only at this level. However, flagellum-kinases and dynein-ATPases need large amounts of ATP to maintain sperm motility all along the flagellum (Cao et al. 2006a,b). It is has, therefore, been suggested that the amount of ATP produced in the mitochondria is not enough to diffuse all along the flagellum to provide enough energy to support motility (Turner 2003, 2006). In addition, studies in mice have demonstrated that defective oxidative phosphorylation does not inhibit sperm motility (Escalier 2006). However, the importance of aerobic or anaerobic metabolic pathways as energy sources for sperm motility apparently varies among species (Marin et al. 2003; Mukai and Okuno 2004; Miki et al. 2004; Rodriguez-Gil 2006). The second important function of the mitochondria in the spermatozoon is the regulation of cell death by the release of cytochrome c and other pro-apoptotic proteins, which induces caspase activation and leads to cell death (Jin and El-Deiry 2005;

Mishra and Shaha 2005; Chan 2006; Lesnefsky and Hoppel 2006; Tsujimoto and Shimizu 2007). Finally, another function has been recently assigned to mitochondria in the spermatozoa in different mammalian species. Until recently, it was widely accepted that spermatozoa were translationally silent, an axiom that was questioned by the demonstration of 55s mitochondrial ribosomes being actively involved in protein translation in human, rat, mouse and bovine spermatozoa (Gur and Breitbart 2006). Moreover, inhibition of protein translation significantly reduced sperm motility, capacitation and *in vitro* fertilization rate in this species (Miller and Ostermeier 2006; Miller 2007). Thus it is likely that protein synthesis occurs in mammalian spermatozoa by means of the biochemical machinery derived from mitochondria. In addition, mitochondria also contribute to the regulation of diverse essential biochemical parameters in the cell, such as redox status, osmotic pressure, and Ca^{2+} homeostasis (Camello-Almaraz et al. 2002, 2006; Chan 2006).

Regardless the importance of mitochondria in motility and other sperm functions, MMP (or $\Delta\Psi_m$) is a major parameter that reflects mitochondrial functionality (Camello-Almaraz et al. 2006; Tsujimoto and Shimizu 2007; Grimm and Brdiczka 2007) and, as previously reviewed, usually is related with other apoptotic changes in the mature, ejaculated spermatozoa (Troiano et al. 1998; Kasai et al. 2002; Paasch et al. 2004b,c; Gallon et al. 2006; Barroso et al. 2006; Martin et al. 2007; Ortega-Ferrusola et al. 2008). The maintenance of a functional MMP relies in the impermeability of the inner mitochondrial membrane for the protons generated inside mitochondria by oxidative phosphorylation. When the permeability of the inner mitochondrial membrane increases for solutes smaller than 1.5 kDa, it is assumed that the permeability transition pore (PTP) is opened and the MMP is disrupted (Camello-Almaraz et al. 2006; Tsujimoto and Shimizu 2007; Grimm and Brdiczka 2007). Subsequently, the cell loses its ability to synthesize ATP with a concomitant blockade of the respiratory chain which leads to the generation of ROS intermediates, most likely *via* direct transfer of electrons to molecular oxygen (Camello-Almaraz et al. 2006; Grimm and Brdiczka 2007). Furthermore, the high concentration of solutes in the mitochondrial matrix also generates an osmotic pressure, which drives the influx of water molecules and the expansion of the inner mitochondrial membrane, eventually disrupting the outer membrane and leading to a massive release of pro-apoptotic factors that finally ends in the demise of the doomed cell (Jin and El-Deiry 2005; Tsujimoto and Shimizu 2007; Youle and Strasser 2008).

Taking into account that mitochondrial physiology critically relies on the MMP, the evaluation of the mitochondrial functionality by means of MMP determination has been proposed by a number of studies as an important factor in predicting sperm fertilizability in ejaculated spermatozoa (Troiano et al. 1998; Kasai et al. 2002; Said et al. 2004; Gallon et al. 2006; Ortega-Ferrusola et al. 2008). In this regard, Gallon et al. (2006) have used flow-cytometric sorting in order to separate two human sperm subpopulations based on MMP and have evaluated some parameters related with fertility in these subpopulations. These authors found that the percentage of morphologically abnormal spermatozoa was significantly lower in the subpopulation displaying high MMP. In addition, in the high MMP subpopulation also were recognized a higher percent of cells showing an intact acrosome and a lower number of spermatozoa showing spontaneous acrosome reaction (Gallon et al. 2006). These results indicate that dysfunctional mitochondria with low MMP were mainly associated with morphologically abnormal spermatozoa displaying impaired ability to undergo the acrosome reaction, thus the maintenance of high MMP is necessary for sperm fertility (Gallon et al. 2006). Contrarily, the motility was not significantly improved in sorted spermatozoa with high MMP as compared with unsorted samples. This could be reinforcing the idea that ATP produced by the mitochondria is not essential for sperm motility, although it can not be discarded an immobilizing effect that has been described in the spermatozoa after cell sorting by flow cytometry (Gallon et al. 2006).

8.5.4 Role of Oxidative Stress and Ca^{2+} Signalling in the Apoptotic Processes in Ejaculated Spermatozoa

In addition to mitochondrial membrane potential, another notable aspect of mitochondria in the context of this chapter is the contribution of these organelles to the production of ROS and to the Ca^{2+} homeostasis, both certainly related with the initiation and development of apoptosis in the cells (Orrenius et al. 2003; Camello-Almaraz et al. 2006; Ball 2008) (See Chapter 1). The mitochondria are the major source of ROS in the cell.

As a consequence of the oxidative phosphorylation, the physiological concentration of $O_2^{-}\bullet$ in the mitochondria is about 5–10-fold higher than in the cytosol or nucleus (Cadenas and Davies 2000). These mitochondria-derived metabolic products can show in somatic cells either physiological or detrimental effects depending on their appearing concentration and on the ability of the cell to remove their excessive production. Similarly to somatic cells, a limited generation of ROS by the spermatozoa has been implicated in the control of normal sperm function. Oxygen species, such as $O_2^{-}\bullet$, H_2O_2 and nitric oxide (NO•), when generated at low and controlled levels, act as second messengers in the mammalian spermatozoa where they could be involved in the regulation of processes such as capacitation and acrosome reaction (de Laminarde et al. 1998; Aitken et al. 1998; Herrero et al. 1999, 2000; Kameshwari et al. 2003; Thundathil et al. 2003; Rodriguez et al. 2005a,b; O'Flaherty et al. 2006; de Laminarde and O'Flaherty 2008). Excessive ROS production, however, is known to impair basic sperm functions, such as motility, mitochondria homeostasis or membrane permeability, especially if neutrophil granulocytes are the source of ROS during chronic inflammation (Barroso et al. 2000; Ricci et al. 2002; Villegas et al. 2005; Aitken and Baker 2006; Tremellen 2008; Shamsi et al. 2008; Koppers et al. 2008).

In somatic cells, the generation of ROS is largely increased when the MMP is disrupted likely by direct transfer of electrons to molecular oxygen (Tsujimoto and Shimizu 2007; Grimm and Brdiczka 2007). Under these circumstances mitochondria might also be a primary target of ROS, leading to an increased mitochondrial dysfunction with concomitant increases in the release of cytochrome *c* and other pro-apoptotic proteins, all of which leading to caspase activation and apoptosis (Jin and El-Deiry 2005; Chan 2006; Tsujimoto and Shimizu 2007; Brum et al. 2008; Koppers et al. 2008). This is particularly interesting in spermatology, since the mitochondria in spermatozoa appear to be preferentially susceptible to agonists of apoptosis and previous studies have found that oxidative stress induce apoptosis in mature spermatozoa via the mitochondria-dependent pathway (Paasch et al. 2004b; Grunewald et al. 2005; Koppers et al. 2008). Furthermore, the cytosolic volume in the spermatozoa is small and contains relatively small amounts of ROS scavengers, which renders these cells especially sensitive to oxidative stress (de Laminarde and O'Flaherty 2008). In view of these data, it is not surprising that the excessive generation of ROS by abnormal spermatozoa with low MMP or by contaminating leukocytes has been recognized as one of the few defined aetiologies for male infertility (Lopes et al. 1998; Barroso et al. 2000; Barroso et al. 2006; Tremellen 2008). Contrarily, a mitochondria-independent impact of ROS in boar sperm motility has also been recently described (Guthrie et al. 2008). Interestingly, in this study H_2O_2 did not affect sperm viability, ATP production, or MMP, likely indicating that elevated ROS can also cause detrimental changes without mitochondria damage or energy disruption in adult spermatozoa (Guthrie et al. 2008). A negative effect in the axoneme involving ATP utilization, or another interference at level of the contractile machinery was proposed to explain the effect of ROS in boar sperm motility (Guthrie et al. 2008).

In addition to induction of apoptosis in a mitochondria-driven manner, ROS are known to trigger the apoptotic pathway by means of a peroxidation-induced loss of plasma membrane function (de Laminarde et al. 1998; Aitken and Baker 2004; de Laminarde and O'Flaherty 2008). This is important in the spermatozoa, since the membranes of these cells are particularly susceptible to oxidative stress due to the relatively high content of polyunsaturated fatty acids in the plasma membrane. Hence, mammalian spermatozoa are highly sensitive to ROS-induced damage mediated by lipid peroxidation (de Laminarde and O'Flaherty 2008; Tremellen 2008).

On the other hand, mitochondria also behave as a high-capacity, low-affinity transient Ca^{2+} store. In somatic cells, changes in intracellular Ca^{2+} concentration ($[Ca^{2+}]_i$) in response to hormones and neurotransmitters induce increases in the mitochondrial matrix Ca^{2+} concentration ($[Ca^{2+}]_m$). Ca^{2+} enter across the inner mitochondrial membrane following its electrochemical gradient when $[Ca^{2+}]_i$ levels reach a submicromolar threshold (Camello-Almaraz et al. 2006). The route is called the Ca^{2+} uniporter, an elusive pathway that remains poorly understood, although recent reports identifies it as a Ca^{2+}-selective channel (Montero et al. 2001). In addition, it has been reported that Ca^{2+} can enter mitochondria through additional mechanism yet poorly understood (Camello-Almaraz et al. 2006). The Ca^{2+} accumulated into mitochondria during the rising phase of the $[Ca^{2+}]_i$ signal is subsequently released during the declining phase of the signals via a Na^+/Ca^{2+} exchanger, although both the Ca^{2+} uniporter and the PTP have also been proposed as efflux routes (Camello-Almaraz et al. 2006).

In a recent study, Bejarano et al. (2008) have shown in human spermatozoa from healthy donors that the activation of caspase 3 and 9 and PS externalization induced by both oxidative stress (H_2O_2) and progesterone, were dependent on elevations in $[Ca^{2+}]_i$. Furthermore, these authors also showed that mitochondrial Ca^{2+} uptake is involved in caspase activation in human spermatozoa, since the blockade of Ca^{2+} uptake into mitochondria by using pharmacological inhibitors was able to decrease apoptosis mediated by both H_2O_2 and progesterone (Bejarano et al. 2008). This report is likely indicating that mitochondria in human spermatozoa, as in somatic cells, are playing a role in the Ca^{2+} homeostasis, as well as can trigger apoptotic events in a Ca^{2+}-dependent manner when the mitochondrial buffering mechanisms for Ca^{2+} are overloaded.

8.6 Concluding Remarks

In this chapter we have compiled relevant information that may help understand the origin and regulation of the apoptotic events that both germ cells and ejaculated spermatozoa experiment. In male germ cells the role of apoptosis in the sustaining of normal spermatogenesis has clearly been defined. There are, however, additional questions that remain to be answered, especially with regard to the involvement of the receptor-death pathway in normal spermatogenesis as well as the apoptotic mechanism involved in the hormonal regulation of spermatogenesis.

In adult mammalian spermatozoa much less in known about the underlying mechanisms that govern the apoptosis and *apoptosis-like* changes detected in a number of species. Reports supporting both abortive and autonomously initiated apoptosis have been published. However, although the mechanism have been reported as independent, it is feasible that abortive and autonomously initiated sperm apoptosis can develop in the spermatozoa under certain circumstances, even simultaneously. This hypothesis might be supported by the fact that the sperm population in a single ejaculate is far to be homogeneous.

Another aspect to consider is the fact that it is yet unknown if all the apoptotic-like changes detected in adult spermatozoa are linked to a detrimental effect or some of those are, indeed, required to develop normal sperm functions such as capacitation. Both, positive and negative roles for apoptosis-like changes have been reported. The answer to this question almost surely will involve research about the intensity, the origin and the temporal regulation of the apoptotic changes developed in the sperm cells. A better understanding of these processes surely will help to improve the fertility of men and animals and will facilitate the improvement of the sperm preservation technology.

Acknowledgments This research was supported by Ministerio de Ciencia e Innovación-FEDER, Madrid, Spain, grants BFU2007-62423 (BFI) and AGL2007-60598 (GAN); and Ministerio of Agricultura Pesca y Alimentación-INIA, Madrid, Spain, grant RZ2008-00018-00-00. JAT further acknowledges support by the program "Ramón y Cajal" from the Ministerio de Ciencia e Innovación, Madrid, Spain.

References

Abou-Haila A, Tulsiani DR (2000) Mammalian sperm acrosome: formation, contents, and function. Arch Biochem Biophys 379:173–182

Adachi M, Suematsu S, Kondo T et al. (1995) Targeted mutation in the Fas gene causes hyperplasia in peripheral lymphoid organs and liver. Nat Genet 11:294–300

Aitken RJ, Baker MA (2004) Oxidative stress and male reproductive biology. Reprod Fertil Dev 16:581–588

Aitken RJ, Baker MA (2006) Oxidative stress, sperm survival and fertility control. Mol Cell Endocrinol 250:66–69

Aitken RJ, Harkiss D, Knox W et al. (1998) A novel signal transduction cascade in capacitating human spermatozoa characterised by a redox-regulated, cAMP-mediated induction of tyrosine phosphorylation. J Cell Sci 111:645–656

Aitken RJ, Nixon B, Lin M et al. (2007) Proteomic changes in mammalian spermatozoa during epididymal maturation. Asian J Androl 9:554–564

Allemand I, Anglo A, Jeantet AY et al. (1999) Testicular wild-type p53 expression in transgenic mice induces spermiogenesis alterations ranging from differentiation defects to apoptosis. Oncogene 18:6521–6530

Almeida C, Cardoso MF, Sousa M et al. (2005) Quantitative study of caspase 3 activity in semen and after swim-up preparation in relation to sperm quality. Hum Reprod 20:1307–1313

Amann RP (2008) The cycle of the seminiferous epithelium in humans: a need to revisit? J Androl 29:469–487

Anzar M, He L, Buhr MM et al. (2002) Sperm apoptosis in fresh and cryopreserved bull semen detected by flow cytometry and its relationship with fertility. Biol Reprod 66:354–360

Aquila S, Middea E, Catalano S et al. (2007) Human sperm express a functional androgen receptor: effects on PI3K/AKT pathway. Hum Reprod 22:2594–2605

Aziz N, Said TM, Paasch U et al. (2007) The relationship between human sperm apoptosis, morphology and the sperm deformity index. Hum Reprod 22:1413–1419

Ball BA (2008) Oxidative stress, osmotic stress and apoptosis: impacts on sperm function and preservation in the horse. Anim Reprod Sci 107:257–267

Barroso G, Morshedi M, Oehninger S (2000) Analysis of DNA fragmentation, plasma membrane translocation of phosphatidylserine and oxidative stress in human spermatozoa. Hum Reprod 15:1338–1344

Barroso G, Taylor S, Morshedi M et al. (2006) Mitochondrial membrane potential integrity and plasma membrane translocation of phosphatidylserine as early apoptotic markers: a comparison of two different sperm subpopulations. Fertil Steril 85:149–154

Beardsley A, O'Donnell L (2003) Characterization of normal spermiation and spermiation failure induced by hormone suppression in adult rats. Biol Reprod 68:1299–1307

Beardsley A, Robertson DM, O'Donnell L (2006) A complex containing alpha6beta1-integrin and phosphorylated focal adhesion kinase between Sertoli cells and elongated spermatids during spermatid release from the seminiferous epithelium. J Endocrinol 190:759–770

Bejarano I, Lozano GM, Ortiz A et al. (2008) Caspase 3 activation in human spermatozoa in response to hydrogen peroxide and progesterone. Fertil Steril 90:1340–1347

Berndtson WE (2008) Comparative reliability and sensitivity of different methods for assessing treatment effects on sperm production. Anim Reprod Sci 105:5–22

Beumer TL, Roepers-Gajadien HL, Gademan IS et al. (1998) The role of the tumor suppressor p53 in spermatogenesis. Cell Death Differ 5:669–677

Blume-Jensen P, Jiang G, Hyman R et al. (2000) Kit/stem cell factor receptor-induced activation of phosphatidylinositol 3′-kinase is essential for male fertility. Nat Genet 24:157–162

Boersma AW, Nooter K, Burger H et al. (1997) Bax upregulation is an early event in cisplatin-induced apoptosis in human testicular germ-cell tumor cell line NT2, as quantitated by flow cytometry. Cytometry 27:275–282

Brum AM, Sabeur K, Ball BA (2008) Apoptotic-like changes in equine spermatozoa separated by density-gradient centrifugation or after cryopreservation. Theriogenology 69:1041–1055

Cadenas E, Davies KJ (2000) Mitochondrial free radical generation, oxidative stress, and aging. Free Radic Biol Med 29:222–230

Camello-Almaraz C, Gomez-Pinilla PJ, Pozo MJ et al. (2006) Mitochondrial reactive oxygen species and Ca^{2+} signaling. Am J Physiol Cell Physiol 291:C1082–C1088

Camello-Almaraz C, Salido GM, Pariente JA et al. (2002) Role of mitochondria in Ca(2+) oscillations and shape of Ca(2+) signals in pancreatic acinar cells. Biochem Pharmacol 63:283–292

Cao W, Gerton GL, Moss SB (2006a) Proteomic profiling of accessory structures from the mouse sperm flagellum. Mol Cell Proteomics 5:801–810

Cao W, Haig-Ladewig L, Gerton GL et al. (2006b) Adenylate kinases 1 and 2 are part of the accessory structures in the mouse sperm flagellum. Biol Reprod 75:492–500

Chaki SP, Ghosh D, Misro MM (2003) Simultaneous increase in germ cell apoptosis and oxidative stress under acute unilateral testicular ischaemia in rats. Int J Androl 26:319–328

Chaki SP, Misro MM, Gautam DK et al. (2006) Estradiol treatment induces testicular oxidative stress and germ cell apoptosis in rats. Apoptosis 11:1427–1437

Chan DC (2006) Mitochondria: dynamic organelles in disease, aging, and development. Cell 125:1241–1252

Choi YJ, Uhm SJ, Song SJ et al. (2008) Cytochrome c upregulation during capacitation and spontaneous acrosome reaction determines the fate of pig sperm cells: linking proteome analysis. J Reprod Dev 54:68–83

Cisternas P, Moreno RD (2006) Comparative analysis of apoptotic pathways in rat, mouse, and hamster spermatozoa. Mol Reprod Dev 73:1318–1325

Coureuil M, Fouchet P, Prat M et al. (2006) Caspase independent death of meiotic and postmeiotic cells overexpressing p53: calpain involvement. Cell Death Differ 13:1927–1937

D'Alessio A, Riccioli A, Lauretti P et al. (2001) Testicular FasL is expressed by sperm cells. Proc Natl Acad Sci USA 98:3316–3321

de Laminarde E, Jiang H, Zini A et al. (1997) Reactive oxygen species and sperm physiology. Rev Reprod 2:48–54

de Laminarde E, O'Flaherty C (2008) Sperm activation: role of reactive oxygen species and kinases. Biochim Biophys Acta 1784:106–115

de Laminarde E, Tsai C, Harakat A et al. (1998) Involvement of reactive oxygen species in human sperm arcosome reaction induced by A23187, lysophosphatidylcholine, and biological fluid ultrafiltrates. J Androl 19:585–594

de Vries KJ, Wiedmer T, Sims PJ et al. (2003) Caspase independent exposure of aminophospholipids and tyrosine phosphorylation in bicarbonate responsive human sperm cells. Biol Reprod 68:2122–2134

De SK, Chen HL, Pace JL et al. (1993) Expression of tumor necrosis factor-alpha in mouse spermatogenic cells. Endocrinology 133:389–396

Ehmcke J, Hubner K, Scholer HR et al. (2006a) Spermatogonia: origin, physiology and prospects for conservation and manipulation of the male germ line. Reprod Fertil Dev 18:7–12

Ehmcke J, Schlatt S (2006) A revised model for spermatogonial expansion in man: lessons from non-human primates. Reproduction 132:673–680

Ehmcke J, Wistuba J, Schlatt S (2006b) Spermatogonial stem cells: questions, models and perspectives. Hum Reprod Update 12:275–282

El-Darwish KS, Parvinen M, Toppari J (2006) Differential expression of members of the E2F family of transcription factors in rodent testes. Reprod Biol Endocrinol 4:63

Embree-Ku M, Venturini D, Boekelheide K (2002) Fas is involved in the p53-dependent apoptotic response to ionizing radiation in mouse testis. Biol Reprod 66:1456–1461

Erenpreiss J, Spano M, Erenpreisa J et al. (2006) Sperm chromatin structure and male fertility: biological and clinical aspects. Asian J Androl 8:11–29

Escalier D (2006) Knockout mouse models of sperm flagellum anomalies. Hum Reprod Update 12:449–461
Falschlehner C, Emmerich CH, Gerlach B et al. (2007) TRAIL signalling: decisions between life and death. Int J Biochem Cell Biol 39:1462–1475
Fatehi AN, Bevers MM, Schoevers E et al. (2006) DNA damage in bovine sperm does not block fertilization and early embryonic development but induces apoptosis after the first cleavages. J Androl 27:176–188
Feng C, Zhang J, Gasana V et al. (2005) Differential expression of protein kinase C alpha and delta in testes of mouse at various stages of development. Cell Biochem Funct 23:415–420
Flesch FM, Gadella BM (2000) Dynamics of the mammalian sperm plasma membrane in the process of fertilization. Biochim Biophys Acta 1469:197–235
Forrest DW, Blanchard TL, Heninger NL et al. (2006) Relationship of germ cell apoptosis to spermatogenic efficiency in stallions with reduced semen quality. Anim Reprod Sci 94:140–143
Franca LR, Avelar GF, Almeida FF (2005) Spermatogenesis and sperm transit through the epididymis in mammals with emphasis on pigs. Theriogenology 63:300–318
French LE, Hahne M, Viard I et al. (1996) Fas and Fas ligand in embryos and adult mice: ligand expression in several immune-privileged tissues and coexpression in adult tissues characterized by apoptotic cell turnover. J Cell Biol 133:335–343
Furuchi T, Masuko K, Nishimune Y et al. (1996) Inhibition of testicular germ cell apoptosis and differentiation in mice misexpressing Bcl-2 in spermatogonia. Development 122:1703–1709
Gadella BM (2008) Sperm membrane physiology and relevance for fertilization. Anim Reprod Sci 107:229–236
Gadella BM, Flesch FM, van Golde LM et al. (1999) Dynamics in the membrane organization of the mammalian sperm cell and functionality in fertilization. Vet Q 21:142–146
Gadella BM, Rathi R, Brouwers JF et al. (2001) Capacitation and the acrosome reaction in equine sperm. Anim Reprod Sci 68:249–265
Gadella BM, Tsai PS, Boerke A et al. (2008) Sperm head membrane reorganisation during capacitation. Int J Dev Biol 52:473–480
Gadella BM, Van Gestel RA (2004) Bicarbonate and its role in mammalian sperm function. Anim Reprod Sci 82–83:307–319
Gallon F, Marchetti C, Jouy N et al. (2006) The functionality of mitochondria differentiates human spermatozoa with high and low fertilizing capability. Fertil Steril 86:1526–1530
Gautam DK, Misro MM, Chaki SP et al. (2007) hCG treatment raises H_2O_2 levels and induces germ cell apoptosis in rat testis. Apoptosis 12:1173–1182
Grataroli R, Vindrieux D, Gougeon A et al. (2002) Expression of tumor necrosis factor-alpha-related apoptosis-inducing ligand and its receptors in rat testis during development. Biol Reprod 66:1707–1715
Grataroli R, Vindrieux D, Selva J et al. (2004) Characterization of tumour necrosis factor-alpha-related apoptosis-inducing ligand and its receptors in the adult human testis. Mol Hum Reprod 10:123–128
Grimm S, Brdiczka D (2007) The permeability transition pore in cell death. Apoptosis 12:841–855
Grunewald S, Baumann T, Paasch U et al. (2006) Capacitation and acrosome reaction in nonapoptotic human spermatozoa. Ann NY Acad Sci 1090:138–146
Grunewald S, Paasch U, Said TM et al. (2005) Caspase activation in human spermatozoa in response to physiological and pathological stimuli. Fertil Steril 83:1106–1112
Gur Y, Breitbart H (2006) Mammalian sperm translate nuclear-encoded proteins by mitochondrial-type ribosomes. Genes Dev 20:411–416
Guthrie HD, Welch GR, Long JA (2008) Mitochondrial function and reactive oxygen species action in relation to boar motility. Theriogenology 70:1209–1215
Harrison RA, Gadella BM (2005) Bicarbonate-induced membrane processing in sperm capacitation. Theriogenology 63:342–351
Hasegawa M, Zhang Y, Niibe H et al. (1998) Resistance of differentiating spermatogonia to radiation-induced apoptosis and loss in p53-deficient mice. Radiat Res 149:263–270

Heninger NL, Staub C, Blanchard TL et al. (2004) Germ cell apoptosis in the testes of normal stallions. Theriogenology 62:283–297
Herrero MB, Chatterjee S, Lefievre L et al. (2000) Nitric oxide interacts with the cAMP pathway to modulate capacitation of human spermatozoa. Free Radic Biol Med 29:522–536
Herrero MB, de Laminarde E, Gagnon C (1999) Nitric oxide regulates human sperm capacitation and protein-tyrosine phosphorylation in vitro. Biol Reprod 61:575–581
Hixon ML, Boekelheide K (2003) Expression and localization of total Akt1 and phosphorylated Akt1 in the rat seminiferous epithelium. J Androl 24:891–898
Ho HC, Suarez SS (2001) Hyperactivation of mammalian spermatozoa: function and regulation. Reproduction 122:519–526
Hockenbery DM, Zutter M, Hickey W et al. (1991) BCL2 protein is topographically restricted in tissues characterized by apoptotic cell death. Proc Natl Acad Sci USA 88:6961–6965
Holmberg C, Helin K, Sehested M et al. (1998) E2F-1-induced p53-independent apoptosis in transgenic mice. Oncogene 17:143–155
Host E, Lindenberg S, Smidt-Jensen S (2000a) The role of DNA strand breaks in human spermatozoa used for IVF and ICSI. Acta Obstet Gynecol Scand 79:559–563
Host E, Lindenberg S, Smidt-Jensen S (2000b) DNA strand breaks in human spermatozoa: correlation with fertilization in vitro in oligozoospermic men and in men with unexplained infertility. Acta Obstet Gynecol Scand 79:189–193
Ikeda M, Kodama H, Fukuda J et al. (1999) Role of radical oxygen species in rat testicular germ cell apoptosis induced by heat stress. Biol Reprod 61:393–399
Inaba K (2003) Molecular architecture of the sperm flagella: molecules for motility and signaling. Zoolog Sci 20:1043–1056
Jahnukainen K, Chrysis D, Hou M et al. (2004) Increased apoptosis occurring during the first wave of spermatogenesis is stage-specific and primarily affects midpachytene spermatocytes in the rat testis. Biol Reprod 70:290–296
Jin Z, El-Deiry WS (2005) Overview of cell death signaling pathways. Cancer Biol Ther 4:139–163
Johnson L, Thompson DL Jr, Varner DD (2008) Role of Sertoli cell number and function on regulation of spermatogenesis. Anim Reprod Sci 105:23–51
Kameshwari DB, Siva AB, Shivaji S (2003) Inhibition of in vitro capacitation of hamster spermatozoa by nitric oxide synthase inhibitors. Cell Mol Biol (Noisy-le-grand) 49:421–428
Kaneko Y, Takano S, Okumura K et al. (1993) Identification of protein tyrosine phosphatases expressed in murine male germ cells. Biochem Biophys Res Commun 197:625–631
Kasai T, Ogawa K, Mizuno K et al. (2002) Relationship between sperm mitochondrial membrane potential, sperm motility, and fertility potential. Asian J Androl 4:97–103
Kierszenbaum AL (2001) Apoptosis during spermatogenesis: the thrill of being alive. Mol Reprod Dev 58:1–3
Kimura M, Itoh N, Takagi S et al. (2003) Balance of apoptosis and proliferation of germ cells related to spermatogenesis in aged men. J Androl 24:185–191
Knudson CM, Tung KS, Tourtellotte WG et al. (1995) Bax-deficient mice with lymphoid hyperplasia and male germ cell death. Science 270:96–99
Kodaira K, Takahashi R, Hirabayashi M et al. (1996) Overexpression of c-myc induces apoptosis at the prophase of meiosis of rat primary spermatocytes. Mol Reprod Dev 45:403–410
Koji T (2001) Male germ cell death in mouse testes: possible involvement of Fas and Fas ligand. Med Electron Microsc 34:213–222
Koji T, Hishikawa Y, Ando H et al. (2001) Expression of Fas and Fas ligand in normal and ischemia-reperfusion testes: involvement of the Fas system in the induction of germ cell apoptosis in the damaged mouse testis. Biol Reprod 64:946–954
Koppers AJ, De Iuliis GN, Finnie JM et al. (2008) Significance of mitochondrial reactive oxygen species in the generation of oxidative stress in spermatozoa. J Clin Endocrinol Metab 93:3199–3207
Lachaud C, Tesarik J, Canadas ML et al. (2004) Apoptosis and necrosis in human ejaculated spermatozoa. Hum Reprod 19:607–610

Lee J, Kanatsu-Shinohara M, Inoue K et al. (2007) Akt mediates self-renewal division of mouse spermatogonial stem cells. Development 134:1853–1859
Lee J, Richburg JH, Younkin SC et al. (1997) The Fas system is a key regulator of germ cell apoptosis in the testis. Endocrinology 138:2081–2088
Lefievre L, Jha KN, de Laminarde E et al. (2002) Activation of protein kinase A during human sperm capacitation and acrosome reaction. J Androl 23:709–716
Lesnefsky EJ, Hoppel CL (2006) Oxidative phosphorylation and aging. Ageing Res Rev 5:402–433
Levy R, Seifer-Aknin I (2001) Apoptosis during spermatogenesis and in ejaculated spermatozoa: importance for fertilization. Ann Biol Clin (Paris) 59:531–545
Liu D, Liao C, Wolgemuth DJ (2000) A role for cyclin A1 in the activation of MPF and G2-M transition during meiosis of male germ cells in mice. Dev Biol 224:388–400
Lopes S, Jurisicova A, Sun JG et al. (1998) Reactive oxygen species: potential cause for DNA fragmentation in human spermatozoa. Hum Reprod 13:896–900
Macias-Garcia B, Morrell JM, Ortega-Ferrusola C et al. (2008) Centrifugation on a single layer of colloid selects improved quality spermatozoa from frozen-thawed stallion semen. Anim Reprod Sci doi: 10.1016/j.anireprosci.2008.08.025
Marchiani S, Tamburrino L, Forti G et al. (2007a) M540 bodies and their impact on flow cytometric analyses of human spermatozoa. Soc Reprod Fertil Suppl 65:509–514
Marchiani S, Tamburrino L, Maoggi A et al. (2007b) Characterization of M540 bodies in human semen: evidence that they are apoptotic bodies. Mol Hum Reprod 13:621–631
Marin S, Chiang K, Bassilian S et al. (2003) Metabolic strategy of boar spermatozoa revealed by a metabolomic characterization. FEBS Lett 554:342–346
Marshall GR, Zorub DS, Plant TM (1995) Follicle-stimulating hormone amplifies the population of differentiated spermatogonia in the hypophysectomized testosterone-replaced adult rhesus monkey (Macaca mulatta). Endocrinology 136:3504–3511
Marti E, Perez-Pe R, Colas C et al. (2008) Study of apoptosis-related markers in ram spermatozoa. Anim Reprod Sci 106:113–132
Marti E, Perez-Pe R, Muino-Blanco T et al. (2006) Comparative study of four different sperm washing methods using apoptotic markers in ram spermatozoa. J Androl 27:746–753
Martin G, Cagnon N, Sabido O et al. (2007) Kinetics of occurrence of some features of apoptosis during the cryopreservation process of bovine spermatozoa. Hum Reprod 22:380–388
Martin G, Sabido O, Durand P et al. (2005) Phosphatidylserine externalization in human sperm induced by calcium ionophore A23187: relationship with apoptosis, membrane scrambling and the acrosome reaction. Hum Reprod 20:3459–3468
Martin G, Sabido O, Durand P et al. (2004) Cryopreservation induces an apoptosis-like mechanism in bull sperm. Biol Reprod 71:28–37
McKee CM, Ye Y, Richburg JH (2006) Testicular germ cell sensitivity to TRAIL-induced apoptosis is dependent upon p53 expression and is synergistically enhanced by DR5 agonistic antibody treatment. Apoptosis 11:2237–2250
McPherson SM, Longo FJ (1992) Localization of DNase I-hypersensitive regions during rat spermatogenesis: stage-dependent patterns and unique sensitivity of elongating spermatids. Mol Reprod Dev 31:268–279
Meehan T, Loveland KL, de KD et al. (2001) Developmental regulation of the bcl-2 family during spermatogenesis: insights into the sterility of bcl-w-/- male mice. Cell Death Differ 8:225–233
Miki K, Qu W, Goulding EH et al. (2004) Glyceraldehyde 3-phosphate dehydrogenase-S, a sperm-specific glycolytic enzyme, is required for sperm motility and male fertility. Proc Natl Acad Sci USA 101:16501–16506
Miller D (2007) Spermatozoal RNA as reservoir, marker and carrier of epigenetic information: implications for cloning. Reprod Domest Anim 42:2–9
Miller D, Ostermeier GC (2006) Towards a better understanding of RNA carriage by ejaculate spermatozoa. Hum Reprod Update 12:757–767
Mishra DP, Pal R, Shaha C (2006) Changes in cytosolic Ca^{2+} levels regulate Bcl-xS and Bcl-xL expression in spermatogenic cells during apoptotic death. J Biol Chem 281:2133–2143

Mishra DP, Shaha C (2005) Estrogen-induced spermatogenic cell apoptosis occurs via the mitochondrial pathway: role of superoxide and nitric oxide. J Biol Chem 280:6181–6196

Miura M, Sasagawa I, Suzuki Y et al. (2002) Apoptosis and expression of apoptosis-related genes in the mouse testis following heat exposure. Fertil Steril 77:787–793

Montero M, Alonso MT, Albillos A et al. (2001) Mitochondrial Ca(2+)-induced Ca(2+) release mediated by the Ca(2+) uniporter. Mol Biol Cell 12:63–71

Moran JM, Madejon L, Ortega FC et al. (2008) Nitric oxide induces caspase activity in boar spermatozoa. Theriogenology 70:91–96

Mortimer ST (1997) A critical review of the physiological importance and analysis of sperm movement in mammals. Hum Reprod Update 3:403–439

Mukai C, Okuno M (2004) Glycolysis plays a major role for adenosine triphosphate supplementation in mouse sperm flagellar movement. Biol Reprod 71:540–547

Muratori M, Maggi M, Spinelli S et al. (2003) Spontaneous DNA fragmentation in swim-up selected human spermatozoa during long term incubation. J Androl 24:253–262

Muratori M, Piomboni P, Baldi E et al. (2000) Functional and ultrastructural features of DNA-fragmented human sperm. J Androl 21:903–912

Nair R, Shaha C (2003) Diethylstilbestrol induces rat spermatogenic cell apoptosis in vivo through increased expression of spermatogenic cell Fas/FasL system. J Biol Chem 278:6470–6481

Nandi S, Banerjee PP, Zirkin BR (1999) Germ cell apoptosis in the testes of Sprague Dawley rats following testosterone withdrawal by ethane 1,2-dimethanesulfonate administration: relationship to Fas? Biol Reprod 61:70–75

Niino YS, Irie T, Takaishi M et al. (2001) PKCtheta II, a new isoform of protein kinase C specifically expressed in the seminiferous tubules of mouse testis. J Biol Chem 276:36711–36717

Nunez-Martinez I, Moran JM, Peña FJ (2007) Identification of sperm morphometric subpopulations in the canine ejaculate: do they reflect different subpopulations in sperm chromatin integrity? Zygote 15:257–266

O'Donnell L, Robertson KM, Jones ME et al. (2001) Estrogen and spermatogenesis. Endocr Rev 22:289–318

O'Flaherty C, de Laminarde E, Gagnon C (2006) Reactive oxygen species modulate independent protein phosphorylation pathways during human sperm capacitation. Free Radic Biol Med 40:1045–1055

Okwun OE, Igboeli G, Ford JJ et al. (1996) Number and function of Sertoli cells, number and yield of spermatogonia, and daily sperm production in three breeds of boar. J Reprod Fertil 107:137–149

Oldereid NB, Angelis PD, Wiger R et al. (2001) Expression of Bcl-2 family proteins and spontaneous apoptosis in normal human testis. Mol Hum Reprod 7:403–408

Orrenius S, Zhivotovsky B, Nicotera P (2003) Regulation of cell death: the calcium-apoptosis link. Nat Rev Mol Cell Biol 4:552–565

Ortega-Ferrusola C, Sotillo-Galan Y, Varela-Fernandez E et al. (2008) Detection of "apoptosis-like" changes during the cryopreservation process in equine sperm. J Androl 29:213–221

Orth JM, Gunsalus GL, Lamperti AA (1988) Evidence from Sertoli cell-depleted rats indicates that spermatid number in adults depends on numbers of Sertoli cells produced during perinatal development. Endocrinology 122:787–794

Otani H, Tanaka O, Kasai K et al. (1988) Development of mitochondrial helical sheath in the middle piece of the mouse spermatid tail: regular dispositions and synchronized changes. Anat Rec 222:26–33

Paasch U, Grunewald S, Agarwal A et al. (2004a) Activation pattern of caspases in human spermatozoa. Fertil Steril 81:802–809

Paasch U, Grunewald S, Dathe S et al. (2004b) Mitochondria of human spermatozoa are preferentially susceptible to apoptosis. Ann NY Acad Sci 1030:403–409

Paasch U, Sharma RK, Gupta AK et al. (2004c) Cryopreservation and thawing is associated with varying extent of activation of apoptotic machinery in subsets of ejaculated human spermatozoa. Biol Reprod 71:1828–1837

Paniagua R, Nistal M, Saez FJ et al. (1991) Ultrastructure of the aging human testis. J Electron Microsc Tech 19:241–260
Pareek TK, Joshi AR, Sanyal A et al. (2007) Insights into male germ cell apoptosis due to depletion of gonadotropins caused by GnRH antagonists. Apoptosis 12:1085–1100
Pena FJ, Johannisson A, Wallgren M et al. (2003) Assessment of fresh and frozen-thawed boar semen using an Annexin-V assay: a new method of evaluating sperm membrane integrity. Theriogenology 60:677–689
Pentikainen V, Erkkila K, Suomalainen L et al. (2001) TNFalpha down-regulates the Fas ligand and inhibits germ cell apoptosis in the human testis. J Clin Endocrinol Metab 86:4480–4488
Pentikainen V, Suomalainen L, Erkkila K et al. (2002) Nuclear factor-kappa B activation in human testicular apoptosis. Am J Pathol 160:205–218
Perdichizzi A, Nicoletti F, La VS et al. (2007) Effects of tumour necrosis factor-alpha on human sperm motility and apoptosis. J Clin Immunol 27:152–162
Perticarari S, Ricci G, Boscolo R et al. (2008) Fas receptor is not present on ejaculated human sperm. Hum Reprod 23:1271–1279
Peter AT, Colenbrander B, Gadella BM (2005) Effect of caspase inhibitors on the post-thaw motility, and integrity of acrosome and plasma membrane of cryopreserved equine spermatozoa. Indian J Exp Biol 43:483–487
Peter AT, Linde-Forsberg C (2003) Efficacy of the anticaspase agent zVAD-fmk on post-thaw viability of canine spermatozoa. Theriogenology 59:1525–1532
Print CG, Loveland KL (2000) Germ cell suicide: new insights into apoptosis during spermatogenesis. Bioessays 22:423–430
Print CG, Loveland KL, Gibson L et al. (1998) Apoptosis regulator bcl-w is essential for spermatogenesis but appears otherwise redundant. Proc Natl Acad Sci USA 95:12424–12431
Ricci G, Perticarari S, Fragonas E et al. (2002) Apoptosis in human sperm: its correlation with semen quality and the presence of leukocytes. Hum Reprod 17:2665–2672
Riccioli A, Salvati L, D'Alessio A et al. (2003) The Fas system in the seminiferous epithelium and its possible extra-testicular role. Andrologia 35:64–70
Riccioli A, Starace D, D'Alessio A et al. (2000) TNF-alpha and IFN-gamma regulate expression and function of the Fas system in the seminiferous epithelium. J Immunol 165:743–749
Rodriguez I, Ody C, Araki K et al. (1997) An early and massive wave of germinal cell apoptosis is required for the development of functional spermatogenesis. EMBO J 16:2262–2270
Rodriguez PC, O'Flaherty CM, Beconi MT et al. (2005a) Nitric oxide-induced capacitation of cryopreserved bull spermatozoa and assessment of participating regulatory pathways. Anim Reprod Sci 85:231–242
Rodriguez PC, O'Flaherty CM, Beconi MT et al. (2005b) Nitric oxide induces acrosome reaction in cryopreserved bovine spermatozoa. Andrologia 37:166–172
Rodriguez-Gil JE (2006) Mammalian sperm energy resources management and survival during conservation in refrigeration. Reprod Domest Anim 41:11–20
Roldan ER, Shi QX (2007) Sperm phospholipases and acrosomal exocytosis. Front Biosci 12:89–104
Roosen-Runge EC (1973) Germinal-cell loss in normal metazoan spermatogenesis. J Reprod Fertil 35:339–348
Rosado JA, Lopez JJ, Gomez-Arteta E et al. (2006) Early caspase 3 activation independent of apoptosis is required for cellular function. J Cell Physiol 209:142–152
Ross AJ, Waymire KG, Moss JE et al. (1998) Testicular degeneration in Bclw-deficient mice. Nat Genet 18:251–256
Rossi P, Sette C, Dolci S et al. (2000) Role of c-kit in mammalian spermatogenesis. J Endocrinol Invest 23:609–615
Royere D, Guerif F, Laurent-Cadoret V et al. (2004) Apoptosis in testicular germ cells. Intern Cong Ser 1266:170–176
Russell LD, Chiarini-Garcia H, Korsmeyer SJ et al. (2002) Bax-dependent spermatogonia apoptosis is required for testicular development and spermatogenesis. Biol Reprod 66:950–958

Russell LD, Ren HP, Sinha Hikim I et al. (1990) A comparative study in twelve mammalian species of volume densities, volumes, and numerical densities of selected testis components, emphasizing those related to the Sertoli cell. Am J Anat 188:21–30

Ruwanpura SM, McLachlan RI, Matthiesson KL et al. (2008a) Gonadotrophins regulate germ cell survival, not proliferation, in normal adult men. Hum Reprod 23:403–411

Ruwanpura SM, McLachlan RI, Stanton PG et al. (2008b) Follicle-stimulating hormone affects spermatogonial survival by regulating the intrinsic apoptotic pathway in adult rats. Biol Reprod 78:705–713

Said TM, Agarwal A, Grunewald S et al. (2006) Selection of nonapoptotic spermatozoa as a new tool for enhancing assisted reproduction outcomes: an in vitro model. Biol Reprod 74:530–537

Said TM, Agarwal A, Zborowski M et al. (2008) Utility of magnetic cell separation as a molecular sperm preparation technique. J Androl 29:134–142

Said TM, Grunewald S, Paasch U et al. (2005a) Advantage of combining magnetic cell separation with sperm preparation techniques. Reprod Biomed Online 10:740–746

Said TM, Grunewald S, Paasch U et al. (2005b) Effects of magnetic-activated cell sorting on sperm motility and cryosurvival rates. Fertil Steril 83:1442–1446

Said TM, Paasch U, Glander HJ et al. (2004) Role of caspases in male infertility. Hum Reprod Update 10:39–51

Sakkas D, Mariethoz E, Manicardi G et al. (1999a) Origin of DNA damage in ejaculated human spermatozoa. Rev Reprod 4:31–37

Sakkas D, Mariethoz E, St John JC (1999b) Abnormal sperm parameters in humans are indicative of an abortive apoptotic mechanism linked to the Fas-mediated pathway. Exp Cell Res 251:350–355

Sakkas D, Seli E, Bizzaro D et al. (2003) Abnormal spermatozoa in the ejaculate: abortive apoptosis and faulty nuclear remodelling during spermatogenesis. Reprod Biomed Online 7:428–432

Sakkas D, Seli E, Manicardi GC et al. (2004) The presence of abnormal spermatozoa in the ejaculate: did apoptosis fail? Hum Fertil (Camb) 7:99–103

Saunders PT, Fisher JS, Sharpe RM et al. (1998) Expression of oestrogen receptor beta (ER beta) occurs in multiple cell types, including some germ cells, in the rat testis. J Endocrinol 156:R13–R17

Schutze S, Tchikov V, Schneider-Brachert W (2008) Regulation of TNFR1 and CD95 signalling by receptor compartmentalization. Nat Rev Mol Cell Biol 9:655–662

Shaha C (2007) Modulators of spermatogenic cell survival. Soc Reprod Fertil Suppl 63:173–186

Shaman JA, Prisztoka R, Ward WS (2006) Topoisomerase IIB and an extracellular nuclease interact to digest sperm DNA in an apoptotic-like manner. Biol Reprod 75:741–748

Shaman JA, Yamauchi Y, Ward WS (2007) Sperm DNA fragmentation: awakening the sleeping genome. Biochem Soc Trans 35:626–628

Shamsi MB, Kumar R, Dada R (2008) Evaluation of nuclear DNA damage in human spermatozoa in men opting for assisted reproduction. Indian J Med Res 127:115–123

Silva PF, Gadella BM (2006) Detection of damage in mammalian sperm cells. Theriogenology 65:958–978

Sinha Hikim AP, Lue Y, Yamamoto CM et al. (2003) Key apoptotic pathways for heat-induced programmed germ cell death in the testis. Endocrinology 144:3167–3175

Sinha Hikim AP, Wang C, Leung A et al. (1995) Involvement of apoptosis in the induction of germ cell degeneration in adult rats after gonadotropin-releasing hormone antagonist treatment. Endocrinology 136:2770–2775

Siu MK, Wong CH, Lee WM et al. (2005) Sertoli-germ cell anchoring junction dynamics in the testis are regulated by an interplay of lipid and protein kinases. J Biol Chem 280:25029–25047

Sofikitis N, Giotitsas N, Tsounapi P et al. (2008) Hormonal regulation of spermatogenesis and spermiogenesis. J Steroid Biochem Mol Biol 109:323–330

Stabile V, Russo M, Chieffi P (2006) 17beta-estradiol induces Akt-1 through estrogen receptor-beta in the frog (Rana esculenta) male germ cells. Reproduction 132:477–484

Suescun MO, Rival C, Theas MS et al. (2003) Involvement of tumor necrosis factor-alpha in the pathogenesis of autoimmune orchitis in rats. Biol Reprod 68:2114–2121

Sugiyama N, Obinata M, Matsui Y (2001) Bcl-2 inhibits apoptosis of spermatogonia and growth of spermatogonial stem cells in a cell-intrinsic manner. Mol Reprod Dev 58:30–38

Sun JG, Jurisicova A, Casper RF (1997) Detection of deoxyribonucleic acid fragmentation in human sperm: correlation with fertilization in vitro. Biol Reprod 56:602–607

Suominen JS, Wang Y, Kaipia A et al. (2004) Tumor necrosis factor-alpha (TNF-alpha) promotes cell survival during spermatogenesis, and this effect can be blocked by infliximab, a TNF-alpha antagonist. Eur J Endocrinol 151:629–640

Tapanainen JS, Tilly JL, Vihko KK et al. (1993) Hormonal control of apoptotic cell death in the testis: gonadotropins and androgens as testicular cell survival factors. Mol Endocrinol 7:643–650

Taylor SL, Weng SL, Fox P et al. (2004) Somatic cell apoptosis markers and pathways in human ejaculated sperm: potential utility as indicators of sperm quality. Mol Hum Reprod 10:825–834

Teng CS, Vilagrasa X (1998) Biphasic c-Myc protein expression during gossypol-induced apoptosis in rat spermatocytes. Contraception 57:117–123

Tesarik J, Martinez F, Rienzi L et al. (2002) In-vitro effects of FSH and testosterone withdrawal on caspase activation and DNA fragmentation in different cell types of human seminiferous epithelium. Hum Reprod 17:1811–1819

Theas MS, Rival C, Dietrich SJ et al. (2006) Death receptor and mitochondrial pathways are involved in germ cell apoptosis in an experimental model of autoimmune orchitis. Hum Reprod 21:1734–1742

Theas MS, Rival C, Jarazo-Dietrich S et al. (2008) Tumour necrosis factor-alpha released by testicular macrophages induces apoptosis of germ cells in autoimmune orchitis. Hum Reprod 23:1865–1872

Theas S, Rival C, Lustig L (2003) Germ cell apoptosis in autoimmune orchitis: involvement of the Fas-FasL system. Am J Reprod Immunol 50:166–176

Thundathil J, de Laminarde E, Gagnon C (2003) Nitric oxide regulates the phosphorylation of the threonine-glutamine-tyrosine motif in proteins of human spermatozoa during capacitation. Biol Reprod 68:1291–1298

Tirado OM, Selva DM, Toran N et al. (2004) Increased expression of estrogen receptor beta in pachytene spermatocytes after short-term methoxyacetic acid administration. J Androl 25:84–94

Tremellen K (2008) Oxidative stress and male infertility: a clinical perspective. Hum Reprod Update 14:243–258

Troiano L, Fustini MF, Lovato E et al. (1994) Apoptosis and spermatogenesis: evidence from an in vivo model of testosterone withdrawal in the adult rat. Biochem Biophys Res Commun 202:1315–1321

Troiano L, Granata AR, Cossarizza A et al. (1998) Mitochondrial membrane potential and DNA stainability in human sperm cells: a flow cytometry analysis with implications for male infertility. Exp Cell Res 241:384–393

Tsujimoto Y, Shimizu S (2007) Role of the mitochondrial membrane permeability transition in cell death. Apoptosis 12:835–840

Turner RM (2003) Tales from the tail: what do we really know about sperm motility? J Androl 24:790–803

Turner RM (2006) Moving to the beat: a review of mammalian sperm motility regulation. Reprod Fertil Dev 18:25–38

Um JY, Choi BM, Kim JS et al. (1995) Expression of protein kinase C delta gene in germ cells. J Urol 154:1237–1240

Veis DJ, Sorenson CM, Shutter JR et al. (1993) Bcl-2-deficient mice demonstrate fulminant lymphoid apoptosis, polycystic kidneys, and hypopigmented hair. Cell 75:229–240

Vera Y, Diaz-Romero M, Rodriguez S et al. (2004) Mitochondria-dependent pathway is involved in heat-induced male germ cell death: lessons from mutant mice. Biol Reprod 70:1534–1540

Villegas J, Schulz M, Soto L et al. (2005) Bacteria induce expression of apoptosis in human spermatozoa. Apoptosis 10:105–110
Wang RA, Nakane PK, Koji T (1998) Autonomous cell death of mouse male germ cells during fetal and postnatal period. Biol Reprod 58:1250–1256
Watanabe M, Shirayoshi Y, Koshimizu U et al. (1997) Gene transfection of mouse primordial germ cells in vitro and analysis of their survival and growth control. Exp Cell Res 230:76–83
Weil M, Jacobson MD, Raff MC (1998) Are caspases involved in the death of cells with a transcriptionally inactive nucleus? Sperm and chicken erythrocytes. J Cell Sci 111(Pt 18):2707–2715
Weng SL, Taylor SL, Morshedi M et al. (2002) Caspase activity and apoptotic markers in ejaculated human sperm. Mol Hum Reprod 8:984–991
Wundrich K, Paasch U, Leicht M et al. (2006) Activation of caspases in human spermatozoa during cryopreservation: an immunoblot study. Cell Tissue Bank 7:81–90
Yan W, Huang JX, Lax AS et al. (2003) Overexpression of Bcl-W in the testis disrupts spermatogenesis: revelation of a role of BCL-W in male germ cell cycle control. Mol Endocrinol 17:1868–1879
Yan W, Suominen J, Samson M et al. (2000) Involvement of Bcl-2 family proteins in germ cell apoptosis during testicular development in the rat and pro-survival effect of stem cell factor on germ cells in vitro. Mol Cell Endocrinol 165:115–129
Yao PL, Lin YC, Sawhney P et al. (2007) Transcriptional regulation of FasL expression and participation of sTNF-alpha in response to sertoli cell injury. J Biol Chem 282:5420–5431
Yin Y, DeWolf WC, Morgentaler A (1998) Experimental cryptorchidism induces testicular germ cell apoptosis by p53-dependent and -independent pathways in mice. Biol Reprod 58:492–496
Yin Y, Stahl BC, DeWolf WC et al. (2002) P53 and Fas are sequential mechanisms of testicular germ cell apoptosis. J Androl 23:64–70
Youle RJ, Strasser A (2008) The BCL-2 protein family: opposing activities that mediate cell death. Nat Rev Mol Cell Biol 9:47–59
Zheng S, Turner TT, Lysiak JJ (2006) Caspase 2 activity contributes to the initial wave of germ cell apoptosis during the first round of spermatogenesis. Biol Reprod 74:1026–1033
Zhou XC, Wei P, Hu ZY et al. (2001) Role of Fas/FasL genes in azoospermia or oligozoospermia induced by testosterone undecanoate in rhesus monkey. Acta Pharmacol Sin 22:1028–1033

Glossary of Terms

Acrosome: a Golgi-derived structure located in the sperm head, forming a cap over the anterior region of the nucleus that is extended up to a posterior domain called equatorial segment. It contains several enzymes such as acid glycohydrolases, proteases, esterases, acid phosphatases, and aryl sulphatases.

Acrosome reaction: a secretory event involving the fusion of the outer acrosomal membrane and the sperm plasma membrane in the anterior region of sperm head. Is a key feature during mammalian fertilization by which the release of hydrolytic enzymes will facilitate the spermatozoa to penetrate the zona pelucida, allowing the fusion of the sperm plasma membrane and the oolemma (oocyte plasma membrane).

Activated sperm motility: sperm movement characterized by a symmetrical beat of the flagellum with low amplitude that drives the sperm in a relatively straight line. Current evidence suggests that the role of activated motility is to aid in propelling the sperm through the female reproductive tract to the oviduct. This movement is seen in freshly ejaculated spermatozoa.

Actin cytoskeleton: cellular "skeleton" contained within the cytoplasm. It is a dynamic structure that maintains cell shape, often protects the cell, enables cellular motion, and plays important roles in both intracellular transport (the movement of vesicles and organelles, for example) and cellular division. Actin filaments are composed of two intertwined actin chains. Microfilaments are most concentrated just beneath the cell membrane, and are responsible for resisting tension and maintaining cellular shape, forming cytoplasmatic protuberances and participation in some cell-to-cell or cell-to-matrix junctions. In association with these latter roles, microfilaments are essential to transduction.

Acute pancreatitis: severe and debilitating inflammatory disease. Some of the effects of pancreatitis are inhibition of pancreatic secretions, an elevation of serum enzyme levels, cytoplasmic vacuolisation, death of exocrine acinar cells, oedema formation and infiltration of inflammatory cells into the pancreas.

Aequorins: Ca^{2+}-sensitive photoproteins which in the presence of molecular oxygen produce bioluminescent light. They possess a number of EF-hand type regions that function as binding sites for Ca^{2+}: when Ca^{2+} occupies such sites, the protein undergoes a conformational change and converts through oxidation its prosthetic group, coelenterazine, into excited coelenteramide and CO_2. As the excited coelenteramide relaxes to the ground state, blue light (wavelength = 469 nm) is emitted.

Alzheimer's disease: progressive and fatal brain disorder characterized by destruction of cells that leads to impairment of memory, thinking and behaviour severe enough to affect work, lifelong hobbies or social life.

Amyloid protein: type of protein that can form insoluble fibrous aggregates sharing specific structural traits. Abnormal accumulation of amyloid protein in the central nervous system may play a role in various neurodegenerative diseases.

Amyotrophic lateral sclerosis: progressive neurodegenerative disease that affects nerve cells in the brain and the spinal cord, characterized by degeneration of the motor neurons that leads to a loss of the ability of the brain to initiate and control muscle movement.

Angiotensin II: is a peptide derived from angiotensin I through removal of two terminal residues by the enzyme angiotensin-converting enzyme, mostly located in the lung capillaries. Angiotensin II belongs to the renin-angiotensin-aldosterone system, which causes the tubules of the kidneys to retain sodium and water and enhanced blood pressure.

Apoptosis: or programmed cell death, is a natural process required for the development and health of multicellular organisms. Cells die in response to a variety of stimuli and during apoptosis they do so in a controlled, regulated fashion. Apoptosis is a process in which cells play an active role in their own death (which is why apoptosis is often referred to as cell suicide). This makes apoptosis distinct from another form of cell death called necrosis in which uncontrolled cell death leads to lysis of cells, inflammatory responses and, potentially, to serious health problems.

Apoptosis-inducing factor: is a flavoprotein involved in the activation of caspase-independent apoptotic events by causing DNA fragmentation and chromatin condensation.

Apoptosome: a large protein complex that comprises cytochrome *c* and apoptotic protease-activating factor-1 (APAF1), and forms in the presence of ATP or dATP. The apoptosome recruits procaspase 9 and results in the allosteric activation of caspase 9.

Apoptotic body: during apoptosis cells collapse into small intact fragments that exclude vital dyes. Such fragments are termed apoptotic bodies.

Astrocyte: also known collectively as astroglia, are a sub-type of the glial cells in the brain and spinal cord. They depict a characteristic star shape and present many processes that envelope synapses made by neurons. Astrocytes are classically identified histologically by the expression of the intermediate filament glial fibrillary acidic protein (GFAP). In the central nervous system this cell type performs many functions, including biochemical support of endothelial cells which form the blood-brain barrier, the provision of nutrients to the nervous tissue, and a principal role in the repair and scarring process in the brain.

Azathioprine: immunosuppressant used in organ transplantation, autoimmune disease, inflammatory bowel disease and multiple sclerosis. It is metabolized into the active metabolites 6-mercaptopurine and 6-thioinosinic acid, which inhibit purine synthesis.

Bile acids: molecules made by the liver from cholesterol that form micelles in the intestine and contribute to the absorption of fats.

Bcl-2 family: Bcl-2 is an oncogene which in follicular lymphoma is frequently linked to an immunoglobulin locus by the chromosome translocation t(14:18). It was the first example of an oncogene that inhibits cell death rather than promoting proliferation. B cells transfected with Bcl-2 were shown to be rendered resistant towards apoptosis induced by IL-3 withdrawal: for the first time it was shown that the pathway toward tumorigenesis depends not only on the ability to escape growth control but also depends on the ability to prevent apoptosis. When several homologues of Bcl-2 had been identified, it became apparent that a Bcl-2 family of proteins could be defined by the presence of conserved sequence motifs known as Bcl-2 homology domains (BH1 to BH4).

BH3 only: Bcl-2 proteins containing the amino acid sequence LXXXGD, in which X represents any amino acid. This motif is conserved between most core Bcl-2 family members and among BH3-only proteins.

Capacitation: a series of biochemical and biophysical modifications that render the ejaculated spermatozoa competent for fertilization of the oocyte. These fundamental processes normally take place in the female genital tract during the migration of spermatozoa to the site of fertilization, although it can also be achieved in vitro.

Caspases: cysteine proteases homologous to *C. elegans* ced-3. They are of central importance in the apoptotic signalling network. The term caspases is derived from cysteine-dependent aspartate-specific proteases: their catalytical activity depends on a critical cysteine-residue within a highly conserved active-site pentapeptide QACRG, and the caspases specifically cleave their substrates after Asp residues. So far, 7 different caspases have been identified in *Drosophila*, and 15 different members of the caspase-family have been described in mammals, with caspase 11 and caspase 12 only identified in the mouse. According to a unified nomenclature, the caspases are referred to in the order of their publication.

Ceramide: lipid composed of sphingosine and a fatty acid found in high concentrations within the cell membranes. Ceramide is a precursor of sphingomyelin, one of the major lipids in the lipid bilayer.

Chaperones: proteins and protein complexes that bind to misfolded or unfolded polypeptide chains and affect the subsequent folding processes of these chains. Chaperones are found in all types of cells and cellular compartments, and have a wide range of binding specificities and functional roles.

Cholangiocytes: epithelial cells of the bile duct. They modify bile by water reabsorption and are involved in angiogenesis, being an important source of vascular endothelial growth factor.

Cholestasis: condition caused by rapidly developing or long-term interruption in the excretion of bile. The obstruction causes bile acids, bilirubin, and lipids to accumulate in the blood stream.

Cirrhosis: chronic degenerative disease in which normal liver cells are damaged and are then replaced by scar tissue. It can result from alcohol abuse, nutritional deprivation, or infection especially by the hepatitis virus.

Cisplatin: platinum-based chemotherapy drug used to treat various types of cancers, such as sarcomas, lymphomas, carcinomas and germ cell tumours. Platinum complexes are formed in cells, which bind and cause cross-linking of DNA and finally apoptosis.

Cyclooxigenase: enzyme responsible for the formation of prostanoids, prostaglandins, prostacyclins, and thromboxanes, involved in the inflammatory response.

Cytochrome *c*: small heme protein associated with the inner membrane of the mitochondrion. It is a highly soluble protein and an essential component of the electron transport chain, where it carries one electron. It is capable of undergoing oxidation and reduction, but does not bind oxygen. It transfers electrons between complexes III and IV of the electron transport chain.

Cytokines: a category of signalling proteins and glycoproteins that, like hormones and neurotransmitters, are used extensively in cellular communication. Cytokines are critical to the development and functioning of immune response, as they are often secreted by immune cells that have encountered a pathogen, thereby activating and recruiting further immune cells to increase the system's response to the pathogen. While hormones are secreted from specific organs to the blood, and neurotransmitters are related to neural activity, the cytokines are a more diverse class of compounds in terms of origin and purpose. They can be produced and released by cells of the central nervous system and can have autocrine, paracrine and endocrine effects, sometimes leading to cytotoxicity.

Death receptors: cell surface receptors that transmit apoptotic signals after ligation with specific ligands. Death receptors belong to the tumour necrosis factor receptor (TNFR) gene superfamily, including TNFR-1, Fas/CD95, and the TRAIL receptors DR4 and DR5. All members of the TNFR family consist of cysteine rich extracellular subdomains that allow them to recognize their ligands with specificity, resulting in the trimerization and activation of the respective death receptor.

Endoplasmic reticulum: organelle found in all eukaryotic cells that is an interconnected network of tubules, vesicles and cisternae. It is responsible for several specialized functions: protein translation, folding and transport of proteins to be used in the cell membrane (e.g. transmembrane receptors and other integral membrane proteins), or to be secreted from the cell; sequestration of calcium; and production and storage of glycogen, steroids, and other macromolecules.

Erythropoiesis: process of red blood cells (erythrocytes) formation or production, which occurs almost exclusively in the red bone marrow. Kidneys release a hormone called erythropoietin in response to low levels of oxygen in the blood; erythropoietin stimulates the red bone marrow to begin red blood cell production.

Excitotoxicity: state caused by excessive accumulation of neurotransmitters at a synapse, that will lead to overstimulation of neurons, and that can end in an impairing of cell function and even cell death.

Exocrine cells: the major functional unit of exocrine glands is the exocrine acinar cell. Typically, exocrine acinar cells are polarised; the infra and paranuclear regions being occupied mainly by rough endoplasmic reticulum, while the Golgi complex and secretory granules are located predominantly at the apical pole of the cell.

Extrinsic apoptotic pathway: upon activation of death receptors, the subsequent apoptotic signalling cascade is mediated by the cytoplasmic part of the death receptor, which contains a conserved sequence termed the death domain (DD). Adapter molecules like FADD or TRADD themselves possess their own DDs by which they are recruited to the DDs of the activated death receptor, thereby forming the so-called death inducing signalling complex (DISC). In addition to its DD, the adaptor FADD also contains a death effector domain (DED) which through homotypic DED-DED interaction sequesters procaspase 8 to the DISC. The local concentration of several procaspase 8 molecules at the DISC leads to their autocatalytic activation and release of active caspase 8. Active caspase 8 then processes downstream effector caspases, which subsequently cleave specific substrates resulting in cell death. Cells harboring the capacity to induce such direct and mainly caspase-dependent apoptosis pathways were classified to belong to the so-called type I cells.

FAD: the flavin adenine dinucleotide (FAD) is a redox cofactor involved in several important reactions in metabolism. FAD can exist in two different redox states and its biochemical role usually involves changing between these two states: FAD can be reduced to the $FADH_2$, whereby it accepts two hydrogen atoms.

FADD: Fas-Associated protein with Death Domain. It is an adaptor molecule that bridges the Fas-receptor, and other death receptors, to caspase 8 through its death domain to form the death inducing signalling complex (DISC) during apoptosis.

Fulminant hepatic failure: liver failure that occurs suddenly in a previously healthy person. The most common causes of FHF are viral hepatitis, and damage from drugs and toxins.

Glial cell: cell type with supportive functions in the central nervous system (both in the brain and in the spinal cord). These cells are characterized because they do not conduct electrical impulses (as opposed to neurons). Nevertheless, glial cells are capable of extensive signalling in response to a diversity of stimuli, establishing a bidirectional communication with neurons and vascular cells. There are three types of glial cells: astrocytes, oligodendrocytes and microglia.

Glial fibrillary acidic protein (GFAP): intermediate filament protein that is found in the cytoskeleton of glial cells such as astrocytes. It is involved in the structure and function of the cell's cytoskeleton, and is thought to help to maintain astrocyte mechanical strength, as well as the shape of cells.

Gliosis: or astrogliosis, represents a remarkably response of astrocytes to all types of injuries of the CNS. It is characterized by an abnormal increase in the number of astrocytes that show enhanced expression of the glial fibrillary acidic protein (GFAP), increase in their membrane processes and increased secretion of neuroactive substances. In the central nervous system, the destruction of nearby neurons can be accompanied by the proliferation of reactive astrocytes, which leads to the formation of the so-called "glial scar".

Granzyme B: a serin protease contained within the secretory granules of cytotoxic lymphocytes and natural killer cells. Granzyme B cleaves its protein substrates after Asp residues, and can promote caspase activation and apoptosis.

Heme group: prosthetic group that consists of an iron atom contained in the centre of a large heterocyclic organic ring called a porphyrin. In haemoglobin, each subunit contains a heme group.

Hepatic stellate cell: or Ito cells or fat-storing cells. They are found in the perisinusoidal space of the liver and are the major cell type involved in liver fibrosis, which is the formation of scar tissue in response to liver damage.

Hepatitis: inflammation of the liver caused by infectious or toxic agents and characterized by jaundice, fever, hepatomegalia, and abdominal pain.

Hepatocarcinoma: or hepatocellular carcinoma. Carcinoma derived from the parenchymal cells of the liver.

Hepatocytes: parenchymal cells that make up 70–80% of the mass of the liver. They are involved in various metabolic functions and also initiate the formation and secretion of bile.

Huntington's disease: inherited disease caused by a faulty gene on chromosome 4. It is a progressive, degenerative disease that causes certain nerve cells in the brain to waste away. As a result, patients show uncontrolled movements, emotional disturbances and mental deterioration.

Hyperactivated sperm motility: sperm movement characterized by an asymmetrical flagellar beating associated with decreased flagellar beat frequency and increased flagellar curvature, which results in circular or figure-eight trajectories. Current evidence suggests that the role of hyperactivated motility is to help sperm detach from the oviductal epithelium, reach the site of fertilization, and penetrate the cumulus and zona pellucida of the oocyte. This movement is associated to sperm capacitation in the spermatozoa of most mammalian species and is seen in most sperm recovered from the site of fertilization.

Immune thrombocytopenic purpura: clinical syndrome associated to a reduced number of circulating platelets (thrombocytopenia) resulting in a bleeding tendency, including easy bruising (purpura). Platelets are coated with autoantibodies to platelet membrane proteins that shorten life span of platelets in circulation by splenic sequestration and phagocytosis by mononuclear cells.

Intrinsic apoptosis pathway: in the so-called type II cells, the signal coming from the activated death receptor does not generate a caspase signalling cascade strong enough for execution of cell death on its own. In this case, the signal needs to be amplified via mitochondria-dependent apoptotic pathways. The link between the caspase signalling cascade and the mitochondria is provided by the Bcl-2 family member Bid. Bid is cleaved by caspase 8 and in its truncated form (tBID) translocates to the mitochondria where it acts in concert with the proapoptotic Bcl-2 family members Bax and Bak to induce the release of cytochrome c and other mitochondrial proapoptotic factors into the cytosol. Cytosolic cytochrome *c* is binding to monomeric Apaf-1, which then, in a dATP-dependent conformational change, oligomerizes to assemble the apoptosome, a complex of wheel-like structure with

7-fold symmetry that triggers the activation of the initiator procaspase 9. Activated caspase 9 subsequently initiates a caspase cascade involving downstream effector caspases such as caspase 3, caspase 7, and caspase 6, ultimately resulting in cell death.

Ischemia: a restriction in blood supply, generally due to factors in the blood vessels, with resultant damage or dysfunction of tissue.

Leukaemia: a number of neoplastic processes of the blood or bone marrow, characterized by an abnormal proliferation of blood cells, usually white blood cells (leukocytes). Displacement of the normal bone marrow cells by immature white blood (normally dysfunctional) cells, results in a number of haemostatic alterations (due to lack of blood platelets) and immunological disorders.

Lysosomes: organelles that contain acid hydrolases to digest excess or worn-out organelles, food particles, and engulfed viruses or bacteria. Lysosomes fuse with vacuoles and dispense their enzymes into the vacuoles, digesting their contents. At pH 4.8, the interior of the lysosomes is more acidic than the cytosol (pH 7.2). The lysosome's single membrane stabilizes the low pH by pumping in protons (H^+) from the cytosol via proton pumps and chloride ion channels.

Liver: large gland in the upper part of the abdomen on the right side. Its functions include secretion of bile, lipid and carbohydrate metabolism, protein synthesis, detoxification, vitamin and metal storage, and many others.

Macrophage system: also known as reticuloendothelial cells, are phagocytic cells involved in the defense of the organism. Cells of the macrophage system derived from precursor cells in the bone marrow, which, in turn, leads to the formation of monocytes released into the bloodstream. Monocytes might remain in the blood circulation, but most of them enter body tissues, where they develop into macrophages.

Microglia: small non-neural cells forming part of the supporting structure of the central nervous system. They depict long processes and ameboid and phagocytic activity at sites of neural damage or inflammation. Normally this cell type is directed to waste products of nerve tissue. However, under certain conditions can lead to neurotoxicity by the release of neuroactive substances.

Mitochondria: double membrane-bound organelles that provide the energy a cell needs to live. While the outer membrane is fairly smooth, the inner membrane is highly convoluted, forming folds called cristae. Mitochondria are the site of eukaryotic oxidative metabolism and contain the enzymes that mediate this process, including pyruvate dehydrogenase, the citric acid cycle enzymes, the enzymes catalyzing fatty acid oxidation, and the enzymes and redox proteins involved in electron transport and oxidative phosphorylation. It is therefore that the mitochondrion is described as the cell's "power plant". In addition, mitochondria play a pivotal role in a variety of biological processes, including steroid hormone synthesis, the urea cycle, lipid and amino acid metabolism, calcium homeostasis and apoptosis.

Mitochondrial outer membrane permeabilization: an event that occurs immediately downstream of Bax and/or Bak activation and is responsible for the release of intermembrane space proteins, such as cytochrome *c*, Smac/ DIABLO, EndoG, Omi/HtrA2 and AIF (but any protein that resides between the outer and inner mitochondrial membranes can be released, depending on its membrane association and solubility). Importantly, the anti-apoptotic Bcl-2 proteins inhibit mitochondrial outer membrane permeabilization.

Mitochondrial permeability transition: MPT, is a sudden increase in the permeability of the mitochondrial inner membrane to molecules of less than 1500 Daltons in molecular weight. This phenomenon is caused by the opening, under certain pathological conditions, of a voltage-dependent, high conductance channel located in the inner membrane that is known as the mitochondrial permeability transition pore or MPTP. The exact molecular nature of the MPT pore is not known and several mechanisms have been put forward to account for MPTP opening. Induction of the permeability transition pore leads to mitochondrial swelling and cell death and may play a role in some types of apoptosis.

Multiple sclerosis: autoimmune condition characterized by physical and cognitive disability. In this illness the immune system attacks the central nervous system and leads to demyelination, which consist of destruction of a fatty layer called the myelin sheath that wraps around nerve fibres and electrically insulates them. When myelin is lost, the axons of neurons can no longer effectively conduct action potentials.

Necrosis: uncontrolled death of cells and living tissues. It begins with cell swelling, chromatin digestion, and disruption of the plasma membrane and organelle membranes. Late necrosis is characterized by extensive DNA hydrolysis, vacuolation of the endoplasmic reticulum, organelle breakdown, and cell lysis. The release of intracellular content after plasma membrane rupture is the cause of inflammation in necrosis.

Neurodegenerative disease: state of illness due to the alteration in the structure and function of the nervous system.

Neuron: cell type, which belongs to the family of cells that compose the nervous system, specialized in sending and receiving electrical signals over long distances within the body.

Neurotoxicity: state caused by the accumulation of certain chemicals in the central nervous system that can lead to damage of nervous cells.

Neurotransmitter: chemical that is used to relay, amplify and modulate signals between a neuron and another cell. For example, in the central nervous system glutamate and ATP act as widespread neurotransmitters between neurons and glial cells.

Non alcoholic steatohepatitis: fatty inflammation of the liver in people who do not abuse alcohol. It is typically a chronic condition that causes no symptoms or very mild symptoms but can sometimes cause progressive scarring and cirrhosis of the liver.

Oxidative stress: metabolic state characterized as a shift of the cellular redox status to a more oxidized state. Such a shift can be due to exposure of cells to environmental oxidants or to the endogenous production of reactive oxygen species (ROS) under pathological conditions such as diseases, or during aging.

Parkinson's disease: chronic and progressive degenerative disorder of the central nervous system as a result of decreased stimulation of the motor cortex by the basal ganglia, normally caused by the insufficient formation and action of dopamine, which is produced in the dopaminergic neurons of the brain. This disease often impairs the sufferer's motor skills, speech, and other functions. It is characterized by muscle rigidity, tremor, a slowing of physical movement and, in extreme cases, a loss of physical movement.

Phosphatidylserine: essential phospholipid of the cell membrane usually kept on the inner-leaflet of cell membranes by an enzyme called translocase.

PKC: protein kinase C is a family of protein kinases consisting of $\sim$10 isoformes. They are divided into three subfamilies, based on their second messenger requirements: conventional, novel, and atypical. Conventional PKCs require Ca^{2+}, diacylglycerol (DAG), and a phospholipid such as phosphatidylcholine for activation. Novel PKCs require DAG, but do not require Ca^{2+} for activation. Thus, conventional and novel PKCs are activated through the same signal transduction pathway as phospholipase C. Atypical PKCs require neither Ca^{2+} nor diacylglycerol for activation.

PLC: phospholipase C is a class of enzymes that cleave phospholipids just before the phosphate group, which plays an important role in eukaryotic cell physiology, particularly signal transduction pathways.

Reactive oxygen species (ROS): molecules or ions formed by the incomplete one-electron reduction of oxygen. These reactive oxygen intermediates include singlet oxygen; superoxides; peroxides; hydroxyl radical and hypochlorous acid. They contribute to the microbicidal activity of phagocytes, regulation of signal transduction and gene expression, and the oxidative damage to nucleic acids, proteins and lipids.

Secretory granules: relatively small intracellular, membrane-enclosed sacs that stores or transports substances. They are a basic tool of the cell for organizing metabolism, transport, buoyancy control, enzyme storage, as well as being chemical reaction chambers. Many in the endoplasmic reticulum, or are made from parts of the plasma membrane.

Stroke: rapidly developing loss of brain functions due to a disturbance in the blood vessels supplying blood to the brain. This can be due to ischemia (lack of blood supply) caused by thrombosis or embolism or due to a haemorrhage.

Syndrome of long or prolonged QT: rare congenital heart condition characterized by prolongation of the QT interval on the electrocardiogram due to delayed repolarization after cardiac excitation. The QT interval on the electrocardiogram, measured from the beginning of the QRS complex to the end of the T wave, represents the duration of activation and recovery of the ventricular myocardium. This syndrome is associated with syncope due to ventricular arrhythmias, which can develop ventricular fibrillation and sudden death.

TRPC channels: a family of transient receptor potential channels those are non-selectively permeable to cations, with a selectivity of Ca^{2+} over Na^{+} variable among the different biomembranes. Many of TRPC channel subunits are able to coassemble. In general, TRPC channels can be activated by phospholipase C stimulation, with some members also activated by diacylglycerol. It has long been proposed that TRPC channels underlie the store-operated channels (SOCs) observed in many cell types. These channels open due to the depletion of intracellular Ca^{2+} stores. Two other proteins, stromal interaction molecules (STIMs) and the ORAIs, however, have more recently been implicated in this process. It should be noted that STIM1, ORAI1 and TRPCs can coassemble.

Thrombin: clotting factor with serine protease activity that converts soluble fibrinogen into insoluble strands of fibrin, as well as catalyze several coagulation-related reactions. In addition, thrombin activates a number of protease-activated receptors (PAR) in the surface of different cell types, including platelets.

Tyrosine kinases: enzymes that can transfer a phosphate group from ATP to a tyrosine residue in a protein. Tyrosine kinases are a subgroup of the larger class of protein kinases. Phosphorylation of proteins by kinases is an important mechanism in signal transduction for regulation of enzyme activity. Most tyrosine kinases have an associated protein tyrosine phosphatase, responsible for protein dephosphorylation.

Vascular leakage: process of transcytosis *via* intracellular vesicles which travel from the luminal to the basal membrane surface of the vascular endothelial cells.

Wolf Parkinson White syndrome: is an electrical abnormality of the heart associated to an accessory pathway, known as the bundle of Kent, which provides an abnormal electrical communication from the atria to the ventricles and induces ventricular pre-excitation. This abnormality might be asymptomatic or accompanied to episodes of tachycardia, dizziness, chest palpitations, fainting or cardiac arrest.

Color Plate Section

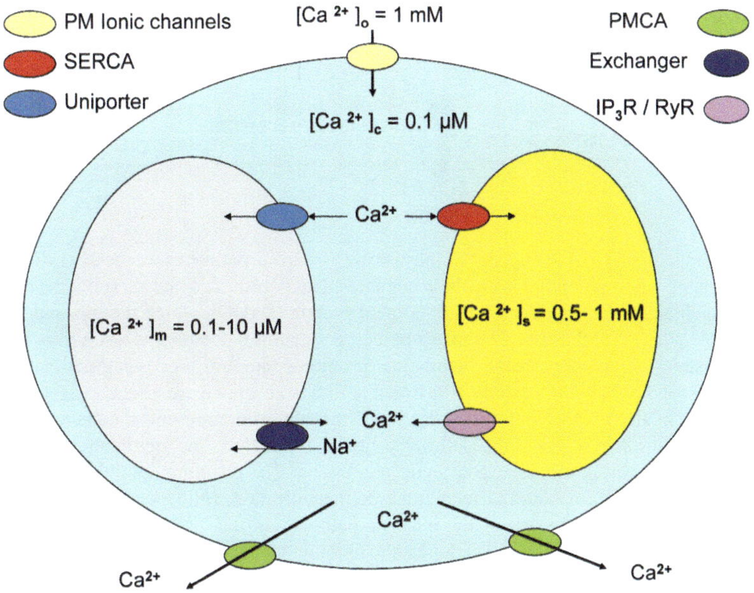

Plate 1 Schematic diagram depicting the major ROS-sensitive Ca^{2+}-handling mechanism (See also Figure 1.1 on page 3)

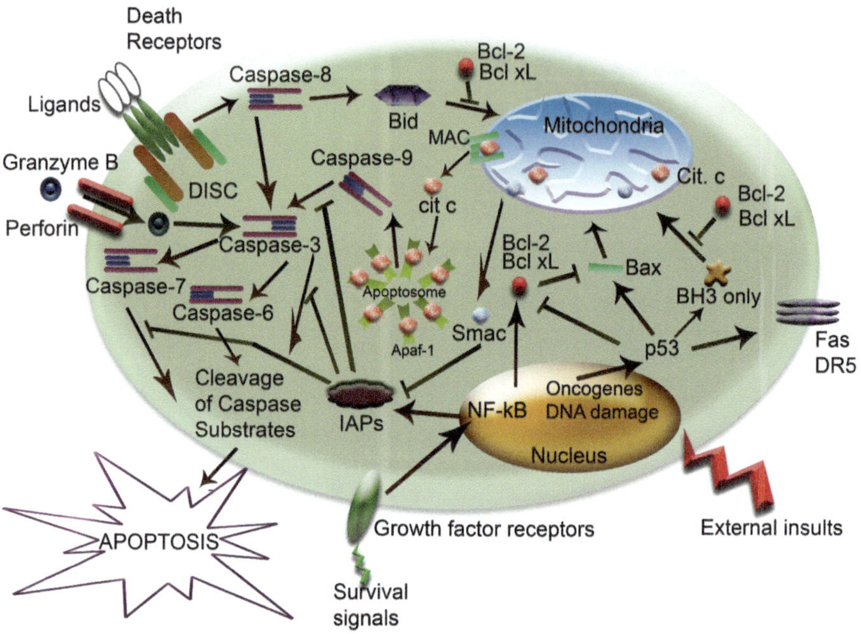

Plate 2 Schematic view of the apoptosis intrinsic and extrinsic pathways. In the extrinsic pathway, engagement of death receptors with their cognate ligands provokes the recruitment of adaptor proteins, such as the Fas associated death domain protein (FADD), which recruits and aggregate several molecules of caspase 8. The death-inducing signalling complex (DISC) thus formed in the cytoplasmic face of the plasma membrane, promotes the auto processing and activation of procaspase 8 (and possibly of caspase 10), which in turn is able to cleave effector caspase 3, 6, and 7. Caspase 8 can also proteolytically activate Bid. Truncated Bid represents the main link between the extrinsic and intrinsic apoptotic pathways, because it favours the aggregation and insertion of Bax and Bak into the mitochondrial outer membrane. The resulting apoptosis induced channel (MAC) promotes mitochondrial membrane permeabilization. Mitochondrial outer membrane permeabilization is the hallmark of the intrinsic pathway. In this pathway, several intracellular signals, including DNA damage and endoplasmic reticulum (ER) stress, converge on mitochondria to induce outer membrane permeabilization, which causes the release of proapoptotic factors from the intermembrane space. Among these, Cyt *c* induces the apoptosis protease-activating factor 1 (APAF-1) and ATP/dATP to assemble the apoptosome, a molecular platform which promotes the proteolytic maturation of caspase 9. Active caspase 9, in turn, cleaves and activates the effector caspases, which finally lead to the apoptotic phenotype. Second mitochondria-derived activator of caspase/direct IAP binding protein with a low pI (Smac/DIABLO) and the Omi stress-regulated endoprotease/high temperature requirement protein A2 (Omi/HtrA2), promote apoptosis indirectly, by binding to and antagonizing members of the IAP (inhibitor of apoptosis protein) family. The granzyme B-dependent route of caspase activation involves the delivery of this protease into the target cell through specialized granules that are released from cytotoxic T lymphocytes or natural killer cells. These granules also contain the pore forming protein perforin, which permits the entry of granzymes. Granzyme B can process Bid as well as caspase 3 and 7 to initiate apoptosis (See also Figure 2.1 on page 24)

Plate 3 Immunocytochemical studies of cultured hippocampal neurons. The figures show the fluorescence image of hippocampal neurons stained with FITC-labelled anti-MAP-2 antibody in brain slices (**A**) and in cell culture (**B**) (See also Figure 5.1 on page 94)

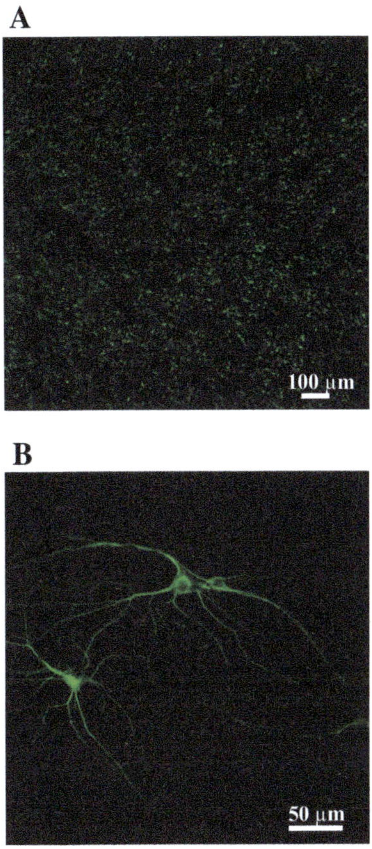

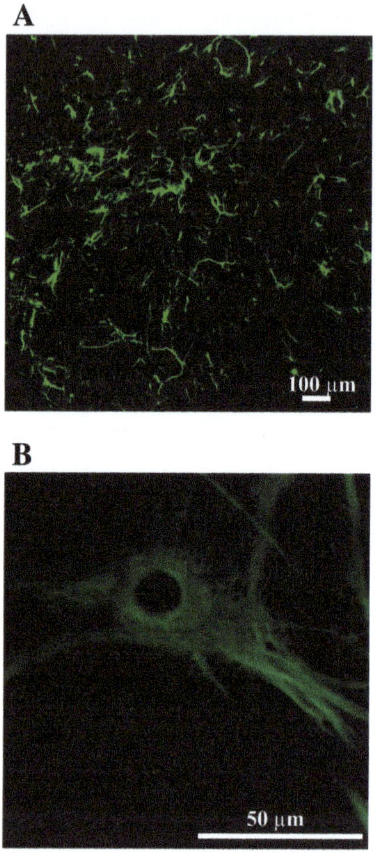

Plate 4 Immunocytochemical studies of cultured hippocampal astrocytes. The figures show the fluorescence image of hippocampal astrocytes stained with FITC-labelled anti-GFAP antibody in brain slices (A) and in cell culture (B)(See also Figure 5.2 on page 95)

Plate 5 Immunocytochemical studies of mixed cultures of hippocampal astrocytes and neurons. The image shows a three-dimensional projection of the spatial distribution of fluorescence of GFAP-positive astrocytes (*red fluorescence*) and MAP-2-positive neurons (*green fluorescence*). The image shows the close association of both cellular types in the cultures, with astrocytic processes engulfing the neuron (See also Figure 5.3 on page 96)

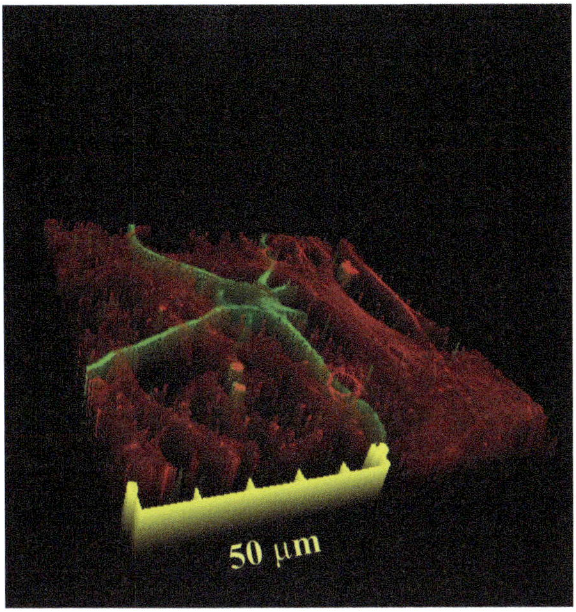

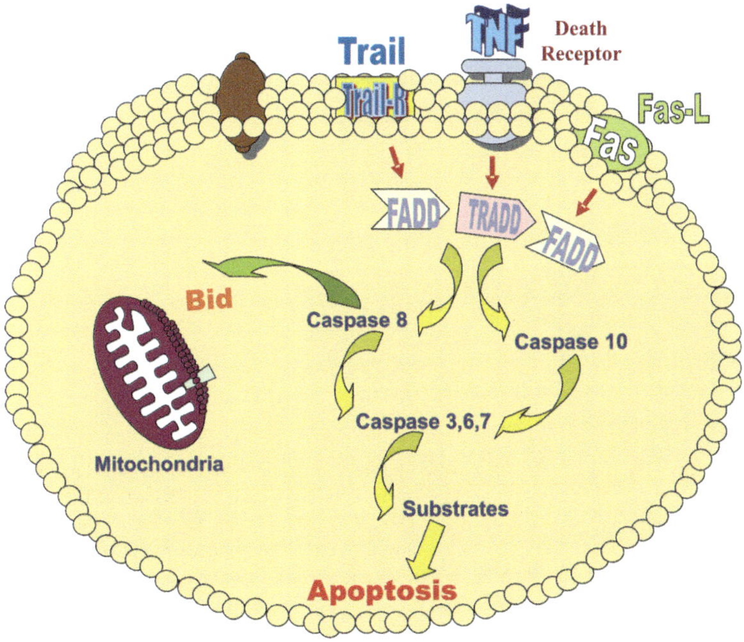

Plate 6 Extrinsic pathways of apoptosis. Brief description of the chain of events activated by death ligands action on their membrane-bound death-receptors and consequent activation of apoptosis (See also Figure 5.4 on page 100)

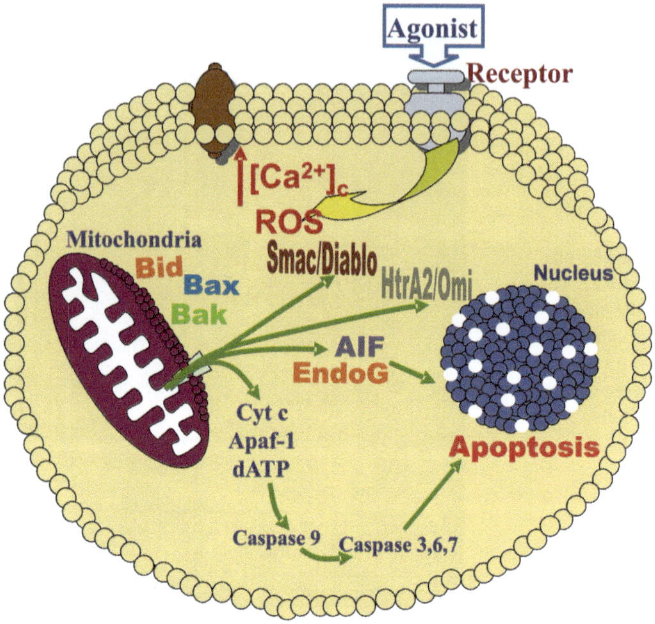

Plate 7 Intrinsic pathways of apoptosis. Brief description of the mitochondrial-linked pathways of apoptosis (See also Figure 5.5 on page 104)

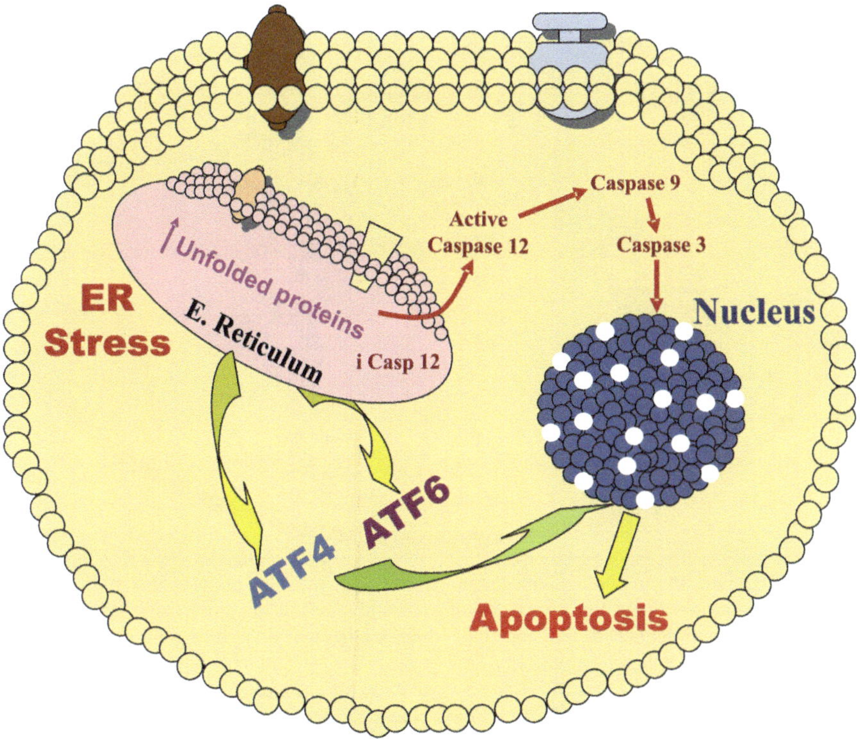

Plate 8 Endoplasmic reticulum pathways of apoptosis. Brief description of apoptosis pathways coupled to endoplasmic reticulum stress (See also Figure 5.6 on page 108)

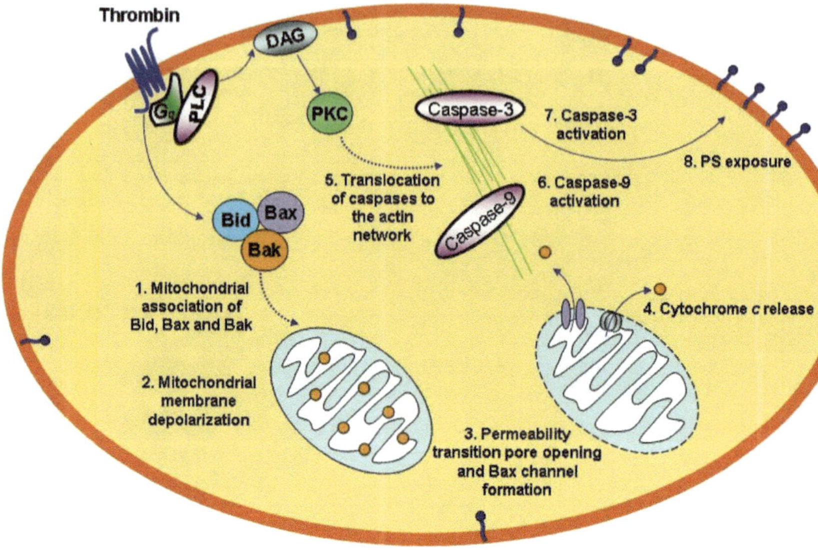

Plate 9 Schematic representation of the apoptotic events activated by thrombin in human platelets. Platelet stimulation with the physiological agonist thrombin results in mitochondrial association of the pro-apoptotic proteins Bid, Bax and Bak, which might lead to the formation and opening of the Bax channel and the permeability transition pore, resulting in mitochondrial membrane depolarization and cytochrome *c* release. Released cytochrome *c* induces the activation of caspases-9, which, in turn, activates the executioner caspase 3, a process that requires association of caspases to the newly polymerized actin cytoskeleton. Caspase activation results in phosphatidylserine externalization and a number of morphological changes characteristics of apoptosis (See also Figure 6.1 on page 141)

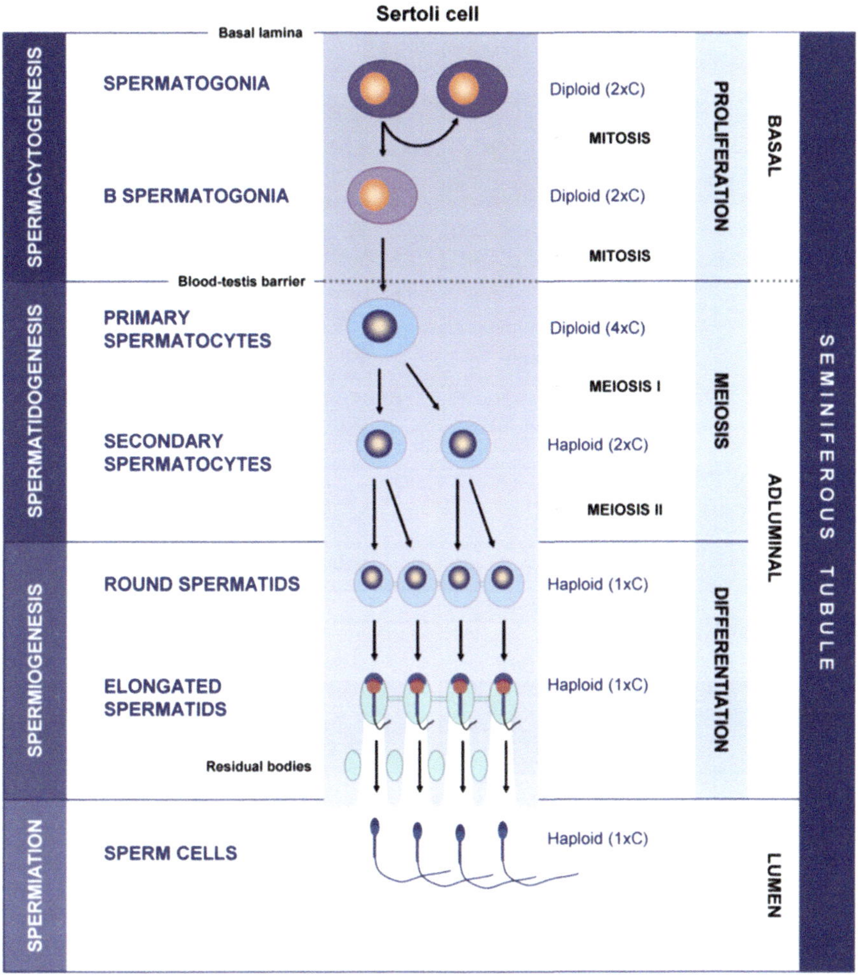

Plate 10 Diagram summarizing the different phases of the spermatogenesis, the characteristics of the germ cells involved in each stage, and the compartments in the seminiferous tubuli. Spermatogenesis is accomplished in the seminiferous tubuli of the testis in four sequential phases, each corresponding with the generation of increasingly differentiated germ cells. The first phase is termed spermacytogenesis, where the spermatogonia, which are the most undifferentiated germ cells, divide by mitosis to renew themselves and to generate B-type spermatogonia. The latter can further divide by mitosis and differentiate forming primary spermatocytes. In the second phase, termed spermatidogenesis, each primary spermatocyte suffer the first meiotic division to generate two secondary spermatocytes which enters second meiotic division to generate four round haploid spermatids. The third phase is termed spermiogenesis and consists in the transformation of a spherical spermatid to a sperm-like mature spermatid. During this transformation the nucleus condenses in size and is stabilized by protamines and most of the cytoplasm of the undifferentiated spermatids is released as residual bodies and phagocyted by the Sertoli cells. The last step, termed spermiation, involves the rupture of the structures and bonds anchoring a mature spermatid to a Sertoli cell, so the spermatozoon is released into the tubule lumen. (See also Figure 8.1 and complete caption on page 169)

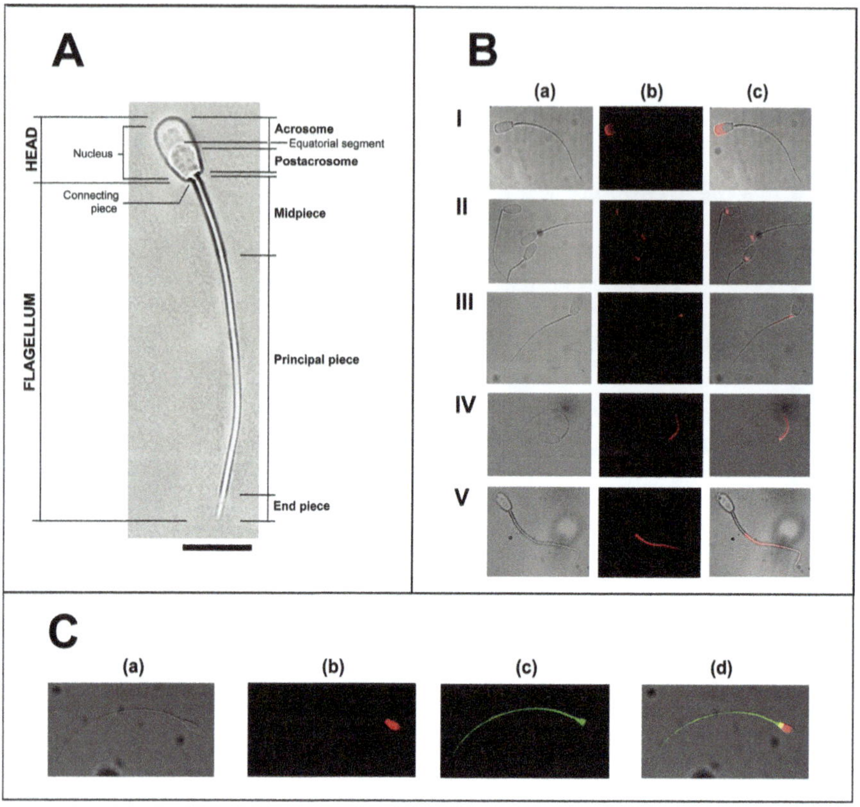

Plate 11 Subcellular structures in a mammalian spermatozoon. *A*. Microphotograph of a boar spermatozoon showing the distribution of the different subcellular structures that can be distinguished in the ejaculated mammalian spermatozoa. The image was obtained with transmission light in an invert microscope (×100). The *black* calibration bar in the *bottom* represents 10 μm. *B*. Fluorescent images of boar spermatozoa labelled with antibodies which specifically recognize proteins located in the acrosome (I), postacrosome (II), connecting piece (III), midpiece (IV), and rest of the tail (V). All images were obtained with a Bio-Rad MRC1024 confocal microscope with a ×60 objective in oil immersion. In each column, panel (a) represent transmission images; (b) is the immunofluorescence obtained in the samples excited at 488 nm with an argon laser and recorded with a 515-nm longpass emission filter; and (c) is the superposition of a and b. *C*. Fluorescent images of a stallion spermatozoon labelled with a nuclear marker and a fluorogenic marker for active caspases. The images were obtained at ×60 magnification in a Bio-Rad MRC1024 confocal microscope equipped with an argon laser. All samples were excited at 488 nm. **Panel (a)** represents a transmission image. **Panel (b)** shows the nuclear fluorescence recorded with a 680/32 nm emission filter. **Panel (c)** is the fluorescence recorded with a 540/30 nm emission filter which corresponds with active caspases. **Panel (d)** is the merge of a + b + c. In this panel, the tail appears as green, the nucleus appears as red, and the postacrosome appears as yellow, as a result of the superposition of *green and red* fluorescence (See also Figure 8.2 on page 183)

Index

GPSR Compliance

The European Union's (EU) General Product Safety Regulation (GPSR) is a set of rules that requires consumer products to be safe and our obligations to ensure this.

If you have any concerns about our products, you can contact us on ProductSafety@springernature.com

In case Publisher is established outside the EU, the EU authorized representative is:

Springer Nature Customer Service Center GmbH
Europaplatz 3
69115 Heidelberg, Germany

Batch number: 10755090

Printed by Printforce, the Netherlands